Ryan,

for your service &
sharing your brother
with us. Let's
bring them all
home.

Ed Sykes (wspr)

The Patch

and

The Stream Where the American Fell

The Patch
and
The Stream Where the
American Fell

Ed Sykes

AIR
CAPITAL
PRESS

The Patch and The Stream Where the American Fell

By Ed Sykes

Copyright © 2020 by Air Capital Press LLC

Wichita, KS

www.aircapitalpress.com

Published in the United States of America

First Edition

This work depicts actual events in the life of the author as recollection permits and/or can be verified by research. Occasionally, dialogue consistent with the nature of the person speaking has been supplemented. All persons within are actual individuals; there are no composite characters.

ISBN 978-1-7352519-0-5

Library of Congress Control Number: 2020942895

Cover design by Carolyn Vaughan (cvaughandesigns.com)

"Last of the Breed" by Dos Gringos, written by Chris Kurek, printed with permission from the copyright owner.

One day you'll see him sitting at a bar.
He's the one drinkin' whiskey and smokin' a cigar.
Pull up a chair and offer him a drink.
It's a good bet that he'll tell you exactly what he thinks
About this country and what it's doing wrong,
And for another drink, he might sing you a song.

He'll tell stories of how we fought the war,
Using words and phrases like you've never heard before.
He'll talk of death as if it were alive,
Then speak of good friends and good times as he looks up to the sky,
Then tell a joke that no man should ever tell.
It don't bother him 'cause he's seen both heaven and Hell.

He's a hard-hearted bastard of a day long gone by,
A mix of emotion and laughter in his eye,
A worn-out junkie on adrenaline and speed.
A fighter pilot, he's the last of the breed.

— "Last of the Breed," Dos Gringos,
written by Chris "Snooze" Kurek

Contents

The Patch

The Stream Where the American Fell

Preface

I have always loved baseball. I think it began when my mother first took me to Wrigley Field to watch her beloved Cubs. I still try to return to Wrigley at least once a year to renew those feelings of contentment I get from visiting the "Friendly Confines." I especially love attending baseball games because it allows the time and atmosphere to visit with the strangers sitting around me and find out a little about their lives. A few years ago at one of those games at Wrigley, I looked around the packed crowd and realized that all 40,000 of those folks had a story to tell—and almost all of them would never document it. I thought about my story and decided it was worth telling. So, like many other times in my life, I am setting out to do something I am ill-equipped to accomplish. I have never liked dealing with research or minutiae, and being outdoors on my little Kansas ranch is much more appealing—but this story is too good, so I'm giving it a shot.

The characters in this story are young men and women who grew up in the 1950s and '60s who found themselves pulled into an event called Vietnam. The story focuses on several young men with no combat flying experience who willingly volunteered to fly one of the most complex machines ever built into an environment where the odds were clearly stacked against them. Most of the young men in our shared situation had similar motivations. Some qualities I'm sure we all shared were a heavy supply of testosterone, a low aversion to risk, and a strong sense of invincibility. High intelligence was not required, but it helped.

It is a story about those who support the soldiers. No man is an island, and this is especially true regarding a soldier. They are surrounded by family and friends who are also affected by the craziness of combat. Even if they never see a literal battlefield, they face enemies of their own.

This is also a story about my quest to make my country respond to a situation that had diminished my confidence in its pledge to support its warriors. I am not alone in this quest, and the people who helped me along the way are all great examples of patriots.

This story is based, for the most part, on my memory. I have also spent some time and exchanged correspondence with many of the characters in trying to "get it right." I have attempted to avoid the temptation to embellish some of the incidents described, but, after all, I am a fighter pilot, and embellishment is a common trait among our fraternity. As explained above, I have a low tolerance for research or minutiae, and I'm certain I will get some of the "facts" wrong, so for those of you who wish to call me out, go write your own damn book!

Ed Sykes, wgfp

Part 1
The Patch

Chapter 1

On Wisconsin

"Sir, my name is Eddie Sykes from Princeton, Kentucky, and I want to be a fighter pilot."

Col. Hosman looked at me in what I perceived was a combination of disbelief and humor. My southern accent probably didn't make the incident any less humorous. He studied me for a few seconds and finally replied, "ROTC orientation is next Monday morning, and you can sign up then."

"No, sir, I don't need to go to orientation. I want to sign up now."

His gaze stiffened a bit, and after a long pause, he extended his right hand and pointed his index finger toward the door, indicating I should go, and replied, "Young man, you can leave my office now, and I will see you Monday at ROTC orientation."

"Yes, sir," I said and left. So, my first official meeting with an Air Force officer resulted in me being thrown out of his office. I don't remember feeling any personal distress from this encounter. Indeed, I remember being very impressed by Col. Hosman.

Col. Hosman was the professor of aerospace studies for the Air Force Reserve Officers' Training Corps (ROTC) program at the University of Wisconsin. This meeting was my first real contact with the Air Force and represented the first objective step toward my goal of becoming a fighter pilot in the U.S. Air Force.

I remember sitting on the swings behind East Side School in Princeton, Kentucky, with my good friend Frank "Butch" Pasteur. We had probably just been to the Capitol Theater for the Saturday morning, 10-cent special shows they ran for kids in the community. These shows were almost always westerns, but sometimes they threw in something different. I especially remember *Sabre Jet*, a Korean War saga featuring F-86s performing some dramatic flying scenes. I never wanted to be a cowboy because I tended to learn about how gravity works when I got on the back of a horse. The mules I sometimes rode around the Kentucky hills were somewhat better, but no self-respecting cowboy would ride a mule. Therefore, *Sabre Jet* gave me a second manly profession to pursue.

As we were swinging, we looked up and watched all the contrails forming in the sky. Somehow the presence of those contrails brought us around to the question, "What are we going to do with our lives?" We spontaneously agreed that we should someday be jet fighter pilots. This seemed like a manly thing to do for two 12-year-olds. Naively, we didn't realize that the contrails were probably not caused by fighters. They were probably made by some huge "trash haulers" (cargo planes), big lumbering bombers, or some other multi-crew airplanes in the U.S. Air Force inventory that we would have had little interest in flying when there were single-seat fighters out there.

Butch and I talked about these goals from time to time as we progressed through our young Kentucky lives, and we left Princeton for college with our paths planned out. I would eventually get there, and Butch came close. His dream came to an end when he was unable to pass the naval aviation physical. However, I was proud to have him as the best man at my wedding, and I know he would have been a great fighter pilot.

My family did not place a high value on education. I had flunked the second grade, so my early academic career got off to a rather shaky start. Neither my mother, father, nor stepfather had finished high school, and I was not aware of anyone from my extended family who had earned a college degree.

Luckily, my high school math teacher, Mrs. Walker, made a huge impact on my life. After a class my freshman year, she pulled me aside and said, "Eddie, you are really good at math." I thought, *Wow, I'm good at something,* and I began to work extra hard at math and the rest of my studies.

With my fighter pilot goal leading the way, I applied for the Air Force Academy in the beginning of my junior year. A friend's mother suggested I do this so I could experience the process and take the required entrance exams in preparation for my senior year. It was a good idea. My test scores in math and science were very good, but the scores revealed a weakness in my English and verbal skills. I was never able to overcome my lack of competence in the verbal skills area (so what the hell am I doing trying to write a book?). In my senior year, the Air Force Academy notified me that I was an alternate, but I never got the awaited phone call.

Not to worry! During my senior year, I applied to Georgia Tech and the University of Wisconsin. I had decided that I would attend a school outside of Kentucky. I knew I wanted to attend Air Force ROTC, and I was sure I needed to be an engineer if I wanted to be a fighter pilot. In retrospect, this was a rather silly assumption, and it cost me a lot of bar time and quality time with some really hot babes.

My scoutmaster, Bill Morgan, greatly influenced my School of Engineering specialty. Bill had been a primary role model in my young life in Princeton. He was a WWII veteran and had completed a tour in Europe as a B-17 turret gunner. His stories about his flying experiences served to heighten my desire to fly and have a military career. Bill also got me involved in ham radio. I would spend hours at his apartment above the Morgan Funeral Home where he worked in the family business. I built several radio kits, transmitters, and receivers, and I built my own ham radio shack beside the old garage in my backyard. All this exposure to electronics moved me to declare electrical engineering as my major.

Why Wisconsin and Georgia Tech? Again, I had some pretty silly reasoning. It was based on their fight songs. My mother always enjoyed sports and had bought a 33-1/3 album of college fight songs, and she played them all the time. I felt Wisconsin's "On, Wisconsin" and Georgia Tech's "Ramblin' Wreck" were the coolest fight songs, plus the schools filled my ROTC and engineering academic needs, so I submitted my requests for admittance.

I was accepted by both Georgia Tech and Wisconsin but didn't decide where I would attend until the late summer of 1962. I'm not sure why I chose Wisconsin, but most likely my decision was influenced by the fact that I had close relatives living in northern Wisconsin (several hours from Madison) whom I might be able to visit. That turned out to be a good decision from the family standpoint but a bad decision from the standpoint of weather. I froze my ass off that first winter in Madison and considered transferring to Tech in sunny Georgia.

Since I worked with a bricklayer/stonemason in the summer of 1962, I was able to put away enough money to cover my entrance expenses at either school. But I hadn't decided how I was going to travel to my new home. My mother finally forced the issue, bluntly asking, "Are you going to go to college?"

"Yep."

"Where are you going?"

"Madison."

"OK, pack your suitcase, and you can leave tomorrow."

I had an old blue metal suitcase, which was about 18″ by 14″ by 8″ (small). I dutifully packed it with a few things to get me started in Madison (nothing in that suitcase was adequate for that first winter!). I went to bed that night assuming all would be well.

The next morning, we got in the car to begin my odyssey. My mother drove east through Dawson Springs and not north toward the ferry at Cave-in-Rock as I had expected. *No problem,* I decided. She was probably taking me to Madisonville to put me on the bus. Not quite. When we intersected U.S. Highway 41 on what I thought was our way to Madisonville, my mother instead turned north on 41. So now I figured, once again, I knew her plan. She was going to drive me to Madison. "Nice!" However, shortly after making the turn north, she stopped the car by the side of the road, hesitated for a moment, and said with her finger pointed down the road, "Madison is that way." There were no hugs or kisses, and I don't even remember an encouraging comment. I got out of the car and waved as my mother did a U-turn and drove off. Chagrined, I stood there collecting myself, then turned and put my thumb in the air.

I arrived in Madison 24 hours later after hitching probably 20 different rides. I arrived in Madison early in the morning, and despite a rather chaotic night, I was filled with energy. Somehow, I ended up on Gilman Street, a few blocks from State Street where all the bars were and, incidentally, a short walk to the campus. I found several buildings with

"Apt For Rent" signs. When I found a man working on one of them, I asked how much the rent was.

"Depends what you want."

"Something cheap."

He took me up two flights of stairs and showed me a small one-room studio with a bunk bed and a bath down the hall shared by six other occupants. He said he had not rented to a second occupant of the room yet but was sure it would fill since school began the next week. I was about to ask how much the rent was when he asked me if I would be willing to work for the rent. Music to my ears!

"Yes, sir."

It turns out he owned two houses next door, and he needed someone to clean the hallways and take out the trash each week. If I would agree to do those chores, in lieu of rent, I had the room. We struck a deal.

As we parted, he told me he had heard that there were some "meal jobs" available at Allen Hall. I didn't know what that meant, but it sounded good. He told me how to get to Allen Hall—a few blocks away—and I found an ad on the front door telling me to proceed to the dining area where I'd meet the "head waiter." He explained the process and told me he needed someone to work in the dishwasher area—I was to be a "stacker." He told me to show up the next day for work as the girls were beginning to arrive and it would be the first day for meals. I would be required to "stack" two meals a day in exchange for three free meals a day. Stacking was essentially breaking down the trays of dirty dishes the waiters brought to the dishwasher area and stacking them in metal trays, which were then loaded into the huge automatic dishwasher. What a deal!

I was elated. I had been in Madison for only a few hours and had a place to stay and what turned out to be really good meals, and it wasn't going to cost me a dime. I was on such a roll I decided to find the Air Force ROTC Office and let them know I was ready to start the process of becoming a fighter pilot. It was a long hike across campus to the College of Engineering buildings where I found the Air Force ROTC Office on the third floor of the Mechanical Engineering Building. I told the sergeant, the first person I encountered upon entering, that I wanted to meet the commander and explain what I wanted.

"Just a second," he replied.

He was gone for a few minutes and came back and told me, "Col. Hosman will see you now—down the hall and second door to your right."

I remember considering, as I walked down the hall, whether I should salute. I'm not sure if I did, but I do remember knocking when I found Col. Hosman bent over some work at his desk. He motioned me in. "What can I do to help you?"

I went into my routine and was summarily dismissed. The following Monday I signed up for AFROTC, beginning my journey.

Leaving my rural Kentucky home and moving to Madison, Wisconsin, in the early 1960s was like taking a trip to another planet. Caldwell County, Kentucky, was a dry county, while Wisconsin was a "drink anything" state with a legal drinking age of 18 years. I had never attended an integrated school, and I was unaware that homosexuality even existed and was bewildered by the concept when one of my new friends pointed a gay man out to me. Finally, the amount of liberal activism against the War and for other liberal causes was beginning to expand, and one could hear and see it on a daily basis on and around the UW campus.

However, despite being on another planet, I adapted rather rapidly because I was here to become a fighter pilot, and not much of the rest of that stuff mattered.

One modification I worked on was shedding my Kentucky accent. I took a speech class during my first semester, and the giggles and smiling glances in the class when I began to speak made me aware that I should start modifying my dialect so that I could sound more like a Yankee. I was fairly successful at this but still find myself slipping into a southern drawl when I'm around some of my old friends from Princeton.

The 18-year-old drinking age was a bit of a nuisance initially because I had never drunk an entire beer before, and now it was a social requirement. I was very bad at it and on many occasions found myself bent over a porcelain stool in full upchuck mode. Even more embarrassing was a "bar date" with one of the lively young Wisconsin coeds. I discovered that Wisconsin girls have hollow legs and could easily drink me under the table—literally. This resulted in having difficulty getting second dates.

However, the 18-year-old legal age ended up being beneficial. It qualified me to drive a beer truck, so beginning in my sophomore year, I worked part-time for Simon Brothers, the local Miller High Life distributor. They paid $3.60 an hour, and I got my beer for half price. I kept that job for the rest of my time at UW and consider it the second-best job I've ever had. Upon my departure from the Air Force to join the Air National Guard in 1972, the Simon Brothers offered me my old job back. I thought about it briefly, but it wasn't really a player.

Every Monday morning, I wore my Air Force uniform to attend AFROTC drill practice. Despite the noise and protests that were so frequent at UW, I never experienced a single incident of disrespect for me or my uniform. The only time I recall being affected by the activities of the protestors was on the occasion when Barry Goldwater, who was running for president in 1964, came to Madison and tried to make a speech on the capitol grounds. The chanting of the protestors completely shut him down.

I was not an overachieving student. In fact, I was a pretty crummy student. I enjoyed learning about electrical theory, but I was not one to dig into the finer points of how things work. Plus, there were too many fun diversions at what *Playboy* magazine designated as "America's Best Party School." I wanted a degree but was not concerned about where I ranked in my class.

A stroke of good fortune occurred during my sophomore year when I discovered Kappa Eta Kappa, an electrical engineering fraternity that came complete with a house, a bar, and a lot of party-wise brothers. I didn't discover the best part until after I had joined—a "gouge file" on the third floor that contained copies of almost all the tests that the electrical engineering professors had given over the past several years. What a great tool! It did much to enhance my "understanding" of electrical engineering.

My primary focus was aimed at AFROTC during those UW years. I enjoyed the drilling and the atmosphere created by the program. The classes were informative and did a good job of preparing me for my future journey. I enjoyed the instructors and their always seemingly upbeat attitudes and presentations. I would later learn that ROTC instructor assignments were undesirable when it came to career progression—especially for pilots. If these men were displeased with their fate, it was never evident to me. One highlight of the program was the check I started getting every month after my sophomore year; although not a lot, it more than covered my beer and fraternity dues.

Between my junior and senior year, I attended AFROTC Summer Camp. This was my first hard-core exposure to active military life and was not entirely a good experience. All the other cadets going to Summer Camp from UW were assigned to active-duty bases in the North. Because my home of record was in Kentucky, I was assigned to Shaw AFB in South Carolina.

I adjusted quickly to the southerners as they were like the kids I had grown up with. However, I was not sure they were as comfortable with this "Yankee" in their midst. I worked hard at doing well, and the others in my flight selected me as their drill instructor for our weekly drill competition, which we won on two occasions.

I thought I was doing great. So when I was informed that my standing in my flight was in the bottom half of the group, I objected, rather forcefully, to my senior officer and camp staff. I was told in no uncertain terms to shut up and "sit down in the boat," which I finally did. It wasn't until my return to UW that I was informed how close I came to being thrown out of the AFROTC program due to my warranted but unwise outburst. So, what did I learn? Not very much, thank you.

During Summer Camp, I experienced my first flight in an AF jet. Each cadet who was identified as a candidate for flying got a ride in a T-33 (a jet trainer derived from the F-80 Lightning, a fighter). The flight was short, but I got a chance to handle the jet and knew I was on the right career path. Once again, I opened my mouth and inserted foot by exclaiming to the pilot in the front seat something like, "I knew you couldn't make me sick." He immediately broke off his approach to the field and taught me how to fill up a barf bag. Continuous aileron rolls and negative g-forces can do that to the unexposed. That was the first and last time I would ever become sick in an airplane.

During my senior year, after passing a flight physical, I entered the Flight Indoctrination Program (FIP). FIP was a flight program that allowed the Air Force to do some limited screening of a cadet's aptitude for flying and also gave the cadets a chance to decide if flying was a good career path. It was conducted as a civilian contract. It allowed a cadet to receive about 30 hours of flying time with a civilian instructor and concluded with a check ride by the Federal Aviation Administration (FAA), who could award the cadet with a private pilot's license as well as a recommendation to the Air Force that the cadet was suitably equipped to enter the Air Force Undergraduate Pilot Training (UPT) program after becoming an officer.

I learned to fly in the Piper Tri-Pacer at Truax Field on the east side of Madison. My instructor did a good job of getting me ready to "solo out," and after a few flights, he turned me loose. The program included several hours of solo flights to prepare the cadet for the FAA check. On these solo flights, I practiced most of the items that were required on the check, but I found the navigation part boring and used much of my navigation time flying low over the Wisconsin River north of Madison—joy riding. Consequently, I did well on all parts of my flight examination except for the navigation part. At one point, I had to admit to the FAA examiner that I was "temporarily disoriented" on a navigation route he had me fly. So, the examiner recommended that I was probably qualified for UPT, but he did not award me an FAA private pilot's license. Once again, the UW AFROTC staff informed me that my continuation in the program was a little tenuous.

My ROTC road wasn't all bumpy, though. I must have demonstrated some leadership skills because I progressed up in rank easily and was elected by my peers to be the commander of the Arnold Air Society (a voluntary professional organization for cadets). I was then elevated to the position of Chairman of the Joint Military Board, which governed all joint activities of the several ROTC units (Air Force, Army, Navy, and Marine) at UW. Upon reflection of my AFROTC experience, it pains me to know that I was slow to understand what it means to be in our nation's military. I just wanted to be a fighter pilot, and this was the hurdle that needed to be crossed. Turns out, I was only a few short years away from learning what being a military warrior really means.

In the summer of 1964, my good buddy Butch and his brother Corky and I decided to drive to Chicago in search of summer jobs. Princeton had little to offer if you didn't want to hoe tobacco or pick strawberries, so we drove north and found a place to live in "Hillbilly Heaven" (many of the folks in the area spoke like us) in Chicago's north side just off of Bryn Mawr Avenue and a few blocks from Lake Michigan. We found jobs that provided cash flow, and we still had some play time. On one occasion, Butch and I took a late afternoon hike to Hollywood Beach, a few blocks from our apartment, in search of whatever drives alpha males to head to the beach. We were both wearing our Princeton University shirts, which were readily available in Princeton, Kentucky. We knew, from time to time, that they were pretty good chick magnets —at least until you had to explain that you didn't really go to *the* Princeton University.

Pay dirt! We noticed two attractive girls sunning on the beach, so we wandered over, and Butch said, "Hi, I'm Butch and this is Eddie. What are you guys up to?"

They checked us out, managed a slight smile, and then locked onto the words "Princeton" on our sweatshirts. Suddenly they became more conversant, and we learned they were Marsha and Mary, who were also in Chicago working summer jobs. Eventually the subject of our shirts came up and we had to confess, but by that time, we had broken the ice. Somehow, Butch gravitated to Marsha and I found myself with Mary. Turns out that she was about to begin her sophomore year at the University of Wisconsin–Milwaukee (UWM) and wanted a career as a science writer. Despite my misrepresentation of an alliance with Princeton University, I guess she felt a verbal résumé of my status—UW engineering student—was worthy of further consideration.

Thus began a relationship with Mary Kathryn Bartz, who would eventually become my wife. Mary grew up on a dairy farm in northern Wisconsin. She was a strikingly beautiful woman who had been the prom queen and valedictorian of her high school class. She was the second of six children in a family with strong religious beliefs and an unbelievable work ethic. I always considered her father, William (Bill) Otto, a "giant," but I'm not sure he shared similar feelings toward this crazy Kentucky kid who was attempting to abscond with one of his innocent young daughters. I didn't understand the level of his concern until a few decades later when my daughters brought home serious suitors.

Marsha dumped Butch within a matter of days, but Mary continued to show interest in my act, and we would meet for ice cream or sandwiches at the local deli. I even took her to Wrigley Field, where we sat in the bleachers. I yelled for the Cubs, and she pretended to show some interest in the game. Just before the summer was over, we took a bus trip to Madison where we walked the campus and recounted experiences. Mary had been a summer science scholar and spent a few weeks at UW while she was in high school and had obviously enjoyed the experience. As our first summer together came to an end, we agreed to meet again after school started.

And meet we did over the course of the school year. The Badger Bus ran every couple hours between Madison and Milwaukee at a total round-trip cost of $3.65. It wasn't long before we were saving our loose change to cover the Badger Bus fare, whether I went to Milwaukee or she came to Madison. We went to Badger football games with a wineskin or flask hidden under our coats and to fraternity parties, which were really crazy. We also went to a formal sponsored by ROTC. One of the most memorable events was a trip I made to Milwaukee to attend a Bob Dylan concert. It turned out to be a huge disappointment when Dylan began singing and then walked off the stage when the sound equipment didn't pick up his harmonica.

I proposed to Mary in the fall of 1965, and after some stupid hiccups and false starts on my part, we married in June 1966. Butch drove all the way from Kentucky to be my best man.

13

That fall, Mary and I set up house in a small bungalow on Madison's east side. It was someone's guest house behind their residence. Mary quit school and worked an 8-to-5 secretarial job to help provide for our living as I finished my last semester in pursuit of my EE degree. Being married proved to have a positive influence on me as I achieved my best GPA of my four and a half years at UW.

In January 1967, I marched in the Engineering School's midterm graduation ceremony. That was a good enough reason to celebrate, but the next day was the ceremony I had been waiting for since that day sitting on the swings in Princeton watching contrails in the sky. My mother pinned second lieutenant bars on my shoulders.

The most remarkable event of that day, though, was a comment from Col. Hosman. He had left the ROTC program at Madison a year or two earlier but had returned to speak at the commissioning ceremony. Following my pinning, he pulled me aside, congratulated me, then smiled and made a comment that I will never forget: "Sykes, you were so dumb you didn't know you couldn't do it."

Chapter 2
My Mother's War

My mother was an interesting character. I would never describe her as a tender or compassionate person but acknowledge that she was tough, quite smart, and somewhat of a rebel. She was born Ruth Elaine Severude in 1925 in Barron County, Wisconsin. Her father, Albin, and her mother, Evelyn, were both from third-generation Norwegian families engaged in farming. Mother was the oldest of six children and at a very young age became a primary caretaker for her siblings. Perhaps as an attempt to escape from these responsibilities, she eloped with my father, Clarence Edwin Sykes, when she was 17 years old. I was born when she was 18, and they quickly followed up with my two sisters, Merry and Sue. So, at age 21, she had three children of her own—some escape!

My grandfather, Albin Severude, had served in World War I and was awarded a Silver Star citation for bravery. He had served as a courier, carrying messages between the trenches on a Harley-Davidson motorcycle. On one of his details, he had encountered a German staff car. He killed the officer and driver and recovered some documents that were taken to his superiors and used as intelligence concerning German plans. Sounds like good qualifications for a Silver Star to me. Following the war, Harley-Davidson used him in many of its magazine ads, showing him with his bike and detailing the reliability of its product. Unfortunately, a few years after the war, he fell from the roof of a barn that he was reshingling and broke his back. He would spend the rest of his life in a wheelchair.

I never saw my grandfather walk, but on those occasions when we visited his home in Dallas, Wisconsin, I would push him to the pool hall every afternoon where he engaged in some heated cribbage games with his old Norwegian buddies and drank "near beer." I would also sit by his bedside and listen to the Cubs game every day they were on, the beginning of my love of baseball. I also watched him work on the leather goods he produced to assist with the family budget. He made beautiful billfolds, purses, and belts using only hand tools. It was at the Severude house that I first began thinking of making the military a career. Out in a shed behind the house, I found several WWI artifacts, including a helmet with a bullet hole in it and a gas mask with an attached canister. The mock battles I played out in my head and with my sisters enriched the fantasy.

My father worked as a farm laborer when I was young and then moved our family to southern Wisconsin, where he took a job at an assembly plant for over-the-road semitrailers. When I was about six, he developed a severe pain in his left knee. He had bone cancer, and the doctors told him the leg would have to be removed. It was removed at mid-thigh in an attempt to prevent the spread of the cancer, but within a short time, it was discovered that it had spread to his lungs. Just before my eighth birthday, he passed.

My mother's family offered to put her up with the rest of the family, but her strong independent streak said "nothing doing," and she took me and my sisters back to southern Wisconsin and got a job in an assembly plant in Fort Atkinson. It was there she met my stepfather, George Sanderson, and after a short engagement, they were married, and he took on our whole bunch. I never called him Dad or Father, a fact I regret to this day. My mother asked us to call him "George," and we always did. George was an over-the-road truck driver, and shortly after marrying my mother, his company relocated him to western Kentucky. We moved to Princeton, and I rapidly settled into my new environment. It was a wonderful place to grow up.

Shortly after moving to Kentucky, my mother took a job with the Princeton Hosiery Mill in the sock-dying area and would remain there throughout my high school years. The best part about my youth was the amount of freedom that was afforded to me by my mother. She kept a pretty good grip on my sisters' activities, but she pretty much let me run wild. The word "love" was never used in our family, and I don't think I ever remember her hugging me. There was never any emotion expressed, but we always had a great deal of mutual respect and rarely argued. Although she never said it, I think she was proud of the ease with which I was moving through adolescence. She never attended the football games, baseball games, or band concerts I played in and never even really talked about them. However, I never felt neglected.

George was just George. He was on the road most of the time, and when he was home, he drank a lot of beer and spent a good deal of time with his truck-driving buddies. He liked guns and hunting. He also made sure I had my first .22-caliber rifle when I was 12 years old, and a 28-gauge shotgun a few years later. In all the years I spent with him, I don't ever remember him raising his voice or scolding me. He was a damn good stepfather considering some of the stories I heard from my friends with similar situations.

My mother attended my high school graduation and seemed impressed that I had won the Joiner Hardware Best Science Student Award. She was mildly interested in my plans to attend college but offered me little guidance about where to go or what to study. College was uncharted territory for both the Sykes and Severude families, and I'm not sure she saw the value of doing it. However, she was supportive enough to drive me to Highway 41 and drop me off so I could hitchhike to Madison. Now that's support. As she drove away, I harbored no ill will or resentment toward her. That was just Mother.

My sisters, however, were a different kettle of fish in her eyes. While I was away at Wisconsin, they both took off on their own paths, and both involved the military. My mother did see value in college for my sisters—primarily as a path to a *Mrs.* degree. When Merry graduated from high school, she entered Morehead State College in eastern Kentucky but soon found that college was not to her liking. With a friend, she moved to Evansville, Indiana, where she found a job at a hospital. Her friend was dating a boxer and asked Merry to double-date with them and fellow boxer Robert Carmody. He was short, thin, and had puffy areas on his face that denoted his boxing experience. She was not impressed at first but soon found Bob to be a fun companion.

Bob Carmody was a tough little guy who grew up on the streets of Brooklyn and, as a smaller guy, had to work hard to maintain his street cred. He dropped out of high school before finishing and joined the Army. It was there he found that his street-fighting skills could be put to good use by the Army. He quickly moved up the ranks in Army boxing and eventually became the number one boxer in the Flyweight division (112 pounds). He was soon representing the Army in all kinds of venues in Europe and even the Pan American Games. Then in early 1964, he competed for and won a place on the U.S. Olympic Team, which was preparing for the Summer Olympics in Tokyo.

In preparation for the Olympics, he trained at Fort Campbell in Kentucky and was assigned to the 101st Airborne Division. Besides boxing, he participated in many of the division's exercises and took great delight in jumping out of perfectly good airplanes with over one half of his body weight strapped to his short little torso. This guy was a real stud.

Merry and Bob hit it off, and she moved with some friends to Hopkinsville, Kentucky, to be closer to their soldier boyfriends. A few weeks before the Olympics, Bob asked Merry to marry him. Merry accepted his proposal without hesitation. They decided they would marry shortly after he returned from the Olympics. A phone call to Mother was in order.

"Mother, I'm getting married. I've met this really great guy in Evansville who is on the Olympic Team and is stationed at Fort Campbell."

Mother was furious!

"Merry, you know he is lying to you. He's just telling you that to take advantage of you. And you know what kind of reputation those Army guys have around Princeton. They just want one thing, and it's not a family!"

"No mother, he's not like that..." she continued. It became obvious that Mother was not convinced. She asked Merry to abandon the relationship. Merry was not about to listen to Mother's rant, and the conversation ended.

As the Olympics approached, it was decided that Mother, Bob, and Merry would get together at a restaurant outside the gates at Fort Campbell. By now, Mother had read some newspaper articles and found out through local TV that there really was a guy named Bob Carmody on the Olympic boxing team. However, she was still not happy about the impending marriage. The meeting did not go well.

"Are you sure you want to marry my daughter?" was mother's overall theme. Bob did what he could to quell her doubts, but he was not too impressed with his future mother-in-law and was not eager to demonstrate any charm. Little bonding occurred as a result of this get-together, and Merry remembered Bob referring to her mother as "your pie-faced mother" as they drove away.

Bob did well at the Olympics, ending up with a bronze medal after losing a split decision to the boxer from Italy. Bob would later express his displeasure with the decision and told me, "He kept biting me in the ear." At any rate, he had an Olympic medal to hang up with his numerous other medals and trophies.

When Bob returned home, he and Merry quickly set a marriage date and discovered one slight hitch as they applied for a marriage license. Because Merry was under 21, she needed Mother's permission to marry. They all met in Hopkinsville to complete the paperwork, and by now, Mother was softening to the idea. They were married at the Fort Campbell Chapel on November 8, 1964. Mother initially said she was not going to attend the wedding but decided better of it. Following the ceremony, there was a reception at a friend's home, and it was here that Mother changed course. She was the mother of the bride in the presence of numerous Olympic boxers and Army brass and suddenly decided that Bob had some redeeming value. He was quickly elevated to pedestal status.

I did not make the wedding, as school had me tied down. My mother, however, was surprisingly expressive when she talked about the wedding reception and all the boxing stars from the U.S. Olympic team who had attended. She was as animated as I can ever remember her being. It was clear she was pleased with her new son-in-law.

Bob and Merry settled in at Fort Campbell, and before long, she announced she was pregnant with my mother's first grandchild. Bob settled in as boxing coach of the Army Boxing team and, because of his position, was safe from being sent to the War that was beginning to worm its way into the nation's consciousness. I was so happy for my sister and mother. She was as happy as I had ever known her to be.

Meanwhile, my youngest sister Sue had decided to give college a try and had enrolled in Southern Illinois University (SIU) with the goal of becoming a teacher. While attending college, she agreed to a blind date, meeting Reinhart Michael Westenrieder. They began to hang out together. The best part for Sue was that he had a car. He was also a member of the SIU swimming team. Reinhart was a bit much for a first name, so everyone called him "Wes."

Wes was a pretty interesting character. His father, although not a member of the Nazi Party, had been part of Hitler's army and had been killed on the Eastern Front in 1944. Wes's mother had been a swimmer and was part of Hitler's dream of an Aryan master race. The defeat of the Germans put an end to that foolishness, and Wes described playing in the rubble of his bombed-out city as a young boy. His mother was in an area occupied by the American army after the war, and it was there that she met and married Wes's stepfather. He was an American serviceman, so eventually Wes ended up in the U.S. It was on his flight to America that he decided he wanted to be a pilot. Once in the U.S., he learned English by watching TV.

He grew up in northern Illinois and became a good freestyle swimmer in high school. He was recruited by SIU and was good enough to compete in many NCAA national swim meets.

Despite being brought up in the U.S., his German roots were obvious to me. Growing up, I had worked a few summers with a German immigrant, Emil Gomalla, who was a bricklayer and stone mason. He was a hard-working and hardheaded character. He paid me well, and we worked 10 hours a day, 6 days a week. He demanded perfection in the performance of my job and was not averse to yelling at me when I screwed up. I guess he established in my mind what a German was supposed to be like, and Wes, except for the lack of an accent, fit the description to a tee. He had an opinion on everything, and he knew he was right. He was not averse to yelling at Sue, but she consistently ignored his outbreaks with a shake of her head and a slight smile. "Now, Wes," was all she generally had to say for him to calm down.

Wes had one other flaw. He wanted to be a pilot. He had joined AFROTC when he arrived at SIU with the intention of doing just that. My sister didn't like the idea, but my mother certainly did. Although he had never met his father, the legacy of his military experience and the experience of growing up in a military family made him comfortable with military life. Upon his graduation from SIU in December 1967, he got his orders to flight school.

Mary and I got married in June 1966, and Sue and Wes were married September of that year. The four of us were all well on our way to completing college, and my mother couldn't have been prouder. Bob Carmody was a superstar with whom she was incredibly pleased, and her son and other son-in-law were both about to become AF officers en route to becoming pilots in the Air Force. What could be better? What could go wrong?

Chapter 3

The Patch

After graduation, Mary and I hung around Madison. She continued working at her secretarial job, and I worked full-time driving my beer truck routes while we awaited my orders that would send us to flight school. When the orders eventually arrived, we anxiously opened the package. We were so excited that we didn't read the name of the town, Lubbock, very carefully. "Where in the hell is Lublock, Texas?" We had never heard of the place or Reese Air Force Base, but the suspense was over. We called friends and relatives to let them know we were headed to "Lublock."

We didn't have to report until early April, so we continued to work. About two weeks prior to our report date, we took a side trip in our old blue Plymouth, "Cognito." We went everywhere in Cognito: through St. Louis, Princeton, and along the Mississippi Gulf Coast. A short stop in St Louis allowed us to marvel at the recently completed Gateway Arch. It wasn't open to the public yet, but wow, it was impressive. We considered this our honeymoon, but I was anxious to get started with my dream.

In Kentucky, it was great visiting the site where it all began four and a half years before. I asked my mother what she was thinking when she dropped me off on Highway 41 and pointed her finger toward Madison. She shrugged and gave me a slight smile and said nothing—nothing needed to be said. Bob Carmody and my sister Merry were now stationed at Fort Campbell where Bob was part of the coaching staff for the Army Boxing team. They drove up to Princeton to see us while we were there, and Bob suggested that I should put on my uniform and come with him to visit Fort Campbell. He gave me some lame excuse for wearing my uniform rather than going in civies (civilian clothes), and I took the bait. Putting on my uniform with my still-gleaming 2Lt. bars, we drove to Fort Campbell.

It was always fun to be with Bob. It was hard to believe the scrawny guy sitting next to me in the car had represented the U.S. in the Olympics and had won a bronze medal. He talked about their new daughter Terri Lynn and mentioned that Merry was pregnant with a second child that was due in the fall. He was obviously happy with his new family.

He also talked about the War and his friends in the 101st who were deploying to the War and how he felt bad about not going with them because of his special status. "I'm thinking about asking them to send me," he stated. "I just don't feel right about not being with my fellow soldiers in this fight." I listened intently but assumed that Bob would remain in the relative security of the boxing world.

Upon our arrival at Fort Campbell, Bob checked in at the gate, and as we were motioned to proceed through the gate, the guard gave me a sharp salute and yelled, "Airborne, sir!" Surprised, I returned his salute. Wearing my officer's bars meant that, even though I was fresh out of ROTC and the lowest-ranking officer, enlisted soldiers were still required to salute me. It was the first time I had been saluted as an officer except for the ceremonial salutes given me by some of the ROTC staff at Madison.

Bob grinned and proceeded to show me some of the facilities used by the Army Boxing team. He then drove to an area where several soldiers were moving around. He stopped the car and said, "Get out, Ed, there's something I want to show you." I jumped out, and we headed along the sidewalk toward a building marked "Mess Hall." Along the way, we met several soldiers, and as we approached, they would all give me a brisk salute and yell, "Airborne, sir." I now began to sense that Bob had set me up and he was trying to wear out my right arm. As we approached the chow hall, I could see the soldiers doing pull-ups on a chinning bar before they entered the hall and yelling "Airborne." These guys were a bunch of studs.

Then Bob pulled the ultimate prank, leading me into the chow hall. As I entered, someone yelled, "Room, ten-hut," and maybe a hundred or so soldiers leapt to attention and yelled, "Airborne, sir." I'm sure those men were pretty annoyed they had to do that for some goofy young lieutenant.

Bob, you asshole, I thought as I stumbled out something like, "At ease," and the soldiers returned to their meal.

Bob, beaming at me with great delight, gave a gleeful little giggle. I stood there for a moment not knowing what to do next, so Bob motioned toward the door, and as we exited, once again came the cry, "Room, ten-hut!" Bob was now laughing out loud as he had demonstrated how easy it is to trick a young second lieutenant. I had been had. We stopped by a liquor store as we left the base and picked up a six-pack of cold beer. We drank it, laughing about his devious trick, as we drove back to Dawson Springs.

Back home, my mother had prepared a great meal, and we feasted on ham and all the fixings while Bob told the group of the trick he had pulled on me. My stepfather, George, joined in on the revelry and concluded that I was not the brightest bulb in the box. As Bob and Merry and their new daughter departed that evening, I saw the delight my mother took in the happiness Bob had brought into Merry's life.

Mary and I then drove south through Tennessee and Alabama to Biloxi, Mississippi, where we checked off that honeymoon square, simply enjoying the beautiful white sand beaches. We also took an excursion on a glass-bottomed boat, where the pilot pulled out a long scoop on a cane pole and asked if anyone would like a raw oyster. No one else volunteered so, wanting to prove my manhood in some small way to my young bride and the other tourists, I raised my hand and said, "I'll try one."

He nodded and dipped the pole into the bay and, after a brief search, retrieved a large gray oyster. Removing it from the hook, he put on some heavy gloves and quickly split the oyster with a large knife, handing me an oyster on the half shell along with a toothpick and some red sauce. Once again, testosterone had gotten me into trouble. Obviously, raw means raw! I spread some of the sauce on the uninviting blob and managed to get a grip on it with the toothpick and put it in my mouth. *Am I supposed to chew it?* I wondered. As the somewhat salty, slimy mass entered my mouth, I realized that all that was required was a big swallow and it was gone. Mary gave me a rather disgusted look as I completed this ritual; apparently, I had not moved up the "manhood scale" in her opinion. When the boat captain asked if I would like another one, I lied, "That was great, but no thanks."

Upon arrival in Lubbock, we quickly found an apartment complex that catered to students in the UPT program at Reese AFB, which was about 10 miles west of Lubbock. Lubbock is located on a huge plateau called the Caprock, and the countryside on top of this plateau is unbelievably flat. The landscape is covered with field after field of cotton along with the frequent dotting of oil-pumping rigs and their accompanying storage tanks. Halfway between our apartment and Reese was a huge cotton gin—the most dominant feature on the horizon.

After depositing our few belongings at the apartment, we drove to Reese and showed the gate guard my orders and proceeded to drive around our new world. There wasn't much to see. Reese was a small base with lots of vintage WWII buildings. Within a few minutes, we had located the commissary, base exchange (BX), theater, and the golf course. However, when we got to the Officers Club, I encountered one of the most beautiful sights ever: next to the parking lot was a ramp full of T-38 Talons. Long and lean and painted white, they looked like they were going 1000 miles an hour just sitting on the ramp. The thought of flying one of those jets was almost more than I could fantasize about—it was surreal. We then drove to the Wing Headquarters where I signed in. I was given a package of material to digest before reporting in the following Monday.

The weekend went by way too slow for me. The excitement of being in flight school was overwhelming. Mary and I went to a "bar" in Lubbock and found that it was not a bar at all—at least not anything we had experienced in Wisconsin. It was really a "club," and you had to join and sign in each time you showed up. We also found out that they did not serve alcohol; you had to bring your own and then were charged for a "set-up" (the ice and mixers). Somehow, I had pictured Texas as a place where bars had swinging doors, cowboys playing poker, and Miss Kitty serving the drinks. Another myth exploded.

We found that all the folks we met who claimed Lubbock as their home were incredibly friendly and ready to help in any way they could.

Another distinguishing feature of the Lubbock natives was front teeth stained slightly brown. The water in Lubbock was high in fluoride content, and folks that grew up there had this feature in varying degrees. We also drove around Texas Tech University to get a feel for the place. Mary had decided to try and finish her degree while we were stationed at Reese, and she, too, had a certain level of excitement to return to her schooling. Driving around the campus, I soon noted that the Red Raider coeds had a different look than the Wisconsin girls. They all wore nice-looking apparel and flawless hair and used a good deal of makeup. Not a single beatnik to be seen.

By Monday morning, we had met some other members of my class within our apartment complex. We found them at the swimming pool and soon determined there were four of us living there. We decided we would carpool to the base, so on Monday morning, I accompanied Garth, Russ, and Jim to the first official active-duty military formation for any of us. There were about 74 students in our class, almost all second lieutenants.

There were a few captains who, for the most part, turned out to be navigators coming from cargo and bomber assignments who had been recommended for upgrade. There was also a small group of first lieutenants who appeared to already know each other and seemed to be more at ease with the situation. Turns out these guys were "Zoomies," U.S. Air Force Academy graduates, who had picked up their silver 1Lt. bars while attending graduate school. "Zoomie" was derived from the term often used by the cadets to describe the Academy—"The Blue Zoo"—because they were always being viewed by tourists who wanted to visit the futuristic facility and frequently watched from high vantage points as students went about their daily routine, feeling like animals in a zoo.

The class, 68-F (the sixth class to graduate in fiscal 1968), was divided into two sections, and each section had the senior captain within the section as a class leader. Most of the first day was filled with paperwork and introductions. We were told we would first spend six weeks flying the T-41 Mescalero (the Air Force version of a Cessna 172), followed by 18 weeks in the T-37 Tweet, which was the primary jet trainer.

If you made it past that, you moved on to the T-38 Talon for the last six months of training.

The best part of the first day was going to Supply and getting four new flight suits, which were often called "goatskins," and our flying boots. *Now we're getting somewhere,* I thought as I tried them on. We dropped our new goatskins off at the dry cleaners to have our name tags and patches sewn on them, and the next day when we picked them up, we were ready to be somebody.

After one of those initial days of training, we decided to stop by the "stag bar." It was a special bar for men only, except for some of the bartenders and an occasional dancing girl. We had heard about it from the bachelors in the class who were giving us married guys a hard time for being "whipped." The stag bar proved to be a large, smoke-filled room with a bunch of type A males in goatskins telling about the day's exploits. I felt at home at once.

The bachelors had also been telling us about Coors beer and how good it was, so we all ordered one. I agreed it was exceptionally good. At that time, Coors was only sold in a few states, Texas being one of them. Its claim to fame was that it was not pasteurized and was always refrigerated, from the brewery to the bar.

As I surveyed the room, it seemed that the instructors were congregated together in one area and the students in another. Their patches telegraphed their status. However, there was a special group of instructors in the corner who somehow stood out from the rest as the most studly of the bunch. They seemed more sure of themselves, spoke louder, and used more hand gestures, recounting different flying scenarios, than the others. As I watched them, there was one distinguishing feature they all had in common. They all wore a red, white, and blue patch on their left arm that read "100 Missions North Vietnam F-105." This patch demonstrated the highest level of achievement in the fighter pilot world. The impact on my testosterone-driven brain was immediate. I had to have "The Patch."

Figure 1 – "The Patch" 100 Missions over North Vietnam in the F-105. Photo by author.

The T-41 program was our first challenge, and we soon found out that it had little to do with flying. Each day was made up of an A.M. and P.M. period, and one section would fly half a day while the other section would have academics. Academics were, for the most part, made up of basic airmanship lessons: navigation, weather, flight planning, etc. Two aspects were introduced that varied from driving a car. Distance was measured in nautical miles, and speed was measured in knots (nautical miles/hour). One knot was about 1.15 miles/hour. Hence, 100 kts (knots) was about 115 mph. Also, fuel was calculated in pounds rather than gallons, so fuel consumption was calculated in pounds/hour.

A few days were devoted to learning Morse code, and I finally found something in the regimen I could excel at. All those years of being a ham operator had given me extensive Morse code skills, and I was pretty proud of myself. However, no one else seemed to care that I was good at something. They all hated the class.

Flying half-day meant loading up on a simple non–air-conditioned bus and making the 25-minute drive to Lubbock Municipal Airport where the Air Force kept a fleet of a couple dozen T-41 aircraft. They also had an operations building that looked more like a machine storage barn. There was a corps of civilian instructors headed up by two Air Force officers. They ran the program and gave most of the check rides, the dreaded Progress Checks, and the even more evil Elimination Checks. These were the rides that were completed by students who did not demonstrate satisfactory progress. A failed progress check followed the next day by a failed elimination ride meant that your hopes of becoming a fighter pilot could be dashed in a matter of two days.

Because of my experience in the Tri-Pacer at Madison, and my rather high opinion of my own flying skills, I assumed this program would be a piece of cake. Not so fast, cowboy! I soon found that flying skills mattered, but this program was all about using the "checklist." The checklist was a little bound folder of normal and emergency procedures, as well as a section concerning the operating limitations of the T-41. It soon became clear that you couldn't take a crap without referring to the checklist. Memorizing the checklist didn't count; you had to actually get it out and read every step and then complete the prescribed action. What a pain in the ass. There was one exception to this process. If you had a critical emergency, you then had to perform the boldfaced procedures (which you had memorized) and then, you guessed it, "refer to checklist."

Every day at Lubbock Muni began the same way—with the most hated ceremony of anyone who ever attended Air Force UPT, the "stand-up." This consisted of a senior officer (aka the Asshole) quizzing about a procedure, almost all in the checklist, that a pilot might need to perform while flying. It was in these sessions that we first discovered the three primary teaching tools used by the Air Force in UPT: fear, sarcasm, and ridicule. The stand-up never seemed to be a win-win situation; it was either a lose-lose or break-even deal. There was no reward for being right when you were called on, but the humiliation of blowing a response was earth-shattering. You always tried to make yourself look small or invisible when the officer conducting the stand-up was looking for someone to call on, and the most dreaded pronouncement you could hear was, "Lt. Sykes," followed by a situational problem and the expectation of an immediate response. If a boldfaced emergency was involved, you had better have it on the tip of your tongue followed by a "refer to checklist," and then you would get out the checklist and read the remaining steps. Screw it up, and the sarcasm and ridicule part kicked in.

One advantage the married guys had was their wives stood in as simulated assholes. Mary became very good at going through all the most probable items the asshole might ask the next day and then making sure my response was correct. The best part about having your wife perform this chore was the lack of sarcasm and ridicule—for the most part. From time to time, Mary would become frustrated with me when I blew a response, but normally it didn't cause me too much pain. Being able to cozy up in bed after her quizzing was much better than having to put up with the harassment that took place after blowing a response during a stand-up.

My T-41 instructor was a guy named Mike who, like most of the instructors, didn't seem to like his job much. About the only time I remember him being happy was when I gave him the compulsory bottle of whiskey after he soloed me out. I was lucky enough to complete the T-41 program before many of my classmates, and I still remember the delight of taking that bus ride from Lubbock Muni back to Reese knowing I would never have to make the trip in the opposite direction. A few of my classmates "disappeared" during the T-41 phase through the elimination process. They rapidly became lost names and faces as the survivors had way too much to deal with.

Near the end of the T-41 program, we completed parachute training. This mostly consisted of learning how to make a parachute landing fall (PLF) and then practicing our PLF while jumping off a 6-foot-high platform and attempting to break the fall by collapsing our legs while rolling to let the thighs and butt absorb the blow. Not much fun.

The parachute training final exam was a parasailing flight behind a Jeep out on the prairie. In a harness with a large parachute canopy, you were hooked by a rope from the harness on your chest to a Jeep that towed you as you ran like hell and were eventually picked up over the ground and elevated to a height of about 75 feet. You then released the rope from your harness and descended at a fairly fast rate to the hard dirt of the Texas prairie. We were told to check our landing area for the prickly pear cacti that dotted the landscape and steer the chute to roll away from them. This process was not fun, but in many ways, it beat sitting in a classroom for 4 hours.

In those last weeks of T-41 training, we were given our first exposure to the altitude chamber. Before this training, we were issued our flight helmets and masks. I proudly took it home and put the helmet on in front of a mirror. I then put on the mask and lowered the visor. I sort of looked like a space alien. Mary also tried it on; I found her pretty cute in the helmet. She didn't hook up the mask. Getting this new equipment made me feel even closer to achieving my goal.

Everyone was a little apprehensive about the altitude chamber after a day and a half of classroom training where we were told about all the bad things that could happen to you if you didn't use your equipment properly. During the flight, you were taken to a simulated high altitude and told to drop your mask in order to recognize the symptoms of oxygen starvation, called hypoxia. Within seconds of dropping my mask, I felt an immediate lightheadedness and tingling sensation in my fingers. It was a sensation I would never forget, and over the course of a long flying career, there were several occasions where this training proved useful. Breathing through an oxygen mask and being exposed to pressure breathing at high altitude was a little uncomfortable for all of us, but we did a pretty good job of "manning up."

I had a couple of days to relax after the T-41s. The heat was off me, and I enjoyed a few days not dealing with the Asshole. However, we were now beginning T-37 academics, and the systems in our first jet were much more complex than anything we had taken on before. There were jet engine components and hydraulic systems and many other new things to learn about. The day we headed to the T-37 flight line was pretty exciting. We were assigned to H ("Hangover") Flight. My new instructor was a young first lieutenant named Jim Lawson, a short-haired Air Force Academy graduate with a subdued manner. There were three other students at our table, and we listened anxiously as he introduced himself. We were his second class and he was still learning the ropes, he explained, but he would give us his best effort to get us through the program. Great!

We soon discovered that we had a new Asshole, and the morning stand-up was still alive and well, except now there was a lot more material to cover. There were many more boldfaced procedures with some long steps, and of course, they must all be committed to memory. More work for Mary.

Lt. Lawson took us to the simulator for several rides where we performed emergency procedures and flew with instruments. The simulator is a good way to prepare, I suppose, but I wanted the real thing. It came soon enough.

The Dollar Ride was our first chance to get used to our new office. It was an environment that really took some getting used to. As Lt. Lawson and I entered Life Support, the area where our flying gear was stored and maintained, my blood pressure must have been pretty elevated and my senses were racing. This was the real deal! I anxiously put on my parachute and grabbed my helmet and put it in my helmet bag; as we walked out the back door to the flight line, I put on my earmuffs.

Getting on the crew van and heading down the flight line toward the T-37 fleet, I could hear the rising crescendo of the shrill scream of the T-37 "Tweety Birds" that were operating in the ramp area. Even with your earmuffs firmly in place, the background noise of the subsonic dog whistle was constant. Arriving at our jet, I placed my chute in the cockpit's left side and got out my checklist to begin going through all the preflight checks. Lt. Lawson had given us a practice walk-around a few days earlier, so I was familiar with what to look for. The walk-around went smoothly.

I climbed into the cockpit, and Lt. Lawson had me read the checklist and accomplish the prestart checks before he turned on the battery and cranked up the left engine. The noise in my left ear was deafening as the engine came up to idle. The chamber flight had prepared me for wearing the helmet and mask, but this strange new environment along with the noise was discomforting. Lt. Lawson cranked up the right engine and called for taxi instructions, then we headed out toward the runway. It wasn't until he had me put down the canopy for takeoff, and the extreme noise subsided, that I began to feel a little more comfortable.

Once cleared for takeoff, Lt. Lawson demonstrated a normal takeoff. The greatest sensation I had was feeling how low to the ground the Tweet was. I felt like my ass was barely an inch off the runway. The little jet accelerated down the runway at a faster rate than anything I had flown before, but it was not overpowering. As Lt. Lawson climbed out, the climb rate was quite a bit faster than the T-41, but it wasn't exactly awe-inspiring. Lt. Lawson took me out to the practice areas that were southwest of Reese, and as he approached a big barn southeast of the AFB, he called "Poppet" (the control agency) and was assigned an area. He then showed me how to get to the assigned area and proceeded to demonstrate all types of air work: stalls, slow flight, and several aerobatic maneuvers. He then let me attempt a few. The new sensations were piling up. The g-forces were pretty strong, and I found myself tightening my legs and abdomen in an attempt to keep the blood in my brain. The mask and oxygen hose were a continual disturbance, but they were bearable. What a strange new environment.

Returning to the field, we came to a point where all Tweets arrived when returning from the assigned area, and as you approached, you called your position and put your head on a swivel as you knew many other Tweets were returning to the same point. Lt. Lawson demonstrated a few touch-and-go landings from normal patterns and let me try a few.

"Watch your airspeed, watch your altitude, trim the aircraft," were all heard, but when you looked inside the cockpit to see what your altitude and airspeed were, he told you to get your head outside and clear the area, watching for other jets and monitoring your ground track. This was not going to be that easy!

Looking back on the whole ordeal, I realized that the glamour of flying jets was not what I had expected. I was sweaty and physically drained. When I got out of the plane, Lt. Lawson came around to my side of the jet and asked, "Well, what do you think?"

"That was really great—better than sex," I lied. Lt. Lawson grinned and simply nodded his head. Looking back, I now realize that the Dollar Ride was as much for Lawson's benefit as it was for mine. Later, in another life, I would become a T-37 instructor and would use the Dollar Ride to do an initial assessment of my new student.

As I talked with my classmates after the Dollar Ride, I found out several of them had become ill on the flight, and the stories about how they managed to get out their barf bag and use it were interesting. A few of these classmates would not become used to the environment; they would eventually declare "uncle" and be allowed to drop out of the program for "manifestation of apprehension (MOA)," a term sort of synonymous with "fear of flying" but not really the same. It is a complex situation in that not only is the motion a problem for some queasy stomachs, but the small cockpit coupled with the helmet and the continuous use of breathing through a tight-fitting mask is hard for some folks to cope with. MOA was one form of self-initiated elimination (SIE), which was a way for someone to say they had chosen the wrong career path and had had enough. Normally SIEs occurred early in the program. There were, however, a small number of men, whom I much admired, who continued to get sick right up until they soloed out and still dutifully got out their barf bag, coughed up their cookies, secured their barf bag, and took the aircraft back from the instructor and continued on with the flight. If this problem continued up until solo, it was MOA/SIE time.

Lt. Lawson was a calm and reserved instructor, and I progressed with average grades through the early instrument rides and then the contact rides. However, I still dreamed of being by myself in the jet. The day came to solo out. I took off and made four acceptable touch-and-goes as well as a full-stop landing, and the words I had been waiting to hear were spoken: "Shut down the right engine, Sykes. Do good, buddy." I shut down the right engine and waited for Lt. Lawson to depart. I watched him walk swiftly to the van where he would be delivered to mobile control to observe my landings and provide words of wisdom, if needed. I got taxi instructions and stopped at the runway, waiting for Lt. Lawson to arrive in the mobile shack. I then received those magic words from mobile control: "Hangover 41, cleared for takeoff."

This was it. I was about to solo a jet paid for by the taxpayers and assigned to the United States Air Force. I ran up the engines and they looked good. I took off in search of my three touch-and-goes and a full-stop landing that would mean I had successfully "soloed out." I turned out of traffic, set up to re-enter the traffic pattern, and had about 4 minutes to drive straight ahead before entering initial. One of the clearest memories of my entire flying career is that of looking at the right seat and seeing it empty. I then looked out at the wings of my little "Hummer" and on one side could see the inscription, "U.S. Air Force," and on the other side the insignia of that same Air Force. Wow, was I excited. *Now I need to get this bitch on the ground,* I thought. I completed some pretty imperfect patterns and landings, but I got it on the ground without any harassment calls from Lt. Lawson. I parked the aircraft; the feeling of getting out of it all by myself was the best.

Getting soloed out is a reason to celebrate, and after getting back to Operations, I was met by many of my classmates who grabbed me and threw me into a large cattle tank filled with water, just as they had done after the T-41. This time, however, Mary was there to watch the festivities and take some photos. She also brought a dry flight suit, and after changing, I made my way back to Hangover and Lt. Lawson. "Nice job, cowboy," he stated, and after giving me a short debriefing, he turned me loose.

One of the other guys in our carpool had also soloed out that day, so we talked Russ into going by the stag bar after work. I bought them all a

Coors, and we talked excitedly about the day's events, but my mind drifted. As I looked around the room, I saw a small group of instructors drinking in the corner, and at least two of them were wearing The Patch. Once again, I knew I had to have The Patch and set my sights on getting there.

Mary and I celebrated with a bottle of Blue Nun that night, and I mentioned I was putting the Thud F-105 high on my wish list. I wanted to gauge her response, which was only lukewarm. As we watched the news every night, it became apparent the Vietnam War was escalating, and the F-105s were right in the middle of it. Every so often, it would be reported that another F-105 had been lost and the status of the pilot was typically unknown. I could sense her uneasiness concerning the Thud, but she didn't say no. End of discussion!

By now, we had been at Reese for nearly three months, and we realized, as two classes graduated, that almost everyone coming out of Reese would end up in Vietnam. If this marriage was going to have any legacy to it, we decided we should begin building a family. *No problem,* I thought. *This doesn't affect me much. I already know what I have to do.*

The T-37 program seemed to fly by. The most exciting parts were the first spin ride, the check rides, and the introduction to formation flying. We had all heard the stories about the T-37 and how violent the spin was, so everyone was a little apprehensive about their first spin ride. It was a wild ride and the recovery was pretty violent and abrupt, but after being demonstrated by Lt. Lawson and then trying it for myself, I thought it was pretty cool. It was certainly the most exciting thing I had done in an airplane up to that moment.

The check rides were pretty exciting too. My first check was an instrument check, which I nailed. I had been told the best way to prepare for any check ride was to "chair fly" the ride over and over while sitting in a chair by yourself and going over the procedures and techniques for each maneuver. It was a pretty boring process, but after a successful first check ride, I developed the discipline to repeat the procedure as each check approached. My T-37 checks all went well, and by the end of the program, I found myself in a high position in my class standings.

Assignments were based on class standing, with the number one student getting first choice and the remaining students getting what was left in numerical order based on their standing. Because 90 percent of our class standing was based on the score from the check rides, I hoped that I might someday be able to finish high enough to get an F-105 and have the opportunity to earn The Patch.

The other exciting part of the T-37 program was the introduction to formation flying. What a hoot. At first it was difficult, but once you got the hang of it, it became a fun challenge. As I finished the Tweet program, I thanked Lt. Lawson for his great instruction and promised I would be back to visit him when I got my assignment. When he asked me what I wanted and I answered the Thud, he gave me a broad grin and said something about me being a dumbass. For now, I was on my way up the street to the T-38 Talon.

As my carpool parked at the T-38 Squadron that first day, we were all excited about getting our hands on the beautiful supersonic jets parked in front of the building. Since our arrival 6 months before, we had driven by those jets on the flight line every day, looking forward to finally getting our hands on one.

Inside, we all grabbed a chair and were introduced to our flight commander and the instructors. I kept looking for a Patch in the group, but there were none to be seen. After explaining the ground rules—more of the same fear, sarcasm, and ridicule—we were assigned to our instructors. "Lt. Sykes, you'll be flying with Lt. Griffin."

Well, crap, I thought. As I went to Lt. Griffin's table, I heard the flight commander make the standard statement, "If you feel like you would like an instructor change, let me know, and we'll move you to another IP." I sat down at Lt. Griffith's table and found out he was a second lieutenant, and we soon learned we were his first class of students. As I listened to him give his story and surveyed the other students at the table, I realized what was happening. For his first class of students, they gave him pretty good performers, but it was not what I wanted.

After a good amount of preparatory stuff, the flight took a break. After a couple minutes of hesitation—*Are you sure you want to do this, dummy?*—I proceeded to the flight commander's office and knocked on his open door.

"What is it, Lieutenant?" he asked.

"Maj. Thompson, sir, I would like an instructor change." I was not surprised by the look I got.

"You just got here, Lt. Sykes. Why do you want an instructor change?"

"Sir, I had a brand-new lieutenant in the T-37 program and didn't get a chance to fly with any experienced IPs, and I especially would like to have a fighter pilot as an instructor."

Maj. Thompson took a few seconds to think about this unusual request. He studied a sheet of paper on his desk. He looked up and studied me before standing up and saying, "Come with me." He led me out of his office and took me to a table in the corner of the room where a senior-looking captain was seated with only two students. "Capt. Corrick, you have a new student at your table," he stated. After I introduced myself to Capt. Corrick, I watched Maj. Thompson approach Lt. Griffith's table, I assume to explain what had happened. Lt. Griffith glanced at me with a quizzical look and then went back to his briefings.

What had I done? I hoped this wouldn't cause some problems for me down the road.

Capt. Corrick proved to be just what I wanted. As it turns out, my abrupt request for an instructor change was a good move. Blaine Corrick was an experienced guy who was easy to fly with. He never raised his voice, and he always had some good tips to help me make my performance look better. After all, I had discovered that more than anything else, you were trying to make your performance shine to get ready for "the show"—the check ride—and I knew those check rides held the key to my chance at getting into a Thud.

If I needed some reassurance, the Dollar Ride provided it. Strapping into the Talon got me fired up. When I lit the afterburners on takeoff, I was amazed at the high rate of acceleration as the Talon quickly achieved takeoff speed and jumped into the air. The rate of climb as we exited the pattern was so impressive, I laughed out loud, and Capt. Corrick laughed with me. I was ready to rock and roll.

Part of the Dollar Ride was an acceleration through the Mach, which was somewhat a nonevent. We climbed to the top of the designated area, and I lit the burners. The jet responded immediately and rapidly accelerated through Mach 1. The only sensation was a little tug on the stick as the nose tried to dip slightly. The wind noise was fairly loud, but there was no magic transformation, and as I came out of afterburner and my little jet quickly slowed below the Mach, I was now a supersonic jet pilot. Yippee!

The T-38, which we called the White Rocket, was a fun little hot rod, and with Capt. Corrick's help, I soloed out without much difficulty. However, I didn't feel the same thrill of accomplishment that I had after soloing out in the Tweet. Could I be experiencing some maturity? Once again, I was thrown into the dunk tank, but this time, it was simply part of the path I was on. After work, the carpool stopped at the stag bar, and as I watched the small group of Patch wearers over in the corner, telling their war stories, I began to get excited about my prospects of eventually wearing The Patch.

Catastrophe! On Oct 23, I got a call from my mother, who was in obvious distress. She was crying and at first said nothing, and then she blurted out, "Bob's dead." I couldn't believe what I had heard. "What?"

Through a continuing torrent of sobs and sniffles, she was able to explain that he had convinced the Army that he wanted to go to Vietnam and serve with his buddies, and they had reluctantly let him go. He had been in Vietnam only 13 days when a sniper's bullet found its mark. My mind flashed back to that day Bob and I had driven to Fort Campbell. He had told me he felt bad because he was not there with his buddies, but I never suspected he would volunteer to go. I obviously did not expect this result. I knew he and Merry were expecting a second child, and I now found out that his son Bobby had been born only a few days before he shipped out.

"Eddie, I want you to fly out to Los Angeles to be with your sister. I can't go right now, and I need you to be there and help her with her kids and all the arrangements."

Oh my God. There is no way I can take leave and walk away from this program. We had been told from the beginning that there would be no leaves granted while we were at Reese, other than the one-week break between Christmas and New Year's. My mother's distress was evident, my dilemma clear.

"I'll get out there as fast as I can, Mother. I should be able to leave tomorrow."

I was not on the schedule at that time, so I went to the Student Squadron where I explained my dilemma to Maj. Freeze. "You know you can't leave, Lt. Sykes. There are no leaves granted during this program."

"I'm leaving, sir. I promised my mother I would go to be with my sister, and I'm going." As I made this declaration, I realized this could be the end of my dream. Going AWOL was not a good way to enhance your chances of becoming a fighter pilot.

"Maj. Freeze, I'm all caught up with my flying and I can make up the academics. I just need a few days to be with my sister and help her through this situation."

Maj. Freeze studied me carefully. I think he realized I was going to go regardless of the consequences, so he finally said, "I can only let you go for three days. Make sure you're back by then."

"Yes, sir. Thank you, sir."

I went back to the flight line and explained to Capt. Corrick and Maj. Thompson what had happened. They did not seem very upset, and Capt. Corrick made me feel like it was not a big deal and that he would make sure I got caught up after my return.

I called Mary and explained what had happened, and she, too, was shocked by the news. We had had no idea that Bob was in Vietnam. I told her I was going to fly to Los Angeles the next day and get to Merry's as fast as I could.

"Do we have enough money for the ticket?" Mary asked.

"I don't know."

Somehow we got the arrangements made, and I called Merry, who sounded distressed. I got directions to where she was living and told her I would be there the next day. The next morning, I was on my way to Los Angeles and my sister.

When I arrived in L.A., I caught a cab to her house. As she opened the door, she calmly said, "Hi, Eddie. Thanks for coming." I realized my sister had a lot of my mother's attributes. I figured she either was in shock or was one tough lady because she was eerily calm and already managing the details that needed to be taken care of. I could sense she was glad to have me there, but her strong internal toughness was going to let her work her way through this situation. I did what I could to help. It was my first experience with changing diapers, but I became pretty good at it.

The Army came by to help her get her goods ready to ship and to fill out some paperwork, but through the whole process, she was as cool as a cucumber. I was impressed. I could only hope that Mary would be as tough if faced with the same situation. I'm not sure I was needed, but I think my presence contributed to her ability to maintain her composure through that time.

I left Los Angeles knowing I had fulfilled my mother's wishes and help was on the way. My mother was now on her way to Los Angeles, and Merry continued to be very Mother-like in the face of this tragedy. Flying back to Lubbock, the reality of my situation began to set in. I was sure the Air Force was not happy about my sudden departure from the flying program. I had all kinds of visions of being eliminated from the program or being washed back a class because of this hiccup. My concerns were not necessary. There was no punishment for my insistence on going to assist my sister, and the instructors in my flight did their best to get me caught up rapidly. A week after my return, I was back to normal progression, and there were never any repercussions for my outburst.

In the middle of the T-38 program, we had the opportunity to update our "Dream Sheet," the Air Force Form 90 that allowed us to list in numerical order our choices for our next assignment. I had been weighing my desires carefully since we had decided to become pregnant, and I wavered a bit in my thinking and put "T-38 Instructor to Reese" as my first choice. All this wobbling stemmed from the prospect that Mary could become pregnant soon, and it would be nice to have the baby at Reese. It also would allow Mary to continue her studies at Texas Tech and become a Red Raider graduate. I then put the Thud as my second choice followed by the other fighters except the F-4. The F-4 pilot slots at the time were only available in the back seat, and I wanted nothing to do with that. I then selected the C-130 as it was as close as you could get to a fighter while still hauling "trash" (cargo). I was fairly confident that I was high enough in my class to avoid anything below the 130, so I didn't study the rest of the choices much.

About that time, a big bomb hit the class. When we had entered training, we were told that our assignments would be based on 90 percent of flying grades and 10 percent academics. At an early morning meeting, it was announced that our class standings would now be based on 50 percent flying and 50 percent academics. I knew at once that this was a stroke of bad luck for me. I had always passed my academic tests but had not focused on them because I knew where my strength lay—flying. I would drop several places in my class standings because of this. I now fell from the top of the class to the high end of the also-rans. I was not happy, but I couldn't change it.

An early winter hit Lubbock, causing us to fall further behind the training timeline. Working on Saturdays became common. As we approached December, Wing Headquarters warned that if we didn't get in better shape on the timeline, our Christmas leave might be canceled. We were supposed to get over a week off between Christmas and New Year's, which everyone was looking forward to—especially the wives. The bad weather continued, and about the middle of December, the word came that our Christmas leaves were canceled. We would be spending the holidays in Lubbock, Texas—ho ho ho!

"All leaves will be canceled until morale improves" became the most popular phrase expressed by me and my classmates. Then, to make matters worse, as the period of our canceled leave began, Lubbock had an unbelievably severe snowstorm, and thus we only flew a small number of sorties. Morale was not improving, and I got a little drunk on New Year's Eve.

From that point until we were scheduled to graduate in early April, we flew almost every day, including weekends. It was trip turns (three flying periods) almost every day. I continued to do well on my flying, but with no more opportunities to improve my academic scores, I was getting a little apprehensive about my chances of getting my "dream sheet" wishes granted.

One highlight of the Christmas weather shutdown was a viewing of a new Air Force documentary, *There Is a Way*, the story of the F-105s flying out of Korat Air Base in Thailand in 1966. It contained some great interviews with many of the pilots and lots of great combat depictions. It included an interview of a young first lieutenant, Karl Richter, who had completed his 100 missions and insisted on being allowed to fly a second 100. *What a stud,* I thought to myself. I also remember it featured a dog named Roscoe who served as mascot of the entire Thud driver fraternity. After the film, it was revealed to us that following the shooting of the film, Karl Richter had shot down a MiG and had been awarded the Air Force Cross (one level below the Medal of Honor). It was also revealed that he had been killed on his 198th mission. That gave me some pause, but it didn't change my mind about earning The Patch.

Right after Christmas, I got a call from my sister Sue. Wes had just gotten his flight school orders, and by coincidence, they were also assigned to Reese AFB. When they showed up in Lubbock, we urged them to join us at the same apartment complex. They did, and just like that, Susie and I were living under the same roof again. Because Susie was pregnant with their first child, they elected to get a two-bedroom suite.

When we showed up for work one day, we were told without warning that our assignments were in and would be announced that day. Maj. Thompson had us gather in the flight room, and he read off the assignments in rank and alphabetical order. Being near the end of the list, I had to sit through the procedure and watch others accept their fates. There were a large number of F-4s being called out. My good friend

Randy Carlson was the first to get an F-105 to SEA (Southeast Asia). As my place on the list approached, I became very nervous. *Help me, Lord!*

"Lt. Sykes, F-105 to SEA."

No shit? I thought as I felt a sense of relief and nervousness simultaneously go through my body. *Could it be that I was really going to get to wear The Patch?*

Following the reading of the assignments, we all walked around the room shaking hands and back slapping one another, while each person remained in their own little world, contemplating what their lives would look like. Almost everyone was destined for a trip to SEA to get up close and personal with the War. I considered waiting until I got home to tell Mary, but I couldn't hold it in that long, so I called her at the apartment where she was busy studying.

"Got our assignments."

"Really? What did you get?"

"Guess."

She began by guessing a T-38, and I let her know she was incorrect. She then guessed the F-4, followed by every other aircraft she could think of except the F-105. When there was nothing left, she finally said, "F-105."

"Yes!" I exclaimed excitedly.

"Oh," she said in a tone that meant she was not exactly happy with the news. She asked me questions that I could not answer, and I told her I would be home in a bit as soon as we were released.

The carpool ride home was interesting. Garth had an F-4 back seat, Russ had a C-141, and Jim had a C-47. Garth was not happy about his F-4 but resigned himself to the possibility of getting a chance to fly in the front seat after a tour in SEA. Jim was just fine, thank you, with his C-47 because he had had enough problems getting through the course, so he was happy to be graduating. Russ was excited about his 141 as that was what he had wanted all along, and obviously, I was excited about my Thud, but I could sense the congratulations of my carpool buddies were mixed with some nervousness concerning my fate.

When I got back to the apartment, Mary met me with a somewhat subdued congratulations. I accepted that as her being concerned but aware that I got what I really wanted. We wandered over to Sue and Wes's apartment and told them of my "good fortune." Sue was rather quiet in her congratulations, but Wes was upbeat. "Way to go, you lucky bastard!"

We went out for dinner that night, and I had a few more beers than I needed. As I lay in bed that night, reality began to settle in. I had gotten myself into a position I really wanted to be in but hadn't been sure I could achieve, so the reality of it was a little hard to imagine. I realized that I had never even seen an F-105, and my only real contact with the Thud was my desire to wear The Patch. *OK, dumbass, now what are you going to do?*

The rest of the T-38 program flew by, and I got a little FIGMO (fuck it, I got my orders). And within a few weeks, I got my orders assigning me to the 23rd Tactical Fighter Wing at McConnell AFB in Wichita, Kansas. En route, I was ordered to Fairchild AFB in Washington for Survival School and to Homestead AFB just south of Miami for Water Survival. My report date to McConnell was June 24. Life was beginning to move rapidly.

I do remember well one of my last T-38 rides with Lt. Griffith, the instructor I had jilted before I even began the program. It was an instrument practice ride, and I was in the back seat practicing instrument approaches, and I wasn't doing very well (FIGMO). As we were completing the ride and returning to base, he made a comment that brought me back to reality: "Keep flying like that, Sykes, and you'll be a Thud smoking hole in the ground." Ouch!

Then, a week before my scheduled graduation from flight school, the evening news threw water on my excitement. On March 31, 1968, President Johnson made an announcement that caused me to doubt that I would ever wear The Patch. The next morning's *Lubbock Avalanche-Journal* summed it up in an article titled "President Orders Halt to Bombing":

> *President Johnson Sunday night ordered an immediate halt to the bombing of 90 percent of the territory of North Vietnam and challenged Ho Chi Minh "to respond positively and favorably to this new step toward peace." The President did not put any time limit on his partial bombing halt.*

The president's order included all air and naval bombardments of North Vietnam "except in the area north of the demilitarized zone where the continuing enemy buildup directly threatens Allied forward positions." President Johnson also stated that he was sending 13,500 more troops to Vietnam and would request further defense expenditures. In closing, Johnson shocked the nation with an announcement that all but conceded that his own presidency had become another wartime casualty: "I shall not seek, and I will not accept, the nomination of my party for another term as your president."

"What now?" we all wondered. There was nothing to do but carry on, but a number of what-ifs emerged in the back of our minds. What if the war will soon be over, or what if they don't need the large fighter force any longer?

We graduated on April 7, 1968, and Mary pinned on my wings. My journey wasn't over, but I had passed the first test. Hallelujah! As part of the graduation ceremony, a number of aircraft had been flown into Reese so parents and friends could see the aircraft their folks had been assigned. There was an F-105 on the ramp, and Mary and I, along with Sue and Wes, walked out to see it. I was immediately struck by its beauty as we approached it, but as I got closer, what really struck me was how big it was. It was huge, and what also grabbed my attention was the fact that it only had one cockpit. Of course, I had known that already, but the visible confirmation that I would fly this big machine by myself was exhilarating.

Following graduation, I had a few days to clear the base and prepare for my trip to Fairchild. Mary was still finishing up the spring semester at Texas Tech and would need to complete a six-week summer school session to get her degree, so we decided she would live with Sue and Wes in their second bedroom while I attended Survival School. Sue was getting close to having her baby, so we thought it might also be a good time for Mary to do a little prep for motherhood. Mary was grateful for the opportunity to stay with them except for having to put up with their dogs.

Survival School was one of the most miserable experiences of my young life. It started off easily enough with some good classroom instruction, especially the part about using survival gear and radios. However, the "trek" was another matter. Several days were spent camping and hiking through the Idaho countryside, and it was cold, wet, and exhausting. Then, for dessert, they captured you and put you through a few days of POW Camp training. The thing that it taught me, more than anything else, was that I did not want to be captured. By the time I had completed the trek, I had lost 15 pounds and my stomach had shrunk so much that the steak dinner they served up at the mess hall was not only hard to eat, but also hard to keep down. Almost everyone had the same experience. I was so glad to get out of Fairchild and back to Lubbock. I spent a week recovering in Lubbock as Mary finished up her degree, and as soon as she had taken her last test, we jumped in Cognito and headed for Florida and Water Survival.

Water Survival was a really fun experience with one exception: when I had to spend the better part of a day in a one-man raft with no one else in sight. That was a little taxing, but I didn't get seasick, and I didn't spot any sharks, so aside from a pretty good sunburn, it went well. At Water Survival, I started to meet some of my F-105 classmates, and we quickly cohered. We were of the same mindset. The excitement was building. Randy Carlson threw a hand grenade on the party by reminding us that we weren't there yet. In a couple of recent classes, some guys with fighter assignments had had their assignment changed to some pretty ugly landings, which put a damper on our party to a certain extent. One thing was clear, however—we all wanted The Patch and had been motivated in similar ways.

As Mary and I left Florida and headed for Kansas, my mind drifted back and forth between my excitement and my mother. I had not seen her since six months before Bob's death, and when I talked to her on the phone, she seemed distraught and unhappy. I'm not sure that my showing up in Kentucky would have helped much—in fact, it probably

The Patch

would have made the situation worse, so we decided to bypass Kentucky. She now knew that her son and son-in-law had a likely visit to Vietnam in their near future, and she did not want another bite of that apple.

Chapter 4

And the Horse We Rode in On

You love a lot of things if you live around them, but there isn't any woman and there isn't any horse, nor any before nor any after, that is as lovely as a great airplane, and men who love them are faithful to them even though they leave them for others. A man has only one virginity to lose in fighters, and if it is a lovely plane he loses it to, there his heart will always be.

— Ernest Hemingway

Goose bumps! Every time I read this, I feel that Hemingway wrote it just for me. This quote was part of an article he wrote in *Collier's* magazine about the defenders of London in the Battle of Britain. The F-105 Thunderchief was that fighter for me, and Hemingway nailed it! Few men on this planet ever get a chance to strap a beautiful fighter to their back and give it life. I lost my "virginity," as described by Hemingway, in the F-105 in 1968 and would go on to fly other fighters, but somehow, the "Thud," as it was called by its pilots and maintainers, will always hold my heart. I spent just over 2,700 hours in its spacious cockpit, and only 13 men who ever walked the face of the earth can claim they logged more. I also logged over 1,000 hours in the F-4 Phantom and another 1,000 in the F-16 Falcon, but I will always consider myself a Thud driver.

This chapter is not a "necessary read" to get the full flavor of my story, but I felt compelled to write it because of my strong affection and respect for this great machine. Much has been written about the Thud, available from many sources. I will talk about this aircraft from my perspective as a Thud driver and perhaps share some details that you won't find in other sources.

Figure 2 - The F-105 Thunderchief ("Thud"). A four-ship of Thuds led by Ed Sykes in the mid-1970s. Photo courtesy of the 184th Tactical Fighter Group, Kansas Air National Guard.

I'm not sure where the Thunderchief label came from, but I suppose it was a follow-on from the earlier Republic aircraft manufactured, the F-84 Thunderstreak. I've also heard stories about why it was called the Thud; from its earliest onset, it had a propensity to blow up in flight during its initial years. However, I always felt it was a reference to *The Howdy Doody Show* that I watched every Saturday morning on our old black-and-white Satchel Carlson TV in the '50s. One of the main characters was Chief Thunderthud (a not-so-native American with a mustache). Political correctness would never allow such a character in today's screwed-up world, but I will always remember the Chief's favorite expression of dismay when things were not going his way: "Cowabunga." It was a made-up "Indian" word, and from time to time when I would screw things up, I would think or utter, "Cowabunga!"

The first issue we need to put behind us when discussing fighters is the concept of what is the best crew setup and how many engines does the craft need. For me, the single-seat, single-engine concept is the way to go. Man has but one heart and one brain. The single-engine route makes a great deal of sense as long as it's a really good engine. The single-seat plane is always better in respect to the ease of decision-making but requires a competent person in the cockpit.

Single engine means that the pilot doesn't have to deal with the complexity of making two or more engines run simultaneously. Also, the additional weight added by a second engine, and the redundancies that come with it, is significant.

The single-seat issue is easier for me to resolve. Having to deal with a fellow human being in a rapidly progressing fight makes decisions much more difficult to resolve and mistakes more likely. No matter how hard a crew works at language and tactics, there always seems to be a difference of opinion when split-second decisions are being made. An argument with yourself is much easier to settle.

The best feature afforded by a second person is the additional set of eyeballs, but technology is rapidly overcoming that advantage. Technology is also overcoming the need for a second member to solve detection of targets and deployment of weapons systems. One of the happiest days of my life was when I flew my first flight in the F-16 and left behind my 7-year association with the F-4. The F-4 with its two engines (pretty good ones) and two crewmembers never seemed to fit me right. In the F-16, I once again had my one heart and brain in sync with a machine with the same characteristics.

A second nice feature of the Thud is the obvious decision to be made when an engine quits and can't be restarted: eject and give the jet back to the taxpayers.

Trying to describe what it was like to operate such a magnificent machine is not easy. Likewise, it's difficult to describe how dangerous a machine it could be if you didn't pay close attention to its limitations. The margins for error were small.

It was designed to accomplish one role but ended up being well suited to another, completely different role. Following WWII, the newly formed Air Force asked Republic Aviation to develop a high-speed, low-level penetrator that would evade the defenses of the Soviet Union and deliver an atomic weapon to the heart of the Motherland. The engineers designed a product that was able to dash at well over supersonic speeds for extensive distances at very low levels, deliver a nuclear weapon, and, on a good day, bring the pilot home (that part was not a guarantee).

To meet these requirements, the engineers developed an airframe and engine to accomplish this high-speed mission. They also gave it some state-of-the-art features like a Doppler navigator, sophisticated ground-mapping radar, and an autopilot to assist the pilot in accomplishing all the associated complex tasks.

The Air Force purchased a total of 833 F-105s. The first 75 were the B model, which was less advanced than the later D and F models. The primary purchase was of the more advanced D model, which contained many state-of-the-art upgrades over the B. As the Air Force brought the D models online, the B models found a home with several Air National Guard and Air Force Reserve units, with a few staying on active duty in a training role. Finally, the Air Force bought 133 F models, which were a two-seat version primarily built as a trainer. Eventually, several were modified to the F-105G configuration whose primary use was as a "Wild Weasel," whose mission was as a weapon system for surface-to-air missile (SAM) suppression.

The earliest models of the F-105 had conventional gauges to display flight information (altitude, airspeed, Mach, etc.). With the advent of the F-105D, these round gauges (Thud drivers referred to them as "steam gauges") were replaced with a state-of-the-art vertical tape system. All flight information was now sensed in traditional ways but sent to a central air data computer (CADC). Now all the flight information was analyzed by the CADC and presented to the pilot by a system of vertical tapes, with the current flight information being displayed on a single horizontal line. With a few more "gee whiz" features, this system made the pilot's cross-check much easier and was the harbinger of modern aviation instrumentation.

Because the Thud was envisioned as a low-altitude penetrator that would operate well behind enemy lines, it became necessary to give the pilot a means to navigate without using traditional ground-based navigational aids. Hence, it was equipped with a semi-autonomous Doppler navigator. This system sent three continuous radar beams at the ground and then, based on the phase shift of the reflected signals, determined the jet's ground speed and drift and supplied this information to the pilot. The accuracy of this system varied from one aircraft to another and was also dependent on the reflectivity of the terrain. However, it got you in the ballpark and would normally get you to a selected destination within an acceptable range. The handiest information was the continuous readout of your ground speed and wind drift angle.

Next problem: What if the Soviets decided to go to war on a bad weather day or at night? To overcome this possibility, the nose cone of the Thud contained a large radar antenna that was primarily provided to map the terrain in front of the aircraft and identify targets that presented a good radar image. Large towns, factories, bridges, railroads, and the like could easily be painted. Also, large bodies of water and mountains were easily distinguished. An additional benefit to the pilot was the ability to "paint" embedded thunderstorms. Best yet was its ability to find and lock on to other aircraft. Looking for a friendly tanker or an enemy MiG was greatly enhanced by this feature.

But the engineers weren't done yet. Once you got to the target using the aforementioned features, you now had to effectively release a nuclear weapon and, hopefully, avoid being destroyed in the process. To facilitate this, a toss bomb computer (TBC) was installed that allowed the pilot to input some target information and the type of weapon he was using before takeoff and depend on the computer to put the bomb on target. The old military adage that "close only counts in horseshoes, hand grenades, and nukes" certainly applied in this situation.

Finally, the Thud came with an autopilot that was so advanced that many of today's modern fighters cannot rival it. The CADC, Doppler navigator, radar, and TBC all made inputs to the autopilot and could be used to perform a myriad of functions. It could maintain an airspeed, altitude, or Mach with ease, but it was not tied to the throttle, so some pilot management was required. It also incorporated an auto-ILS (instrument landing system) that assisted the pilot in bad weather situations. Many of today's modern fighters do not incorporate an auto-ILS.

Some of the most unique functions had to do with delivering nuclear weapons. The system allowed the pilot, while making a very high-speed, low-altitude penetration, to find a target on his radar and designate it by selecting "Autoss" on the autopilot panel. This allowed the airplane to take over the delivery. The high-speed dash would continue until the TBC determined it was time to pull up as it approached the target. The autopilot would then begin a hard wings-level pull-up and continue the loop until the aircraft was headed back toward the entry route and then roll the wings level. (Aviation buffs will recognize this as an Immelmann turn.) During this pull-up, the TBC would determine when it was time to open the bomb bay and eject the weapon (normally when headed straight up). The bomb would then be tossed toward the target and normally had a parachute to give the pilot and aircraft more time to escape the blast.

Stories are told about Thuds and B-52s sitting nuclear alert being stationed at the same bases in the Pacific in the early '60s. The B-52 crews proudly wore jackets with their crew position displayed boldly on the back, i.e, Pilot, Co-Pilot, Navigator, Bombardier, WSO (weapon systems officer), or Tail Gunner. In response, the Thud drivers decided to follow suit and came up with a similar jacket that included all the B-52 positions posted on one jacket ("Tail Gunner" was replaced by "Nose Gunner"). A certain level of arrogance is not uncommon among the fighter pilot fraternity.

These systems were really cool, but they required a high level of study and concentration on the part of the pilot. In today's world, where kids are adept at computer usage and video games, young people would quickly adapt to the use of these systems. However, in the early 1960s, computers were only something you read about in science magazines and sci-fi novels. It was a significant "learning leap" for the not-very-high-tech airmen of the era.

Although none of this "magic" would be used to deliver a nuclear weapon, the availability of many of these systems would greatly enhance the Thud's capabilities in Southeast Asia. The F-105 was never called upon to penetrate Soviet airspace and deliver a nuclear weapon. However, as the Air Force began to contemplate attacking targets in North Vietnam, it occurred to them that the Soviet-made defenses used in North Vietnam looked a lot like the weapon systems the Thud was designed to defeat. Hence, the F-105 would evolve from a high-speed penetrating bomber to a tactical fighter/bomber.

It was recognized pretty early on that nukes would not be an option in Vietnam, so the bomb bay that was originally the nest for a nuclear weapon was fitted with a large fuel tank—best move ever, as you could always use more gas. They also put pins in the bomb bay actuator to make sure some dumbass didn't inadvertently open it. For the pilot, however, the best part was gaining a little respect. Non-Thud fighter pilots had given Thud drivers crap for years: "Can't be a fighter; only bombers have bomb bays!" You could always counter by letting them know how much additional gas you had.

The immediate response everyone has when they first walk up to the Thud is, "That's one big sumbitch." It's the largest single-engine, single-seat jet fighter ever built. Its size was intimidating to a young buck like me, but once you strap it on and get it up to speed, it begins to feel like a smaller aircraft, and the faster you go, the better it feels. A lot of F-100 Super Sabre pilots transitioned into the Thud in the '60s, and I often heard the comparison, "It was like going from a sports car to a Cadillac."

It's a basic rule of physics: you can get anything airborne if you get it going fast enough. This certainly applies to the F-105. Its 65-foot length with only a 35-foot wingspan meant it was heavily wing loaded; that is, every square inch of wing had to carry a large amount of weight. Therefore, you had to generate a large amount of lift over the wing to generate enough force to get all its weight off the ground. And how do you do that? Go fast as hell on takeoff. Hence, nicknames like the "lead sled" and the "world's fastest tricycle" were often applied to the Thud. It was often said that if you built a runway around the world, Republic would build an airplane that needed all of it to get airborne.

If high speed is required, then the Pratt J-75 engine was just what was needed. It was a large turbojet engine with afterburner and the possibility of water injection. When the throttle was full forward thrust—at "military" (100% RPM)—it generated about 16,100 pounds of thrust. Afterburner kicked that up to 24,500, and water injection would give additional 2,000 pounds, bringing the total up to 26,500 pounds at sea level on a normal day.

To monitor engine performance, the pilot had several gauges to let him know how his engine was doing. The RPM gauge let him know that the engine speed was matched by his throttle. The exhaust gas temperature (EGT) gauge told him the amount of heat that was being sensed as the gases exited the engine. The exhaust pressure ratio (EPR) gauge let the pilot know the ratio between inlet and exhaust pressure of the flowing gases. A fuel flow gauge measured the amount of non-afterburner fuel that was flowing to the engine. (A rule of thumb was to multiply fuel flow by four to approximate fuel flow in afterburner.) The last gauge having to do with the engine, but the one that most got your attention, was the oil pressure gauge. Loss of oil pressure told you to get the beast on the ground in short order or you would soon find yourself in a nylon letdown (ejection) situation. The J-75 had a low tolerance for operating without oil, and 5 to 10 minutes was about the most you could expect.

The engine could be started using an auxiliary power unit (APU) to supply air pressure to initiate rotation of the turbine section of the engine. But in the early days of the F-105, and in all starts in SEA, the cartridge start was used, which was a large canister containing several pounds of black powder and a starter cap. The crew chief installs the cartridge in the starter breech before start, and when the pilot hits the Cartridge Start button, electrical power is applied to the cap, and the black powder begins to burn with exhaust gases that begin to turn the engine over. In the cockpit, the pilot can hear the cartridge fire and, if the wind is blowing in the right direction, can smell the manly odor of carbonite. It is a pure delight!

Starting the Thud was simple. No key needed. Simply turn on the battery, press the Cartridge Start button, press the Air Start button at 8% RPM, and move the throttle to Idle, then the rest was pretty much automatic.

You can hear the engine light as the gas ignites, and as it comes to life, there is an initial purr that grows into a delightful vibration. If the pilot listens closely, when he presses the Air Start button, he can hear the igniters crackling in his headset. Now all the pilot has to do is monitor the EGT as the temperature rises to make sure the engine doesn't over-temp and then start up the air turbine motor (ATM) as the engine approaches idle. The ATM supplies power to the utility hydraulics and AC power, and in a matter of a minute or two, the Thud is ready to roll. All that is left is turning on the avionics and performing some checks.

The Thud driver experienced two exhilarating occasions on each flight into combat. One, of course, was the period in which he is engaged with the enemy, and the second is takeoff. Takeoff with a fully loaded Thud on a hot day was not to be taken lightly.

Water injection was only used for takeoff. A 40-gallon tank of demineralized water was loaded prior to starting the engine on days when conditions warranted such use—always with combat loads in SEA. The water was sprayed directly into the compressor section of the engine. The pilot selected water injection right after the afterburner was successfully engaged. A somewhat reliable blue light in the cockpit indicated that water injection had been initiated, but the pilot always verified the system was working by feeling a slight boost in thrust and a pretty significant rise in EGT. In SEA, the use of water on most takeoffs was considered essential. A pilot on duty in a mobile shack beside the runway always monitored the takeoffs and would notify the pilot to "check water" if the afterburner flame did not turn from blue to orange as takeoff was initiated. Once airborne with the gear and flaps up, the pilot would select DUMP on the water injection switch, and the remaining water would be dumped. Failure to dump the water could result in the remaining water freezing at high altitude, which could cause damage to the system.

Just before takeoff, the pilot would ask the tower for the temperature and, using his checklist, compute a minimum EPR for military power. Before takeoff, the pilot would run the engine up to military power and check to make sure it was producing minimum exhaust pressure. Also prior to takeoff, the pilots would compute the minimum line speeds that would be expected at 2,000 and 3,000 feet down the runway. If you were much below these computed speeds, it was wise to abort the takeoff. Shortly after passing the 3,000-foot marker, you entered "no man's land" where, if your engine failed or something else bad happened, you didn't have enough runway to stop or enough airspeed to take off.

Before takeoff, especially with combat loads, the pilot would always locate the Jettison Stores (aka the Panic) button located right in front of the throttle. If the engine quit or you had a fire on takeoff, that button jettisoned all your fuel tanks and bomb load, instantaneously reducing your weight by thousands of pounds, allowing you to possibly get the aircraft airborne and eject or continue the takeoff to sort out your problem. (The bomb fuses were not armed, so the bombs should not explode, but there were no guarantees. If the bombs did explode, it simply was not your day.)

Landings were not as tenuous as takeoffs but did demand a high level of attention on the part of the pilot. Of course, the biggest problem with landing the Thud was the high approach speeds. Just prior to landing, the jet would be going well over 200 miles per hour and, depending on weight, crosswinds, and configuration, could be much higher. A strong headwind was always of great benefit. It seemed like no two landings in the Thud were the same, but getting on the ground was not your biggest problem—it was now time to stop this big hot dog.

The first trick was to land as slow as you could without stalling and with as much runway in front of you as possible (commonly referred to as "landing on the first brick"). Once you were on the ground, you pulled the drag chute handle and waited for the welcome tug on your shoulders as the 20-foot nylon chute deployed. If you weren't sure it deployed, you could check your cockpit rearview mirror. Also, if you didn't get your chute out right away or if it failed to deploy, you would get a gentle reminder from the mobile control officer, e.g., "Drag chute 2." Once you had cleared the runway, you would jettison the chute to be retrieved by maintenance. Landing in strong crosswinds could be tricky as the drag chute would cause the aircraft to weathervane into the wind, and in some instances, it had to be jettisoned to avoid running off the runway.

While the drag chute was being deployed, you would gently raise the nose until it was even with the horizon to provide aerodynamic braking. This needed to be done with care because if you raised the nose too high, you could get the dreaded "piece of tail" as the tail of the Thud dragged on the runway. This action normally destroyed the lower speed brake at great cost to the taxpayers. It also generated a source of heavy, good-natured ribbing from your fellow pilots, e.g., "Heard you got a piece of tail. Was it a moaner or a screamer?" More on the speed brakes later.

One of the worst nonfeelings you could get was the drag chute not deploying. Without it, applying the brakes at high speeds became necessary. This was a delicate situation. The trick was to apply brakes with gently increasing pressure to avoid the brakes from entering anti-skid, an automatic releasing of pressure in a rather abrupt, on/off progression. This normally occurred with the two wheels out of sync— very uncomfortable. It was also important to avoid laying on the brakes too aggressively as you could slip past anti-skid and blow a tire or two. This was unacceptable because the tires would be torn away rapidly, and you would now be riding on the rims of the wheels.

The last resort for stopping the Thud was the tailhook. If selected by the pilot, a tailhook would be blown down near the aft of the plane and would be held on the runway by gas pressure. All runways where Thuds operated had one or more barriers near the end of the runway. These barriers were marked by large signs with a yellow "meatball" inside a black background and consisted of cables held off the runway by rubber donuts. About 500 feet from the cable, if there was any doubt about your ability to stop, you would extend the hook. The tailhook would grasp the cable as you passed over it and, through a braking system, would bring the aircraft to a gradual stop. The Navy uses a similar but more sophisticated system on their boats to stop their aviators and their toys.

The Thud tailhook system required that the aircraft be slowed to around 150 mph (depending on weight) before engaging the cable. If you engaged the cable at too high a speed, the hook was stronger than the aircraft structure and the tail end of the aircraft would be ripped away, resulting in a runaway tricycle with no means to steer or stop it. Another problem was if you blew one or both tires, it was possible for the rims of the tire to cut the cable, and now the level of excitement would really pick up. There was one last hope. At the end of the runway was a "rabbit catcher." This was another form of barrier that had a cable and netting that extended about 3 feet above the runway. The cable and its braking system would catch the landing gear of the aircraft and slow it down and stop it before it ran off the overrun.

If you missed the rabbit catcher, you were in the hands of the Lord. Ejection was risky because the Thud ejection system did not have any guarantees if you were on the ground going less than 85 knots. More than one Thud driver has died after unsuccessfully stopping the tricycle, and one actually drowned as his plane ended up submerged in a water hole past the end of the runway.

The Thud, especially the way it was used in SEA, had some rather serious flaws. When the aircraft was designed as a high-speed, low-altitude penetrator, it was not envisioned that it might be subjected to intense anti-aircraft artillery (AAA) at higher altitudes. At high altitudes, the Thud was subjected to a multitude of different sizes of AAA, and the majority of hits ended up in the aft bomb-bay area of the jet. It seemed that enemy gunners rarely led the aircraft by enough to hit the cockpit area.

The designers of the aircraft used the bomb-bay area as the primary conductor of hydraulic pressure to the aircraft. The two primary flight control sources of pressure were P1 and P2. All the actuators for the flight controls had dual hydraulic systems so if one was lost, the second could allow for a safe recovery of the aircraft. However, the P1 and P2 pumps were located at the rear of the engine and were driven by the 15th stage of the engine's turbine. The pressure from these pumps was then channeled to the flight controls through high-pressure lines that ran next to each other, many of them in the bomb-bay area. An AAA hit would frequently be taken in the bomb-bay area and would oftentimes take out both systems simultaneously. Bad deal! If both systems were lost, the controls would freeze up within a matter of seconds and the pilot would be forced into an immediate ejection, oftentimes in an uncontrolled state. Numerous injuries (and worse) occurred because of this situation.

Early in the War, it was recognized that this was a serious problem, and a fix was added to allow the pilot to get out of the combat area before he bailed out in a safe area. The engineers designed a stabilizer lock (stab lock) system. As soon as a pilot recognized he had been hit, he would roll wings level on a good escape heading and bring the nose of the aircraft above the horizon. As he slowed down, he would engage the Stab Lock switch and move the stick slightly until he felt the stabilizer lock up. Now the pilot was required to maintain a fairly slow airspeed (below 325 kts), which was not an airspeed you wanted to maintain while departing a high-threat area. However, it was your only shot. If the pilot attempted to increase the airspeed above 325 kts, the nose would come down, and if the nose got too low, it could not be brought up. This condition resulted in a rapid increase in airspeed and negative g-forces. In this situation, the pilot had to eject immediately or be faced with a high-speed ejection with negative g-forces (resulting in a lot of self-inflicted pain and possible broken bones).

In addition to allowing the pilot to lock up the stabilizer, the trailing edge flap system was also modified. The Thud had originally been designed with a large cable that interconnected the left and right trailing edge flaps. This cable made sure the two trailing edge flaps came down together. This cable was removed and a switch was added to the cockpit that allowed the pilot to lower either the left or right flap to keep the wings level or complete a slight turn. This modification was great if you needed to utilize the stab lock system, but more than a few Thuds and a few Thud drivers would terminate their existence because of a previously nonexistent problem—split flaps.

The stab lock system did successfully allow some pilots to get out of the target area and successfully eject. It also allowed some to escape the target area, determine that they had some primary hydraulic pressure available, and unlock the stab and successfully recover the aircraft. (The aircraft could not be landed with the stab locked.) A running argument among the Thud community debated the total benefit realized by this modification.

Another serious design flaw of the Thud was the ATM system. The ATM was a small turbine motor driven by air pressure and was used to power the AC electric and utility hydraulic systems. Much of the aircraft's avionics and weapons delivery systems were powered by AC power, and the utility hydraulic system powered such essential items as normal gear extension, afterburner ignition, speed brakes, air refueling, and the gun. Without the ATM, the aircraft was severely crippled. You could return it to the field for landing, but it was certainly not suitable for combat.

The source of the air for the ATM was very hot air taken off the 15th stage of the turbine, but the ATM itself was located in the nose area well forward of the pilot. It had a high-pitched whine that was very loud. If you lost the ATM in-flight, the noise reduction was so great that the first thing you suspected was that your engine had quit. The hot air was delivered from the aft of the aircraft to the front of the jet by a large steel tube covered with a protective shroud. Along the entire distance of this tube was a heat-sensing system that told the pilot if the system had developed a leak. Fire and Overheat lights told the pilot things were not going well, and if the situation was not corrected, bad things could happen. A switch was provided to shut down the main air line. Turning off this switch also shut down the ATM, so life became more complicated. This hot air line also ran through the bomb-bay area, so a hit taken in the bomb-bay area could cause additional havoc if the main air line was severed.

Every fighter I've ever been around had some method of slowing the vehicle down. In combat situations, you never outgrow your need for speed, but there are times when you need to slow down really fast. Speed brakes are the answer, and the Thud had some great ones. Using a switch on the throttle, the pilot could extend four (two if the gear was extended) huge pedals attached to the back of the aircraft. When extended at very high speeds, the pilot was thrown forward abruptly. They were effective but had one "gotcha": when they were extended, it was hard for the pilot to detect they were out at lower speeds. In most fighters, you could sense a pitch change or additional noise, and some fighters had a light on the instrument panel that let you know the brakes were out. The only indications, which were not obvious, that they were out in the Thud were a higher than normal throttle setting and higher fuel consumption to maintain desired speed. Every Thud driver would experience the embarrassment of his wingman letting him know his speed brakes were inadvertently extended.

Another problem with the speed brakes was the possibility of getting a "piece of tail"—in this case, not a good thing. On landing, the pilot would typically raise the nose once on the ground to "aero brake," using aerodynamic drag to slow the aircraft down. It worked well as long as you did not bring the nose up too high. If you got the nose too high, the bottom speed brake would contact the runway, grinding off part of the metal surface. You could also do damage to the bottom speed brake if you landed too slow. You were not aware you had gotten a piece of tail in the cockpit, but it was always one of the first things the crew chief checked before the pilot shut down.

"You got a piece of tail, sir," he would exclaim, normally with a big grin on his face. It was not what you wanted to hear when the crew chief climbed the ladder to give you your helmet bag and help you unstrap.

If the damage was very extensive, the crew chief had to change out the pedal and the pilot owed him a case of beer. A side note: Pilots learned to check the bottom speed brake pedal while they were doing their preflight. If there was any minimal damage to the pedal, caused by a previous flyer, you brought it to the crew chief's attention. If you missed it, you would end up paying for a case of beer for someone else's piece of tail. Crew chiefs were devious bastards and would try to get as many beers as they could from pilots who didn't pay attention to the pedal on preflight.

So, enough about the routine mechanical stuff. Let's talk about the business end of this great war machine. The manliest way to shoot down an enemy fighter was to gun down the son of a bitch. Missiles were OK, but if you really wanted to savor the event, you got in close and put lead into the MiG's airframe. Early in the War, the F-105 was the best gunfighter in town. The early F-4 C and D models did not have an internal cannon but sometimes carried an external gun pod outside the aircraft. The pods were not as accurate as internal cannons and created a lot of aerodynamic drag. (In the case of the F-4, who cared? It was a draggy machine.) The Air Force eventually got wise and put an internal cannon in the F-4E.

The M-61 cannon on the F-105 was both accurate and manly. It carried just over 1000 rounds of 20-millimeter shells and fired them out at 100 rounds per second. It was mounted on the left-hand side of the nose, and when you shot it, the sound was exhilarating. Also, the smoke from the cannon would go down the engine intake with some of it ending up in the air conditioning system and eventually in the cockpit. I delighted in smelling the cordite gases!

The early Thuds required the pilot to go through several switch settings to go from bombing mode to air-to-air guns, but it was determined early on that the pilot needed a rapid way to select guns in case some MiG driver ventured into view. Because these opportunities lasted only a few seconds, a MiG Killer modification was added, which allowed the pilot to select air-to-air guns mode by hitting a single button on the stick. When MiG Killer was selected, the gun was set to fire and the gunsight became a lead computing sight. You simply put the target near the center of the gunsight, and the radar would automatically "lock on." The gunsight system indicated when you were locked on and told you when you were in gun range. When the range gauge signaled you were in gun range, you squeezed the trigger, turning on your gun camera, and the gun began to fire. "Take that, Red Baron."

Thud drivers had 23 confirmed gun kills in SEA with another four and a half kills using Sidewinder missiles. (Half kills happened when two pilots claimed the same kill and it was split between them.) The way to shut up an F-4 Phantom driver when they gave you crap about being a bomber was to ask who had the most gun kills.

An interesting aspect of the Thud was when it came to aerial refueling. Refueling could be accomplished using a receptacle located about 6 feet in front of the pilot on the aircraft's nose or via a probe located just to the left of the receptacle. Both the receptacle and probe were enabled by a handle located below the instrument panel.

Refueling using the receptacle was pretty easy once you became comfortable flying very close to a big airplane. Once you were stabilized behind the tanker, the tanker boomer would insert the refueling boom in the receptacle, at which time the Thud system would lock on to the boom. The boomer would give a tug to make sure the receptacle was locked and commence refueling. Once you had your gas, the pilot unlocked the jaws of the receptacle. Receptacle refueling was so easy in the Thud that oftentimes the pilot would simply fly the boom into the receptacle without an assist from the boom operator.

Refueling using the probe was a much different challenge. This task was carried out primarily by the pilot maneuvering to aim the probe at the basket. However, as you approached the basket, the air flow off the nose would move the basket away from the probe, and the rodeo began as the pilot chased the basket. Chasing the basket was commonly referred to as the "dick dance" or "J.C. (Jesus Christ!) maneuver." A number of Thud drivers can lay claim to having ripped a basket off a tanker, and almost all Thud Squadron bars had one on display.

Over time, the pilot could become fairly good at it, but the receptacle was always preferred. The worst thing that can happen is coming back from a strike, low on gas and pulling up behind a tanker dragging around a basket. The most useful feature of the probe was a Thud driver's ability to give the finger—or, "shoot the bird"—at some rival squadron or F-4 driver as you taxied by them by extending the probe.

The cruelest system in the cockpit was the one-quart water bottle paired with a one-pint pilot relief bottle (a pisser). Attentive crew chiefs would normally fill the water bottle with cold ice water before flight, and there was a flexible straw attached to allow the pilot to suck its contents out—normally right after successfully getting the takeoff behind him. However, the pisser was less user-friendly. The one-pint bottle had a funnel on top and a trigger that had to be depressed to allow fluid to enter it. Using it was a real goat rope. Unzipping your flight suit was never easy as the g-suit and parachute harness interfered with the process. Next, you had to find your little shriveled-up tool and expose as much length as you could. (Flying combat doesn't do much to enhance the size of old Willy.) Then you grabbed the pisser and tried to place it in the right position and, while depressing the trigger, let it fly.

In the meantime, you had to fly your jet, and if you were on the wing, you tried to plan for an opportunity to get far enough from the lead so you didn't hit him while making sure there were no big clouds up ahead. I'm certain there is not a Thud driver alive who has not missed the funnel on at least one occasion. Plus, it always seemed to fill up before you were ready to quit, and you'd leave some liquid in the funnel. The left switch console could become a little messy if you had to do much maneuvering before landing. After landing, you never wanted to tell the crew chief you had a full pisser, so you would try to dump it over the side as you were taxiing to parking.

I would be remiss in this short tour of the Thud if we didn't discuss "giving it back to the taxpayers"—that is, ejection. The cockpit of a fighter, for the most part, is a pretty comfortable environment. Most of the time, you have good temperature control and the visibility is awesome. A form-fitted helmet shelters you from the noise of the air rushing past your craft and the noise created by the engine and all its moving parts. (In the Thud, the ATM was an exception, but you soon blocked its high-pitched whine from your consciousness.) By the time you checked out in your first fighter, you were well acclimated to wearing an oxygen mask, and if your butt got sore, you could bring some relief by blowing up your g-suit with the Test button. In short, one felt relatively secure and happy. The prospect of leaving this comfortable nest for the harsh environment on the other side of the plexiglass canopy was daunting.

Early versions of the Thud did not have any capability to launch a pilot's tender body from the jet until the aircraft was airborne and going pretty fast. With the advent of Vietnam and the Thud being used in some tense takeoff situations, the ejection system needed improvement. A rocket seat was added along with an explosive charge in the parachute that provided a rapid deployment of the parachute when a "butt slapper" pushed the pilot away from the seat. These enhancements allowed the pilot to eject on the runway after he had exceeded 85 knots, giving him a fair chance at survival. Once you were past 100 knots on takeoff, you knew you had a fighting chance.

On the other side of the spectrum was the very high speeds at which Thuds operated when engaged in battle. Taking a catastrophic hit at high speeds without any ability to bring the aircraft under control left the pilot with only one option for survival—eject at once and take what fate brought. Conditions would only get worse if the pilot delayed. The prospect of broken bones and spine injuries was not pleasant, especially if you were about to become a prisoner of war. However, it was all you had left. Raise the ejection handles, pull the triggers, and recite, "Our Father, who art in…!"

This was a basic synopsis of what I best remember about the Thud. I never had to give one back to the taxpayers but came close on a few occasions. As I sit on the back deck of my small Kansas ranch, drinking a beer and watching the sun fade into an unbelievably spectacular sunset, I sometimes pretend I can feel the gentle rumble of the J-75 engine winding up and urging me on to another adventure in this great machine. I also grin as I consider the fact that the newest fighter in the Air Force inventory, the F-35, is a single-seat, single-engine fighter with an internal gun. And yes, it has a bomb bay!

Chapter 5
Putting on Our War Paint

Graduating from UPT with an F-105 in hand was as good as it gets. My biggest fear was that my assignment might be changed before I reported to Kansas. I'd heard of several assignments being changed between UPT and pilots reporting to their training units. I didn't want to be sent to the back seat of an F-4.

Cognito, our old 1956 Plymouth, passed its first test in Kansas as we departed Kansas City and headed southwest toward Wichita. We had never seen an 80 mph speed limit before, and since we liked to drive 10 mph over the limit, we asked Cognito for 90. She responded with ease. Cognito's lack of air conditioning didn't bother us much as we had plenty of circulation on this warm, sunny Kansas day. We were about 60 miles from Wichita when two dots caught my eye above the horizon to my left. At first the dots appeared to be moving slowly, but as they came closer, their speed accelerated. I then picked up two more dots following the first two across the prairie. As the dots became distinguishable, I recognized the long, lean profiles of the Thuds roaring along at high speed and low altitude. The four Thuds passed in front of us and soon became dots again before disappearing to the north. What a magnificent welcome to my new life. Mary and I looked at each other, and I expressed my extreme delight with some loud profane comment. She was somewhat more reserved. "Those were 105s, weren't they?"

"Damn straight!" I answered.

I was bubbling with excitement as we hit the east side of Wichita. We had our road atlas ready to get us to McConnell AFB. However, we soon realized that we could find the base with little difficulty because there was a frequent parade of Thuds proceeding south down the initial approach at McConnell. Four ships, two ships, and single flights of Thuds all pointed in the same direction, and as we looked further to the south, we could see them circling the base and extending the gear and flaps as they prepared to land. We drove down Rock Road to the East Gate of McConnell, and I asked the guard for directions to Base Headquarters so I could sign in. He directed us to turn right and head along Salina Drive, which would eventually take us to Building 1, which was the Headquarters building.

Salina Drive is the perimeter road that crosses the two main runways on the north side of the base. As we approached the east runway, we noticed a red light with a sign below it that said, "Stop on Red for Low-Flying Aircraft." We dutifully stopped as a flight of two Thuds passed very low in front of our car and proceeded to land on the west runway. Meanwhile, four Thuds had taxied into position on the east runway and were running their engines up for takeoff as another four positioned themselves in front of a large earthen bunker short of the runway, preparing to have their weapons armed and to obtain a "last-chance" inspection. The four aircraft on the runway lit their afterburners. The booms of the afterburners coupled with the sounds of the aircraft in the last-chance area gave me a rush of excitement as we proceeded along Salina Drive. Is it possible that in a few short weeks, I will be part of this phenomenal process? I mentally pinched myself.

As we passed the west runway, I became aware of a large ramp filled with F-100 Super Sabres. At the time, I didn't know what this flock of beautiful jets was, but it turns out it was the 184th Tactical Fighter Group of the Kansas Air National Guard (KANG). (I had no way of knowing that in another four years I would become a member of this group, and in 1986, I would become its commander, a position I held for six and a half years.)

We arrived at Building 1. It was a beautiful old building that had been the original Wichita Airport. It was now the Headquarters for the 23rd Tactical Fighter Wing (TFW). The 23rd was charged with training most of the pilots that were headed to Thailand to fuel the fight. I checked in and was told to show up for class the following Monday. Over the weekend, Mary and I rented a small house near the West Gate of McConnell. I had told Mary that I wanted to rent a house rather than an apartment because "I wanted us to live in a house before I died." Not sure what motivated me to make such a decision, but I suppose it could be the enormity of the situation I had gotten myself into. Now we were ready to begin our new adventure. Sleep did not come easy for me as I stewed over what I was about to do.

On Monday morning, I drove east across Salina Drive with the sun in my eyes reflecting off the silver F-100s of the KANG and eventually encountered the beautiful ramp of Thuds. I was sure there was over one hundred of them. I was due in class at 8 A.M., and already the noise of Thuds taxiing and taking off was music to my ears. When I arrived at my destination, the 4519th Combat Crew Training Squadron, I was happy to see Randy Carlson standing by the entrance smoking a cigarette. We greeted each other and shook hands nervously, both a little overwhelmed that we were here but ready to face this somewhat scary unknown.

I motioned at Randy's cigarette. "Better stop that shit."

Randy nodded and grinned. "You know I've never been an optimist, don't you?"

I looked at him questioningly, and he explained, "An optimist is a Thud driver who smokes and is concerned that he is going to die of lung cancer."

We chuckled and headed in the door.

Students and instructors were beginning to gather to our right, and we could hear the sound of a Ping-Pong game being played in that direction. Heading through an open door, we found a large briefing room with large cushioned leather-bound chairs facing a raised stage with a podium. Behind the podium was a blackboard with the words, "Welcome Class 69-CR." Guess that's us. In the back of the room were two guys in goatskins playing a pretty heated game of Ping-Pong. One short, thin player was beating the hell out of a student I recognized from Water Survival. Several of us watched intently as the short, thin player easily dispatched his victim. The winner was wearing The Patch, so I was sure he was an instructor. He taunted his beaten prey a bit and then stated, "I bet I can beat you left-handed." Sensing an opportunity to regain some self-pride, the student indicated he was up for that, so the instructor changed hands and again dominated the student. Mercifully, the game was interrupted by a short (my height) older man in a goatskin. It was time to get down to business.

Lt. Col. Harry Schur moved to behind the podium and waited for a few minutes until the students took seats in the nice chairs. A few of the instructors took chairs, but most of them stood around the room listening to Lt. Col. Schur and switching their gaze between Schur and the assembled students.

In a deep commanding voice, he began, "You all know why you're here and where your next assignment will be. The instructors in this room have all been there and know what you are about to face, and damn it, we're going to get you as ready as we can. You are the second F-105 class made up completely of recent UPT graduates. The first class left McConnell a few weeks ago and, by now, are flying their first combat missions out of Thailand.

"That class started with 27 students and graduated 26. Unfortunately, one of their students was killed on a ground attack ride. There are 22 of you, and it is our intent to graduate every one of you and not lose any aircraft in the process. We will attempt to make this exercise as safe as we can, but you can expect to be asked to do some things that will really test your skills and check out the size of your balls. We expect you to work and study really hard and adhere to the schedule 100 percent. Any questions?"

Holy shit! No one was going to ask a question in this setting.

Hearing no questions, Lt. Col. Schur turned the meeting over to the squadron operations officer, Lt. Col. Billy Joe Dulin, who began by giving us some insight as to what we were facing. He spent a good deal of time talking about things we could do to really piss him off, like not being ready to fly, being late to briefings, or busting a boldfaced exam. We were comfortable with these rules as they were much like the ground rules we had faced in UPT. He then announced our instructor pilot (IP) assignments. As assignments were given, the concerned IP would raise his hand and identify his students. Lt. DeVoss and Lt. Sykes, you'll be flying with Capt. Sheehy." I looked around the room and soon found a grinning face that belonged to the Ping-Pong master. As the meeting ended, Jim DeVoss and I gravitated to Capt. Paul Sheehy.

"You kids ready to learn anything?" He grinned.

Kids? Hell, Paul couldn't have been much older than us. Like us, he had been picked to fly Thuds out of UPT, had flown his 100 missions, and had recently returned to McConnell. Turns out he was only a year older than me but several years wiser.

"Let's get some coffee and find a briefing room, and we can talk," he exclaimed. He showed us where the coffee pot was and explained the honor system rules that required you to put a dollar in the coffee can each month. We wandered down a long hall with doors spaced every 6 feet or so. Some of the rooms already had some of our classmates being briefed by other IPs. We found an empty briefing room and followed Paul in. The room was pretty small with six chairs and a table with a couple of ashtrays on it. There was a shelf containing several loose-leaf books and a bank of briefing boards that contained numerous maps and depictions of all types of ordnance delivery patterns and air refueling procedures. Paul took a seat in front of the briefing boards, and Jim and I sat across from each other.

"You guys ready for this?" Paul asked.

Jim and I looked at each other for a second then simultaneously turned to Paul and enthusiastically answered in the affirmative.

"Shit hot!" Paul exclaimed.

Shit hot! I thought. I had heard the term used by some of the IPs during the Ping-Pong match and assumed it was a way to express approval. Growing up in Kentucky, we used the term "shit fire" when things didn't work out right. One of the first cuss phrases we learned from our older friends was, "Well…shit fire." I used to like to squirrel hunt with my .22, and when I would miss a fairly easy shot, my most likely expression would have been, "Well…shit fire." "Fire" and "hot" were close in meaning, but the two phrases were clearly aimed in different directions. Jim and I would soon learn that the "attaboy"s we were always looking for from Paul usually included the term "shit hot."

Paul pulled down one of the books off the shelf and explained it was the syllabus we would be using to get us ready for our TAC (tactical) Check—the ride that certified we were ready to go to war. The first two weeks would consist of a lot of academics and simulators. Once we had our F-105 Systems course out of the way and had completed our instrument and emergency procedure simulators, we would begin flying.

"These first few weeks come at you fast, so don't get behind. Get through this, and we'll go have some fun." Paul grinned. "One other thing—no whining or weak excuses for screwing up. If you've screwed up and I say you've screwed up, the first thing out of your mouth should be 'I fucked up' and then continue on with the conversation."

Lucky for me, I was paying attention when he said this. I would use that phrase many times over the next few months, and it always calmed the waters. I also watched some of my classmates, who were reluctant to admit error, suffer through some uncomfortable moments until they realized that not admitting you screwed up led you down a dangerous path.

That afternoon we found the Academic building and met our F-105 Systems instructor, Capt. Bill Ardern. He looked like he was ready for central casting for a Hollywood thriller. He was a handsome, smooth-talking guy who didn't seem to have any rough edges. It didn't take long for him to turn on the fire hose that we all expected. After a little bit of housekeeping, we were handed our F-105 Dash Ones (pilot's flight manual) and noted immediately that it was much bigger than the ones we had used for the T-37 and T-38. Doing a quick, random search through the manual revealed that my vocabulary was about to expand a bunch. We were also issued a thick In-Flight Checklist and a few other publications. Several years of reading only engineering texts had slowed down my reading speed a lot, and I could see some long study sessions ahead.

Bill said, "I know you are all recent UPT grads, and there will be a lot of new types of systems. Over the next couple weeks, I will cover the most important stuff, but it's up to you to soak up as much as you can on your own or working together. Besides systems operation, you need to put emphasis on chapters two and three in your Dash One. Chapter two is Normal Operations, and your IPs will expect you to be familiar with normal ops from the get-go. Chapter three is Emergency Procedures, and you need to hit that chapter hard beginning today. There are a number of boldfaced procedures you will be expected to know verbatim. You guys are about to become single-seat fighter pilots, and your working knowledge of these procedures may determine if you live or die. Also, you will begin flying simulator missions in the next few days, and your instructors will know if you are slacking off, and you can expect some pretty severe ass kickings if you get behind."

With that, he launched off on the first session of his Systems class, "The J-75 Engine." After showing us a couple of slide depictions of the heart of the F-105, he took us to a room down the hall that contained an uninstalled engine. It was pointed out that the J-75 weighed more than an entire empty T-38. After he pointed out the major components of the massive engineering feat, it occurred to me that jet engines had only been around for a few decades, and this wonderful machine had to be at the forefront of engine technology. I was amazed to think that in a few short weeks, this monster would be hurling my pink body through space with unbelievable power. And so it went for the rest of the afternoon and for the next several afternoons that first week. The "fire hose" was turned on at full volume and deeply inserted in my mouth. I was drinking as fast as I could, but I found myself attempting to sort what was nice to know versus what was essential.

Classes typically met for 4 hours each day, and the other half day was spent with Paul as he talked through some of the procedures we would need to know for our first flight. We would get two instrument flights in the back seat of a two-seat F-105 (the "family model"), and if we did OK on those two rides, we would be soloed out. We also began our three simulator rides where we had our first chance to use the vertical tape flight instrument system—how cool! The normal and emergency procedures were repeated on these simulator rides. The simulator was a 1960's technology machine, bolted to the floor with zero visual presentations, but the seriousness of our endeavor made us attentive and appreciative of the opportunity to get somewhat close to reality. The emphasis placed on ejection training was a little disconcerting.

We were also allowed to go out to the flight line and sit in an empty cockpit to go through our checklist procedures.

"Don't even think about moving any switches," Paul cautioned.

This was my first opportunity to climb the ladder and sit in a real cockpit of the F-105. What a great view of the world. One difficulty with this excellent view of the world was that when the ladder was removed, it was a long drop to the ground. During a ground egress practice, I found that I could hang from the canopy rails and still have 5 feet to drop to the concrete below. Also, because of the bend of the fuselage, you would hit the ground falling backward. Don't take off your helmet before attempting this! During egress training, a ground crewman was always available to ensure a "soft" landing. Paul also showed us how to enter the cockpit without stepping on the ejection seat—stepping on the seat was bad form and, on hot days, could leave some black tar footprints on the seat. But most of all, stepping on the seat did not look very manly.

Paul also did a practice preflight walk-around with me and Jim. A lot of items needed to be checked before you strapped this mother on, and you wanted to make sure all the gauges and gadgets on the exterior of the plane were at the appropriate settings and positions. Removing the gear safety pins was part of the walk-around, and I normally had to stand on a wheel chock to reach and remove the pin. *This thing is a big mother.* As the first week went by, the anticipation of actually flying was building. I had not flown in nearly three months, so I was afraid there might be a little rust to knock off, but bring it on.

As the first flight approached, I realized how lucky I was to have been assigned Paul Sheehy as my instructor. He was clearly the guy I had been looking for in UPT. He had a great upbeat attitude about Jim and me and his job. You knew he sincerely wanted to make us as good as we could be. He was brutally honest when we screwed up but did it with a smile and chuckle and an "OK, let's try that again, and let's do it this way." And when you did it right, you knew it at once: "Shit hot!" You never had to guess where Paul Sheehy stood.

The first week flew by rapidly and was an exhausting blur. By the end of the week, I was ready for a root beer and some relaxation at the Officers Club. Up until now, I had gone straight home after work each day and spent evenings studying and having Mary grill me on my boldfaced emergency procedures. I wasn't sure she enjoyed the process, but she didn't complain. Some of the bachelors had drifted over to the Club during the week to eat and have a beer or two, and they expressed a good deal of excitement about what they had found. (I had packed a lunch each day and studied over the noon break.) However, I was not prepared for what I found. Unlike Reese and the other few O'Clubs I had visited, McConnell did not have a stag (men-only) bar. There were two bars: the casual bar and the formal bar. Translated, this meant "Fighter Pilots and Friends" bar and "Everybody Else" bar.

Friday night at the McConnell casual bar was something to behold. As you walked across the dining room area and approached the bar, you could hear the noise of many people laughing and yelling along with loud music. As you approached the bar, there was an enormous block of cheese about 2 feet across and 18 inches high with an immense meat cleaver next to it so you could grab a hunk. Bowls and saltines by the cheese allowed you to really load up. On Friday night, draft beer was free from 1700 to 1800. The smoke in the room was thick. Outside the bar stood some tables filled with folks who wanted to momentarily escape the craziness of the place. Inside were tables filled with lots of boisterous fighter pilots, many with their hands in the air describing some air battle or maneuver (most fighter pilots cannot talk well if their hands are not engaged). Several women were scattered throughout the bar, and none of them seemed to be bored. All were engaged in conversation or sizing up some pilot as he described his exploits or manliness.

Looking around, I discovered a table with some of my classmates and our IPs. Paul was not among them. The table was pretty full, but I found an unused chair from another table and drug it over, squeezing in between two of my classmates. "Hey, it's Happy Hour and beer is free— go get one," I heard someone yell above the noise, so I made a beeline to the bar and looked for an open space to squeeze in. I found a slight opening between two pilots talking to each other and two young women also talking. I squeezed in, and the pilot on my right ignored my slight push for position. The lady on my left turned to see who was moving into her space. Upon sizing me up, she broke into a pleasant smile.

"Hi," she exclaimed. "Are you in the new class?"

"Yep, fresh meat. New guy on the block," I said and noticed while I was talking that she glanced at my left hand to see if it had a ring on it. It didn't. Mary and I had been pretty poor when we married and didn't buy rings. She borrowed my mother's and I borrowed her cousin's ring to use at the ceremony. She was happy with her engagement ring, and I liked not wearing a ring. I had carelessly lost my high school class ring within the first few months of college because I was always taking it off because it was so uncomfortable.

"Great! You'll catch on quick," she exclaimed with a smile before turning back to her friend.

I had been motioning for a bartender and caught the attention of a short, busty, curly-haired woman who was bantering with many of the fighter pilot patrons. When she hustled over, I noticed her nametag had "Elsie" printed on it. Before she could ask, I requested, "Elsie, could you snag me a draft beer please?" She nodded and retrieved it quickly, so I returned to my chair. I took a long swig of my beer and was soon joined by an IP who I did not know who knelt by my chair.

After exchanging introductions, he got right to the point. "What did you think of that woman you were just talking to at the bar?"

"She was pretty nice. Do you know her?"

He hesitated. "That's Darlene. She comes in a lot. She's looking for her third Thud driver. Her first two bought the farm, and she's looking for another one."

"Ouch!" I exclaimed as I glanced over at the bar where she was still standing with her back to us. She sure didn't appear to be grieving and didn't match my image of what a twice-widowed woman might look like, so I surmised why my newfound friend might be giving me the scoop on Darlene. I let him know I had all the wife I could handle, thank you, and Darlene was not a threat to my young ass. He laughed and departed.

I didn't want to hang around the bar too long. I knew Mary wanted to go out to eat, and I was still not swallowing the contents of the fire hose fast enough, so I needed some study time. However, the beer was good, and this had been my first chance to relax, so I hung around for another half hour or so. During that time, I picked up some great insights concerning this wonderful playpen. It was pointed out that the bell over the middle of the bar, complete with a pull cord, was a special tool, and when it was rung, someone was going to buy drinks for everyone in the bar. Of course, if you wanted to buy a round to celebrate some great event, you could ring it and tell Elsie to start a tab. If you walked into the bar with your hat on, the first to observe this faux pax would yell, "Hat on in the bar," and anyone close to the bell could ring it in celebration of the offender buying a round. The worst offense, however, was if your wife called you on the bar phone, at which time the bartender would announce your name along with "your wife is on the phone." Again, someone close to the bar would ring the bell, and the offender would be harassed unmercifully after he started a tab and while he tried to carry on a conversation with his wife. *Note to self—tell Mary never to call me at the bar!*

I also got some insight from my tablemates into some of the legends who inhabited this place. "See that scrawny little guy over there—that's Terrible Ted Tollman, one of the guys who strafed the Russian trawler in Haiphong Harbor," or "That big guy with the cigar at the bar—that's Black Matt—he's the guy that boomed the Air Force Academy last month and blew out all those windows."

Others were pointed out who had been shot down and picked up or had shot down MiGs. I was fascinated by this collection of guys and couldn't wait to buy some of them a beer and get the full scoop, but I decided to get out before I had my third beer. As I stood up to leave the bar through the rear entrance, I noticed a table near the door full of women seeming to have a good time.

"Who are the honeys at that table over there?" I asked.

"Those are the missile wives that come over to watch the animal show."

McConnell had a ring of 18 Titan II intercontinental ballistic missiles (ICBMs) located on desolate parts of the prairie that surrounded the base. The missile crews would go on 24-hour alert for several days at a time, and some of the wives would pack up and wander over to see the show.

"Do they ever feed the animals?" I asked.

"Been known to happen" was the reply.

Driving home, I concluded that the McConnell bar was a dangerous place if I wanted to succeed in this endeavor. I had gone through a period of excessive bar time in college, and it almost cost me the opportunity to get a degree and commission. This place was more luring than any of those college bars, but it needed to be treated like a hot stove. Little did I know that once I started flying, especially when we got to the later part of the program, almost all the instructors debriefed their afternoon flights at the bar.

Mary and I went out to eat that night and celebrated our first week in Wichita. I excitedly described my visit to the bar, leaving out the part about Darlene and the missile wives. I told her about the penalty for her calling me at the bar, and she accepted that...sort of.

"So how do I get a hold of you?"

No answer for that, so I allowed that I would make sure I got home early. I suggested she could visit the bar—there was no penalty for bringing your wife to the bar. I later found out that although there was no penalty for your wife showing up "no notice," you could catch a lot of crap if she did.

Mary said she had been looking in the local rag, *The Wichita Eagle*, and had found a job opportunity working for a professor at Wichita State University (WSU). She had called him, and he needed a research assistant for a book he was writing. It didn't pay much, but it sounded interesting, and she could catch the city bus to WSU as it went near our house. It had been almost a year since we had decided we would start a family and nothing had come of it so far, so getting a job sounded like a good idea. She was going up to WSU Monday for an interview, and I had no doubt she would get the job. And, of course, she did.

The next week passed in a busy blur, and by the end of the week, I had completed a couple of simulators, life support training, and the F-105 Systems course. The first two flights would be in the back seat of the family model, two-seat F-105F. Paul would be in the front seat, and he would watch me practice instrument approaches and then some landing pattern work from the back seat. As I strapped into the back seat, I understood why the instructors hated flying in the back seat with students. You couldn't see anything outside looking forward.

As Paul started the engine, I experienced a feeling I would never grow tired of, the gentle vibration of the engine as it came to life and then an acceleration followed by an increased intensity of the vibration into a rumble as the heart of the Thud settled in to idle power. You knew your horse had a really big heart as your body became a part of the vibration. So cool! Paul completed the ground checks, and we taxied out to the arming area, where the ground crews performed a last-chance inspection, then we took the runway. Released for takeoff, Paul and I checked our engine gauges, and finding all was well, he released the brakes and lit the afterburner (AB). Having heard the deafening boom of the AB from the outside, I was a little disappointed by what happened inside the plane. Although there was no doubt the AB had lit, I experienced no immediate sense of rapid acceleration. This was no T-38.

As we approached the 2,000- and 3,000-foot line checks, I realized that the T-38 would already be airborne by now, but our takeoff roll was just beginning. Finally, somewhere around 180 knots, Paul rotated the nose, and we skipped on the runway a few times and became airborne. Paul came out of afterburner at 300 knots and accelerated to 350 knots, which was the normal climb speed for the Thud.

"You got it, cowboy," he said.

"I've got the jet," I replied and forcefully took control of the stick and throttle. As I shook the stick to acknowledge that I had the jet, I knew right away I had a big machine strapped to my back. The meaning behind the expression, "The Thud is not a sports car, it's a Cadillac," became clear.

I would later find that you could make the Thud feel like a T-38, but you had to be going about 550 knots.

From my simulator rides, I had discovered I liked the vertical tape flight instrument system. I was also impressed by how stable the aircraft was when you made corrections with the flying instruments. I was pretty proud of myself after we had completed the instrument work, and we returned to McConnell to practice some landing patterns. The landing routine was the same as I had flown in flight school except the speeds were much higher. It also became clear from Paul's coaching that speed control was essential—both ways. You couldn't be too slow or too fast. We practiced several approaches, but I always had to relinquish control to Paul as we neared the part of the pattern where I could no longer see the runway. The Thud was also not authorized to make touch-and-goes, so the only landing I would see would be the full stop.

Paul's landing was a little firm, and he used some aero braking as he deployed the drag chute and we taxied in. Paul's debriefing was comforting as he didn't slap my wrist for any major screwups, and he let me know we would repeat the same procedure the next day. When we did, it went well.

"You ready to solo out?" he asked.

"Damn straight!"

"Shit hot, we'll get your cherry Monday, cowboy!"

I wished I didn't have to take two days off before soloing out, but what the hell, it was a Friday and time to revisit the casual bar. As I joined several of my classmates, they were all abuzz about two students who had just run into each other but bailed out safely. Since they were from another class, we didn't know them, but they were senior officers and had run into each other during formation practice.

"The two son-of-a-bitches just hit each other," was the call made by the instructor who observed the two collide. As we talked with the instructors about the incident, they were more casual about it than we were. Cost of doing business. These two aircraft losses brought us to four for the year at McConnell. Prior to our arrival, one of the lieutenants from the first all-UPT class had hit the ground and was killed, and one of the instructors had jumped out when he put a jet out of control and into a spin.

Given that I had to solo out on Monday, I decided to get home early and again go over the things I needed to know for the big day.

"Two jets crashed today," I announced as I entered the front door.

"I know, I heard it on the news at work. I assumed it wasn't you."

Note to self: Anytime there is something like this on the news, give Mary a call and say something like "Did the Cubs win today?" That would be code for "I'm OK." We had been briefed that if accidents occurred, we were not allowed to confirm it directly by a phone call. Once I explained that the two guys that hit each other were from another class and were both OK, she didn't seem to be very interested in the accident and did not express any concern about my welfare. Perfect!

The weekend dragged by. I went over the Dash One normal and emergency procedures many times and had Mary drill me on the boldfaced procedures. I slept reasonably well. However, this was a BIG deal. I had been dreaming about it since Butch and I had fantasized about it on the swings at East Side School many years ago. Jet fighter pilot—only one step left!

Monday morning as I was driving to work across the flight line, I watched the single-seat Thuds taking off and realized that today I would be one of those. The meeting with Paul and the briefing occurred like nothing special was happening. I felt good about the "big boy" approach and soon fell into the same mindset. This was just another day at work. At my jet, Paul climbed up the ladder behind me and watched me strap in. He gave me a short "blindfold" cockpit check where I had to close my eyes and identify things like the drag chute handle, the air start button, and the panic button. It was all pretty silly, but it was required by the syllabus, so we got it done.

"Go ahead and start up and get your ground checks completed, then check in on the radio in 12 minutes." Then he was gone, leaving me all alone with this big high-speed tricycle strapped to my back. Very cool! The feel of the engine coming to life in a jet that belonged only to me was a great feeling. I completed the ground checks carefully and checked in with Paul at the appointed time. I led Paul out to the arming area and then onto the runway, where he assumed the "chase" position. I ran up my engine and checked back to see that Paul was ready to go. He nodded that he was ready, and we were off.

Somehow the anxiety that I had experienced when I soloed out in flight school was gone. This first solo flight was not filled with the giddy exhilaration I'd felt on previous adventures. I had taken a giant step toward maturity as an Air Force fighter pilot, and I would never look back. This was what I was born to do.

The flight went well, and I soon forgot that Paul was behind me playing babysitter as I got things done as briefed. My touchdown at landing was a little firm, and I bounced a bit. The long distance between my eyes and the main landing gear was a little hard to discern. As I raised the nose to aero brake and pull the drag chute handle, Paul flew over me, low and loud, and dipped a wing before gracefully entering a closed pattern to land behind me. I waited for Paul in the arming area, and we taxied back to park together. It was here that I uttered Paul's favorite expression with a smile on my face, "Shit hot!" Already on the taxi as I exited the aircraft, Paul greeted me with a big smile and, "Shit hot! Nice job, cowboy!"

I felt a little out of place going to the bar on a Monday, but this was a time to celebrate. A few of my classmates were there who had also soloed out that day, and I would say we all had a similar experience. We were all excited but also subdued about our accomplishments. Much more was to be learned. We also discussed some rumors that a few of our classmates were having some difficulties and were under some additional scrutiny. I was a little surprised to find this out, knowing that all my classmates had finished high in their UPT classes. However, I had complete confidence in Paul and knew he would let me know if I was screwing up, and so far, not a peep.

The next few days went by in a blur. I took more rides, learning about how the aircraft handled—everything from stalls and slow flight to acrobatics. Ten days after the mid-air, we got word that Capt. Ardern, our movie star classroom instructor, had jumped out of a two-seat family model after his engine seized up with loss of oil pressure.

"Did the Cubs win today?"

We soon discovered that two of our classmates had disappeared. We did not talk about it or ask questions. They hadn't said goodbye. The system simply gobbled them up, and we were happy to be part of the survivors.

In mid-July, I got a call from my brother-in-law, Wes, who was about to finish up his stint in the T-37 at Reese. He told me that he was planning his T-37 cross country and would like to stop and spend the night at McConnell on a Saturday later that month.

"Damn right, Wes. I'll check with Base Ops and find out your inbound and meet you there."

I greeted Wes and his IP, Capt. "Bear" Stallings, at Base Ops, and we all jumped in Cognito and drove to the Club for a beer. There were only a few other pilots in the casual bar as not much was going on that Saturday afternoon. Once we had some beers in hand, I let them know that they were out of uniform for a fighter bar. They gave me a questioning look and I pointed at their knife pockets. Knife pockets, a small pocket inside the left leg designed to hold a survival knife, were on all issued AF flight suits. The knife was attached by a string and primarily used to cut parachute cords during an ejection to make the chute more steerable.

"You can let me 'scalp' you, or it will get done later by a bunch of rather aggressive Thud drivers—your choice."

Because fighter pilots wore g-suits, which already had knife pockets, the knife pockets on our flight suits were unnecessary, so the first thing you did when issued a new flight suit was cut it off. Wearing a knife pocket in the bar was a violation of etiquette and had to be remedied by immediate removal. Many fighter pilot bars displayed a collection of all the scalps taken—often many hundred.

I chose Wes as my first victim and had him remove the knife, then I got a firm grip on the top flap of his knife pocket and pulled down abruptly. His flight suit was pretty new and offered a good deal of resistance, but I managed to take the scalp with one hard yank. I then looked at Capt. Stallings and noted he was not really into our little game, but with a grudging nod, he approved of the ritual. Bear's flight suit was much older, so I concluded the pocket would come off easily. I yanked too hard and found myself holding the entire inside of his flight suit leg. By now, a number of spectators had gathered, and they were laughing wildly, both at me and Bear. "You're incompetent, Sykes," or "Nice legs, buddy," were the typical comments.

I couldn't get up from my knees I was laughing so hard, until I looked up and saw the scowl on Bear's face. "Uh-oh." Wes, too, was laughing at his somewhat exposed IP, but Bear was not amused.

"Do you have another flight suit with you, sir?" I queried, and he shook his head with a continued look of disgust. Time to remedy this situation. I snuck Bear and Wes out the back door of the Club, jumped into Cognito, and drove to Clothing Issue. It was still open, and Bear exchanged his old tarnished flight suit for a brand-new issue. Luckily, they had one close to his size. Bear had been aptly nicknamed, and there weren't many fighter pilots that came in his shape. Looking back, I suppose I'm glad Bear didn't decide to beat the crap out of me as there is little penalty for captains kicking the shit out of second lieutenants. We were soon back at the bar and enjoying life. What we could not have known was that Bear had only a short time to live as he would be killed in a T-37 accident a little over a year later.

***

In early August, as I was getting ready to fly, there was a buzz around the dispatch desk. I figured out from the talk that someone from my class had experienced a mishap. One of my classmates had jumped out of a jet near the Hutchinson Airport, some 50 miles northwest of McConnell.

"Did the Cubs win today?"

After I landed from my flight, the facts were coming in that my classmate had run out of gas and was trying to make an emergency landing at the Hutchinson Airport. I assumed it must have been some type of mechanical fuel malfunction, but as the whole thing unraveled, we learned that he had inadvertently put out his speed brakes, and when told by his instructor to rejoin, he could not catch him and finally selected afterburner to speed the rejoin up. Using speed brakes and afterburner at the same time was a procedure used to dump fuel if you needed to get rid of fuel in a hurry, and it was quite effective. In this case, however, he did not realize he was low on gas until his Fuel Low warning light illuminated. At that point, he was between a rock and a hard place. The airplane ended up flaming out a few miles short of the runway, and he had no option other than to give it back to the taxpayers.

"What a dumbass" became the chorus echoed by his classmates, but deep down, we all knew that we could have easily committed the same error. But now, the event was much closer to home. I think we all sensed that the instructors were beginning to take a more serious look at us. "How many more fuckups are there in this group of kids?" You could sense there was a heightened sense of tension in the workplace. Good news for me—Paul seemed unaffected by the commotion.

Calm would not be restored easily, however, because three days later as we were preparing for our morning flights, we got the word that all flying was canceled for the day. A student from another class had crashed on takeoff just south of McConnell. We were told there would be a mandatory meeting for all available pilots at the base theater in a couple hours. We knew nothing about the accident, but we were concerned about our future given that we were at the end of the whip and probably the most expendable of all the students undergoing training. Maybe they were about to send us to the dreaded F-4 back seat or turn all of us into trash haulers. The fact that two of our classmates had disappeared and the guy who had run out of gas was not flying again and in some type of holding pattern made us suspicious that we could be next.

We dutifully hiked across the street to the base theater a half hour before the appointed time and began to find out what had happened earlier that morning. A student who was a previous T-38 instructor had attempted to take off without afterburner and found out why the Dash One prohibited this procedure. As he struggled to get airborne and raised the gear, the engine was unable to sustain climbing flight, began to stall, and the pilot jumped out, picked up his chute, and walked away. We were told that he was making a formation takeoff, and we immediately branded the guy a dumbass. How could he possibly not know he didn't have an afterburner as his instructor rapidly accelerated away from him?

Someone made the comment that we were losing more airplanes at McConnell than they were in combat. True or not, the record of students at McConnell over that month was not sterling. A mid-air collision, running out of gas, and attempting a non-afterburner takeoff were all preventable. Getting hosed by anti-aircraft gunners, a missile, or a MiG were certainly more justifiable—and a lot more manly when you talked about it at the bar.

If I ever have to jump out of one of those machines, and live to tell about it, I hope I have a manly story to explain my exploits at the bar, I concluded.

As we entered the theater, we noticed that the wing commander was sitting in a chair on the small stage in the front, and we assumed he was waiting for the meeting start time. About 10 minutes before the start time, I heard some commotion in the room followed by some applause and cheering. As I looked toward the source of the commotion, I saw a young captain walking down the aisle getting handshakes and pats on the back from other pilots who were seated in a back corner of the theater.

"Who the hell is that?" I said out loud.

Someone responded, "That's the guy who jumped out."

I was a little surprised that a dumbass would be getting such a hearty reception. With the commotion ongoing, I looked toward the stage and noticed the wing commander was slumping forward in his chair with his hand shielding his eyes. His wing had now lost five Thuds in a little over a month, and he had little reason to rejoice. I suspected we were here to listen to a lecture from the wing commander about cleaning up our act and stopping the bleeding.

Suddenly, we were called to attention as two more men entered the stage opposite the wing commander. We were told to take our seats, and some Grand Poobah—I assumed he was a general officer—walked up to the microphone and in a short statement let us know there was a new shirt in town. As of this moment, our new wing commander was Col.

James Hartinger. As Col. Hartinger was introduced, I feared that his message might not be good news for me and my classmates.

He began by letting us know that he was here to clean up the mess, and it started today. First, he chastised the instructors for being too loose with their students, and from now on, they were going to be more attentive to identifying the weak dicks and taking action to correct the situation.

"We can't carry students," he stated.

The first question that entered my mind: *Is Paul carrying me?*

Now finished addressing the instructors, he let the students know we had to work harder to make sure we were doing our best. To help us progress, he was going to bring in several T-33s to give us a little "seasoning" if we demonstrated that we were not keeping up with the program.

Oh, shit, I thought. Is my life about to go to hell in a handbasket?

A somber group of young guys departed the theater and watched the administration of Air Force "justice." But our concerns were not for the previous wing commander; our doubts lingered on the future of our young asses. Within a few minutes, we had branded Col. Hartinger's T-33 program as "Purgatory" and figured one was more likely to go to Hell than Heaven if you ended up in Purgatory.

Back at the Squadron, nervous chatter about the morning's accident and our future abounded. I saw Paul enter the room and soon concluded that he wanted to pull Jim DeVoss and me aside, so we followed him back to one of the briefing rooms. He closed the door behind us, and I feared I was about to hear some bad news. He looked at us with a big grin and said, "Whoa, what did you think of that shit?" He hesitated briefly then continued, "You guys are not in trouble, and as long as you don't do something really stupid, you won't be. Got any questions?"

What was there to say. I felt like I had been given new life. I shook my head in the negative, as did Jim.

"Shit hot! See you guys tomorrow."

Chapter 6

The War Dance

Having witnessed a swift change of command and a promise from new management to fix "our problem," I was interested in seeing how that was going to be done. Was I part of the problem? By now, two of my classmates had been eliminated from the program, and Col. Hartinger was about to begin his "seasoning" program with the T-33s. I didn't know the specifics of these actions and really didn't care much. I was too busy keeping my head above water. The good news was that Paul had given me no indication I was in trouble, and trusting Paul completely, I felt pretty comfortable with my status.

One immediate effect of the new sheriff in town's T-33 program was the disappearance of our classmate who had ejected after running out of gas. He was sent to Purgatory, where he was supposed to receive some seasoning and be put back in the Thud. That's not the way it worked out, I guess, because we never saw him again. I did see one of my classmates who had been sent to Purgatory later on my way to Thailand. He was stationed in South Korea flying C-47s. Purgatory my ass—more like hell!

It was now time to learn how to kick ass. We had gotten through the basics and were now ready to hone our war fighting skills. I had just started feeling comfortable with my environment when I found myself with a fire hose in my mouth once again. Due to graduate and go to war in less than four months, we now had to learn how to air refuel, kill MiGs, drop bombs, and shoot rockets, missiles, and the gun.

We returned to academics where we were exposed to the Dash 34. When we entered the classroom, we found a bound manual somewhat bigger than the Dash One. Doing a quick peak through its contents, I discovered a myriad of charts and tables for almost every conceivable weapon and delivery available to the Thud.

Our lifestyle at McConnell left us with little time to work out or exercise. We also found ourselves at the bar almost every day, and I consumed more than a few beers on each occasion. By this time, I noticed my "goatskin" (flight suit) feeling a little tighter, and when I had an occasion to jump on a scale, I found I had gained over 25 pounds from my 160-pound fighting weight at UPT. For some reason I had also grown a mustache—probably because so many others had grown them as we knew they were fashionable in SEA where grooming regulations for pilots were very relaxed. One of my classmates decided I looked like a no-necked character in the James Bond movie *Goldfinger* named "Oddjob," and he hung that nickname on me and it stuck. Paul Sheehy always grinned a little when he called me Oddjob. Lucky for me, it didn't stick beyond McConnell, but years later when I would run into my classmates, I was always greeted with "Hi, Oddjob."

Understanding the fundamentals of math and physics from my days as a Badger engineer helped me understand the concepts of how bombs and bullets work. Quite simply, you release or shoot them from your jet, and they will eventually reach the ground. And if you did it right, the round would connect with the head or hardware you wished to destroy. Get it wrong and the taxpayers would be pissed because you just wasted a lot of their money.

The first trick is to get all your switches set up right. The Thud weapons delivery panel takes up much of the left side of the front cockpit display and contains many switches, knobs, and buttons. It affords the pilot many options for execution. It also affords the pilot the opportunity to fuck it up.

All the options we needed to know were covered in class, including the use of the manually set "pipper." The pipper is a red dot surrounded by two concentric red circles superimposed on a combining glass located above the cockpit front panel. It is adjusted by a knob at the top of the front panel and allows the pilot to depress the pipper to a prescribed "mil" setting. During a bomb or gun delivery, the pilot uses the pipper as his aim point and maneuvers the Thud to put the pipper where he wants it. That's about as much as the reader needs to know about this exercise.

This entire system is now considered antique. In today's world, computer-generated aim points and smart bombs give the pilot great "stand-off" capability. He or she no longer has to venture very close to the target and the associated defenses. The bombing system in the Thud had the opposite effect. Experience soon taught the pilot that the closer you could get to the target, the better the chance of success. Thus, pilots would "press" (get closer and lower) in an attempt to be more accurate. Of course, this meant a closer exposure to the enemy defenses surrounding the target. This is an issue that each pilot had to sort out for himself. It usually boiled down to three things: (1) How important is this target? (2) How big are my balls? (3) I'm pretty sure I'm invincible, so what the hell!

By late July, we were getting ready to begin dropping bombs and shooting the gun, so the excitement level was pretty high. We were also beginning to do some air-to-air training and learning about high-speed and low-speed yo-yos, breaks, and hard turns. You could use these maneuvers to shoot down a MiG or to avoid being shot down by the same. We also got a chance to use the air-to-air gun tracking system in the Thud, which was pretty impressive.

At this time, another of our merry band was sent to Purgatory for an undisclosed reason and never seen again. We were now down to 18 from our original 22.

We began practicing in-flight refueling. Refueling using the receptacle proved to be simple once you got the hang of it, and it was rather fun. The probe, however, was a different story. Getting your probe to settle into the basket required a good amount of skill and a bit of luck. As the basket approached the probe, the jet wash off the nose of the aircraft began to push the basket away from the probe, and the temptation was to chase the basket, resulting in what we called a "J.C. (Jesus Christ!) maneuver" in which the aircraft would porpoise up and down and, if not stabilized, could rip the basket off its hose. Most Thud squadrons would have a basket hanging from the ceiling somewhere that had been stolen from a tanker. Refueling missions also gave me and my classmates a chance to poke fun at one another at the bar as it was hard to hide your inadequacies while your flight mates watched your every move behind the tanker. But with practice, in-flight refueling became a pretty simple procedure.

The real war dance began when we started to drop practice bombs and shoot the gun on the gunnery range. The first bombing runs were made using a small fat-nosed bomblet used to simulate dropping high-drag bombs or napalm. The pass was made at 450 knots from 100 feet above the ground. As your pipper tracked along the ground, you had to hit the "pickle" button at the right time to get the bomblet to hit inside a small sandbox. The first time you did it, the rush of the ground speeding below your aircraft was a little unnerving.

We also began to drop bombs from higher dive angles, up to 45 degrees of dive, and release altitudes, techniques designed to give the pilot adequate escape distance yet also ensure the aircraft was close enough to the target for a good chance of hitting the target. During these passes, the pilot had to compensate for all manner of factors: wrong airspeed or wrong dive angle, not to mention the effect of the wind (a significant factor in Kansas). By repeating flights to the gunnery range day after day, your scores eventually improved, and when you met certain criteria, you were qualified in each event. Qualification in an event for me meant a trip to the bar to celebrate. Each pilot had to be qualified in all events before they could be given a Tactical Check and be certified as ready for combat.

The event that was everyone's favorite was low-angle strafing. This was accomplished with a slight dive angle at 450 knots while attempting to place the pipper in the right position to hit the "rag" (a large plywood board). You could begin shooting from as far away as you wanted, but you had to cease fire at 1200 feet from the rag. If you fired past the 1200-foot foul line, you were given a foul and penalized for that pass. Two fouls on a mission got you kicked off the range. A little experience taught you that your scores improved a lot the closer you were to the target, but shooting too far past the foul line could result in damage to your aircraft or possibly shooting yourself down from a bullet that skipped off the ground and ricocheted up and hit your jet.

It was a fun exercise that allowed you to demonstrate your superior skill and cunning. Each pilot had 150 rounds of 20-millimeter cannon ammo available and five passes at the target to qualify (25 percent of your bullets through the rag). Since the gun shot 100 rounds per second, you had to judiciously use the trigger in short bursts until the final pass. Once you figured out how the winds were affecting you that day and where your gun was harmonized, normally on the fifth pass, you pressed the foul line and held the trigger down hard until the gun stopped shooting. When you ran out of bullets, you were "Winchester," and the fun was over.

Does this sound like an activity a bunch of alpha males might wager on? You bet your sweet ass it was! After the first few rides to the range, the most anticipated part of the briefing was settling on the bet. There were several ways to define the bet, but once it was negotiated, you could compute what the reward for a good day would be and the penalty for screwing it up. Some of the instructors would not take part in the betting, but most would lick their chops and join in. The instructors who bet normally won most of the money. A real sign of progress was having an instructor hand you money after the results were in. However, that didn't happen often.

With all this excitement going on, it was hard for me to think about anything else, but Mary changed my focus by letting me know she had taken a pregnancy test at the base hospital and the results would be available in a few days. She didn't say what had prompted her to take the test, but I didn't ask. I had little interest in what made that part of a woman function. I understood my role in the process and that was enough. She called me at work a few days later and told me the results of the test were in and asked if I would like to meet her at the hospital. I wasn't on the morning schedule, so I hiked over to the hospital and met her as she checked in. In those days, the idea of a partnership in such matters had not taken hold, so Mary, when called, went into the doctor's office by herself. A few minutes later, she returned with a slight smile on her face.

"Did the rabbit die?" I asked (Google it). She nodded affirmatively, and we were officially about to become a family. I wasn't sure how this would impact my life, but I knew it was supposed to happen. I gave her a big hug and said something stupid like, "Good job, girl." I had to get back to the Squadron and prepare for the afternoon flight. That evening, we went out to eat to celebrate and bought a bottle of our favorite wine, Blue Nun Liebfraumilch. This was well before the days when mothers were told not to drink while they were pregnant.

It was early August, and life was good. My new career was going well, and I felt safe from Purgatory and likely to make it to the next step, earning The Patch. Then things took a turn for the worse. We were being introduced to a new weapon—rockets. Rockets were carried on the outboard wing station of the Thud and offered a little better accuracy than dumb bombs. However, they didn't carry as much explosive capability. We were told that we would carry them in combat from time to time when targets were smaller and required better accuracy.

We were not required to qualify in rockets, but we were being familiarized with them in case we ended up carrying them in combat. As we were briefed about shooting rockets, the warning was always about not continuing to watch them as they left the aircraft. Because of the trailing flame from a rocket, there is a temptation to watch it to see where it hits the ground. The common warning was, "If you see your rocket hit the ground, it will be the last thing you ever see. You and your aircraft will end up unscoreable at 12 long (in a smoking hole well past the target).

On August 13, one of my classmates, Jerry Zoeller, who was one of the three captains in my class, probably succumbed to this lure and ended up crashing long of the target. He tried to eject once he realized he could not recover but had no chance of survival. It was his first rocket familiarization ride. The disappearance of our first four classmates because of flying deficiencies had made us all a little nervous, but the actual death of a classmate was beyond sobering.

"Did the Cubs win today?"

About half of our class were married, and the wives became even more anxious. Jerry was married and had an 8-month-old son. As I broke the news to Mary, she displayed little emotion and seemed to not be stressed by the event. I think from the beginning of this adventure, she had been aware of the risks and would accept whatever happened as part of God's plan.

A few days later, I had my first rocket familiarization (FAM) ride, being quite aware of the dangers. We would never be certain that our Jerry had fallen into the rocket-watching trap, but we assumed that was the cause. I knew what I was going to do: shoot the rocket and begin an immediate pullout. 'nuff said. As I set up for my first rocket pass, I didn't feel especially nervous. I had my switches all set up for a rocket delivery, and I was on a "ground track" to roll in on the target.

As I began to roll in on the target, I did what you always do when you roll in on a target—push the microphone button to let the range officer know I would like clearance to shoot a rocket. The only problem was I pushed the wrong button. By mistake, I pushed the pickle button, and as I watched the rocket leave the aircraft, I knew I had really fucked up. From where I was, it appeared the rocket was headed for the small Kansas town of Falun, which was located a few miles south of the range. I wanted so bad to reach out and grab that rocket and retrieve it, but that was obviously not possible. *Oh, crap.*

"Two is in rockets, with an inadvertent release," was the call I made to the range officer.

"Roger, two cleared in dry [don't shoot]." There was a short delay. "Was it pilot induced?"

Here was a way out. I could blame it on the aircraft weapons delivery system. But then I remembered Paul's initial briefing—always admit when you fucked up!

"Roger," I replied, and the range officer told me to continue my first pass dry and then continue with the mission. All I could think of as I climbed to make another rocket pass was how many people I might have killed in Falun and how this would surely land me in Purgatory. I was relieved to see that there was no smoke or fire emanating from Falun as I set up for my next pass. Very carefully, I pushed the correct button and rolled in and shot my first deliberate rocket. Right away, I was able to understand the instructor's warnings. The rocket was an enticing object as it streaked off toward the ground. As I had planned, I began an immediate pull-off. At least my second pass was successful.

On the return flight to McConnell, the thought went through my mind that this could be my last flight in the F-105. How could I have been so stupid? Perhaps I was more nervous than I thought as I prepared for that first rocket run. Maybe I still had my classmate in the back of my mind. Upon landing, there was a car marked "Safety" waiting for me at my parking spot. A young captain with a clipboard met me as I came down the stairs and asked me what happened. I explained what I had done and let him know the last time I saw the rocket it was headed for Falun.

"Have you had any reports of damage?" I asked hesitantly.

"Not that I'm aware," he replied. "We'll let you know if we hear anything."

The flight debrief was rather embarrassing as I told the others what I had done. The instructor didn't seem to be all that upset, but I wondered what was going through the minds of the other two students in the flight. "What a dumbass" was most likely. However, one comforting result of the debrief was that the instructor let me know, as I described the location of the incident, that the rocket undoubtedly landed well short of the town of Falun and might have scared the hell out of a cow or deer but not much else. *God, I hope he's right,* I thought as I breathed a little easier. I was surprised that I didn't bust the ride, although I wasn't given a flattering grade. As I read the grade sheet, I was relieved to see there was no recommendation for elimination or corrective action. Perhaps Purgatory was not in my future.

I decided to go to the bar and face the abuse of my fellow classmates, and there was plenty of it. Most of it was geared toward teaching me the difference between the mic and the pickle button. One classmate also said he heard on the radio that the town of Falun was burning to the ground for some unknown reason. I assumed he was just giving me crap, but it did give me a start. After a couple beers, the conversation moved on from "let's kick the hell out of Oddjob" to other topics. Thank God!

As I got up to head to the pisser, I was met by an older full-bird colonel who asked if I was the lieutenant who made the inadvertent rocket release that afternoon. *Oh shit, here it comes, my introduction to Purgatory.*

When the colonel introduced himself as the wing safety officer, I braced myself for the worst. What I heard was unexpected: "Thanks, Lieutenant, for fessing up about your little miscue. If you hadn't done that, the maintenance guys would be tearing that airplane apart right now for a malfunction that didn't exist, and it may have been off the schedule for several days."

Holy crap, I'm a hero, I thought in disbelief. This called for another beer or two.

This incident would have a huge impact on the rest of my career. Over the 20 years I was a fighter pilot instructor, I would never tolerate an excuse for an obvious mistake by one of my students. "I fucked up, sir" was the only acceptable answer.

The rest of August and all of September slipped by rapidly as we honed our skills on the gunnery range. We now progressed into the most current tactics that were being used in SEA. Tactics on how to avoid MiGs and surface-to-air missiles (SAMs) were introduced, and we eagerly absorbed any information that might help us survive. Our instructors became more enlivened as we entered this phase of our training, backing up their instruction with personal experiences. It was exciting.

Every ride provided a new challenge, and we were beginning to have solid feelings of competence and confidence. In late October, we would deploy for two weeks to George AFB in California for a "graduation exercise" of sorts as we took advantage of the large tactical gunnery ranges and supersonic areas afforded by the vast expanses of the desolate desert in the Death Valley area. We also would soon be receiving our orders assigning us to our new fighter squadron in Thailand, and in a few months, we would be departing for the War.

With all this excitement in my life, I suppose I was not paying as much attention to Mary as I should have. She had begun experiencing morning sickness shortly after we'd discovered she was pregnant and had left her job at Wichita State. One morning when I was at work and getting ready to go out and fly, she called me, and I could tell she was upset.

"I just passed out in the bathroom and woke up on the bathroom floor," she let me know.

"Do you want me to come home?" I asked, knowing I shouldn't even ask. I should just let the Squadron know I had an emergency and leave.

There was a good deal of hesitation. "No, I think I'll be alright," she said in a soft, shaken voice. It was the answer I wanted to hear, but this was a scene that would be repeated several times over our relationship. Was I married to her or was I married to these jets and this crazy lifestyle? I know the answer, but I won't put it in writing.

McConnell had gone two months without losing an airplane, and we were beginning to think we had broken the curse. Then on October 10, we lost another Thud and its pilot. This time, it was a popular instructor, "Foxy" Flowers, who had experienced split flaps after takeoff and was unable to control the aircraft, crashing a few miles northwest of the base. The split flap situation was a result of a modification that had been made to the Thud to enable pilots with severe combat damage to lock up the horizontal stabilizer and use the flaps to guide the aircraft out of the combat area. It was a standing argument among the older Thud drivers as to whether the modification saved or destroyed lives. The arguments at the bar that night were pretty graphic, but all mourned the loss of Foxy, a driver who would certainly have been there drinking with us that night if it had been anyone else.

A short four days later, another McConnell Thud was lost due to an engine failure. This loss happened at George AFB. The pilot ejected successfully. We knew little about the accident, but it made us acutely aware that we were still involved in a dangerous occupation. Since I had been around Air Force flying, I had heard the phrase "accidents happen in threes," and we wondered if the third hammer would drop.

The answer wasn't long in coming. On October 18, I was flying a gunnery mission to Smoky Hill Range. We first flew a low level, which was uneventful. As we approached it, I noticed a thick black smoke rising from the center of the Range, and when we contacted the range officer, we were told that the Range was closed because of an aircraft mishap. I knew that the flight on the Range ahead of us was made up of an instructor and three of my classmates; a sickening feeling swept through me. After landing, we found out almost at once that another member of my class had unsuccessfully attempted to eject and was killed during a high-angle strafe pass. Capt. Wayne Wright was the second student from my class to "buy the farm," and our class strength was reduced by six from the original 22.

The mood in the Squadron was somber. We had all begun to believe that we had matured enough as pilots that it was unlikely we would lose another classmate. Especially affected was the sole remaining captain, Jerry Modolo. He had been a member of both flights during which his fellow captains were killed. Three captains in our class, accidents happen in threes, and he was the last one standing.

Also, our deployment to George AFB was supposed to be in three days, but we wondered if it might be delayed or canceled. Later that afternoon at a solemn Squadron meeting, we were told that the deployment would go as planned. Nobody had canceled the War, so we were still needed to feed the fight and we had better strap in for the ride.

One fortunate situation was that Mary was out of town when all this happened. She was with her family in Wisconsin celebrating her parents' 25th wedding anniversary. She would not return until after my deployment to George. I decided not to let her know about Wright's death until after she returned. I was glad I didn't have to remind her, yet again, that her husband had chosen a dangerous occupation.

Wright's death was soon pushed out of our thoughts by the George deployment. The first day of the deployment simulated a large combat strike force attacking a tactical target in southern California. All 16 students and several instructors would take off over a period of a few minutes and join up with tankers over western Kansas who would top off our gas and escort us toward the target. We would then fly to the target and roll in on it using some of the tactical formations we had been taught over the last few weeks. Sitting in my jet as we were being armed, a nervousness I had never felt before came over me. This was as close to an actual combat mission as I had experienced to date, and even though there would be no enemy to contend with, this was beginning to look like the real thing.

I had gotten comfortable at refueling, but this time it felt a little harder because now the gas was essential to getting me to the target. In this case, the target was a full-up surface-to-air missile site. Flying to the target in tactical formation, and then rolling in on the target simultaneously with three other jets, was so exciting. Attempting to join up with the whole gaggle of jets required my full attention. Landing at George, I had a real sense of accomplishment. As far as I knew, I hadn't screwed anything up and my bombs had landed pretty well.

As our remaining merry band of 16 joined up at the bar a little later, the excitement over our adventure grew amid the stories of our new accomplishments. The death of Wright only three days earlier was quickly moving into the background. We were all beginning to develop that first layer of "crustiness" that we had seen demonstrated by our instructors. The war must go on.

George went by in a blur, packed with all kinds of new, essential survival information. We had been unable to use the supersonic capabilities of the Thud in Kansas but now were free of that burden. As the Thud entered supersonic flight, there was a significant pull on the stick, and if the conditions were right, you could see shock waves emanating from your canopy bow and from other aircraft in your flight. In supersonic flight, the Thud felt like a T-38 with great responsiveness.

The most exciting event was a practice combat egress from a high-threat area. Following an event, our instructor pointed our flight path toward the desert floor, and he continued to descend to an altitude that was very low—probably 50 feet or so. He then called, "Afterburner now," and as the burner lit, I was pushed back in my seat with a noticeable acceleration. Now very low with my eyes fixated on the lead, I could see the ground rush by in a blur. I could also tell I was supersonic by the pronounced tug on the stick as we went through the Mach. The instructor then called, "Check altitude." I took a quick glance at my altimeter and was surprised to see I was 200 feet below sea level. We were now racing across Death Valley at breakneck speed and would be some of the last men allowed to do so. The practice was banned a few years later for environmental reasons. I am certain that many iguanas and other desert creatures had suffered from broken eardrums during that era. Oh well. Few men who ever walked the face of this planet can claim they have been supersonic below sea level, and I appreciate those critters' sacrifice on my behalf.

We were also introduced to some new weapons systems and got to use our magnificent gun in an air-to-air situation. On two missions, we got to strafe a "dart," a large aluminum-covered target that was dragged behind another Thud by a long cable. The dart was towed around a large racetrack circuit, and the students took turns rolling in on it. Using the air-to-air sight, you locked on the dart, drove into firing range as denoted by the pipper analog, and opened fire. Sometimes you could see pieces fall off the dart, but after each round, an instructor would move in close to the dart and let the student know if he had a "hit" or "miss." It was great sport, and you soon had your confidence in the gun reinforced.

We also experienced shooting an AGM-12 air-to-ground missile known commonly as the "Bullpup." It was a large missile that you fired from a wing station. It sounded and felt like a big freight train as it came off the rail. A bright burning flare was attached to the back of the missile that allowed the pilot to observe it until it hit the ground. A little joystick attached to the left control panel allowed the pilot to guide the missile to the target. This system was one of the Air Force's first attempts to give the pilot greater stand-off capability. Shooting the Bullpup was conducted from 20,000 feet or more (the stratosphere for Thud drivers). Of course, the time between missile launch and impact required the pilot to fly a predictable course over the target. In low-threat areas, this was a great weapon, but in high-threat areas with large anti-aircraft artillery (AAA) sites, this predictability was deadly. In combat, more than a few Thud drivers dived or were forced to jump out of their jets as they attempted to guide their missile to its target.

We were introduced also to the AIM-9B Sidewinder. This was an air-to-air missile that every Thud would carry into combat and, on rare occasions, use. It was an infrared, heat-seeking missile that had a movable seeker head that was "slewed" to the pilot's pipper. To practice shooting this missile, the pilot carried a target missile and a Sidewinder. You would first shoot the target missile and then move the pipper to the plume of fire emanating from the target. The Sidewinder seeker head would acquire the heat source and then give the pilot a manly "growl" in the pilot's headset. Once you got a solid growl, you could shoot the missile and watch it track down the heat source and explode upon contact. Unless a MiG driver got in front of you by accident and was showing you his tail, it was pretty unlikely you would ever employ the Sidewinder. However, it was available, and we all dreamed of getting to use it.

After a few days at George, word got back to us that not everything back at McConnell was normal. Apparently, someone in the chain of command, way above our pay grade, had decided to try and analyze what was going on with our class. After Wright's death, it was decided to assemble a study group made up of specialists in human behavior and interview our wives while we were gone. I got reports from my other married classmates that they were asking wives about their husband's mood, diet, drinking, and whether we were getting laid enough, or too much. We thought it was pretty funny, but I don't think the wives did. I was so glad Mary was out of town and not part of this clown act. We would never be affected by or hear anything about the results of this survey, but I was always curious about what their findings were. However, if one of their conclusions was that we were just a bunch of dumbass lieutenants, they would have gotten that right.

The last weekend of our deployment, a bunch of us drove to San Bernardino to celebrate our success. The married guys got drunk while watching the bachelors get drunk and make fools of themselves. Just because you can fly jets doesn't mean you know how to dance. Despite our losses and numerous screwups, we were all feeling like we were up to the challenge facing us. We were no longer afraid of going to Purgatory, and we were anxiously awaiting our orders to Thailand and the War.

Flying back to McConnell was also exhilarating. On this beautiful fall day, the base was clearly visible from 50 miles away. Making the descent into the base, I knew I was about to leave her, but it was the place that had elevated me to a new sense of maturity. It had taken me from being a young pup to an accomplished pilot and someone with a sense of comfort for my new world. I was now anxious to move to the next level.

The following day, Mary returned from Wisconsin. I excitedly told her about my George adventure before finally getting to the admission that we had lost another classmate. Again, the response was stoic and without emotion. She had long ago accepted the perils of being a fighter pilot's wife. I only hoped she never had to face the experience of an Air Force staff car pulling up to the front door and letting her know she was a widow. Or even worse, that she was the wife of a prisoner of war or that I had been declared missing in action.

It was now late October with only a few weeks left before our graduation December 9. My orders to Thailand were waiting for me when I returned from George; I was being assigned to the 34th Tactical Fighter Squadron at Korat. As the class compared assignments, I discovered only two others were being assigned to Korat. The other 13 were headed for Takhli. The instructors told me I was one of the lucky ones because Korat was the nicest base in Thailand, bar none. However, I was somewhat disappointed that I would no longer be interacting with all my classmates.

One of my most memorable flights at McConnell took place right after our return from George. I was scheduled to fly #4 in a flight led by a new instructor to the Squadron, Dan Cherry. I had never met him before but liked him at once. The flight was to be a standard four-ship to the Range with #2 leading a low level before we got to the Range. Because I was #4, I didn't pay much attention to the low-level route to be flown as it wasn't my responsibility to get the flight to the Range. However, #3 aborted in the chocks after startup and #2 aborted on the runway with some type of malfunction. It was now just me and Dan.

Oh shit. I sure hope Lead doesn't make me lead the low level. Sure as hell, right after takeoff, Dan gave me the lead, expecting me to lead him to the Range. I quickly pulled out my low-level map and headed east toward where I thought the low level was to start. As I passed a small town with a railroad going through it, which could possibly be the start point, I called "Hack," indicating I was hacking my clock and the low level was started. Because of my dumb assumption that I did not have to worry about the low level, I hadn't even set up the coordinates in my Doppler navigator. I was screwed. There was nothing left to do but continue on and use the headings and times on my chart to fake it while looking for a landmark I could identify. I found nothing but made my turns based on the chart data, and luckily, as I got close to the Range, I spotted the most prominent landmark around the Range, Kanopolis Lake. Somehow, I had headed in the right direction, and as we approached the Range, I gave the lead back to Dan, knowing I had busted the low-level portion of the ride—I didn't have a clue!

In the debrief following the ride, Dan said, "Good job on the low level, Oddjob."

Holy shit! He was as lost as I was. I admitted nothing. I always laugh to myself as I remember this episode. Dan Cherry would go on to lead the Thunderbirds, the Air Force demonstration team. On several occasions as I watched him lead the team, I would laugh to myself and wonder if he ever got them lost—I trust not.

One final hurdle had to be faced before graduation. I still had to take my TAC Check. This was a two-ship low-level to the Range with bombing and strafing that also included a wing takeoff and wing landing. In the past, I always dreaded check rides, but this one was different. My level of confidence was quite high, and I could fear no evil. The day before my check, I reviewed the next day's schedule and found I was to take my check ride with "Lucky" Ekman. Lucky was one of the most respected Thud drivers on base, and I wanted to impress him. The check ride went well, and as Lucky filled out the form certifying I was "Combat Ready," I felt confident that I *was* ready for what lay ahead. I felt honored to be endorsed by Lucky Ekman!

To the bar, by God. Many of my classmates were assembled there, and we were all telling fighter pilot lies and boasts when someone turned up the volume on the TV and told everybody to shut up. It was Thursday, October 31, and President Johnson was giving a speech to the American people:

> I have now ordered that all air, naval, and artillery bombardment of North Vietnam cease as of 8 a.m., Washington time, Friday morning. I have reached this decision on the basis of the developments in the Paris talks. And I have reached it in the belief that this action can lead to progress toward a peaceful settlement of the Vietnamese war.

You could have heard a pin drop. My classmates and I looked at each other in disbelief. We never talked about it, but all of us had been motivated by the prospect of someday wearing The Patch, and this did not bode well for our visions of manliness. No North Vietnam, no Patch.

Now what? Would our orders be canceled, or would we be assigned to other aircraft? Numerous exclamations of disbelief rose from the bar, but eventually our instructors calmed us down. They let us know that just because the war was over in Vietnam, there was still plenty going on in Laos and along the Ho Chi Minh Trail. Many of them also expressed the opinion that it was only a matter of time before we returned to missions over North Vietnam. It turned out they were correct about that, but it would not happen until 1972 under President Nixon. I went to bed that night still in a state of disbelief, but the next morning as I read the newspaper, I knew it was for real.

An article discussed the reasons for the complete halt of bombing, including the agreement that the people of South Vietnam's government will participate in the Paris peace talks, but, once again, I stewed over what this decision might do to my future. It was now unlikely I would ever wear The Patch, and I might never be called upon to use my recently acquired skills. Would I even be needed in the role of fighter pilot?

A few weeks later, all the students and instructors attended a graduation dinner. With the War Dance over, we were happy to be among the survivors.

Our instructors told us we were ready for what we were about to face but also said we still had a lot to learn. As I thanked Paul Sheehy, he gave me an infectious grin and his standard, "Shit hot! Do good, Oddjob."

I did not win the award for "Top Gun" or "Outstanding Graduate," but I was elated that Mary and I had survived this most important test. "Thank you, Lord!"

Figure 3 - Ready to be Somebody: The 16 Graduates of the author's F-105 class just before they left for Thailand, 1968. Photo courtesy of the 23rd TFW

Now off to face the real deal. Putting Wichita in the rearview mirror in early December 1968 was filled with mixed emotions. The F-105 schoolhouse had been one of the most difficult yet satisfying experiences of my life. The excitement of learning to fly this beautiful and powerful machine was a real high. However, I now had orders to proceed on January 1 to a port of call at Travis AFB with an eventual destination of Korat Royal Thai Air Force Base (RTAFB).

The orders specified that I would be assigned to the 388th Tactical Fighter Wing (TFW) and the 34th Tactical Fighter Squadron (TFS). Those orders also specified that I would first proceed to Clark AFB in the Philippines to attend Jungle Survival School, which was referred to as "Snake School" by all who had attended. These orders were a sobering realization of the eventual end result of the path I had chosen—very sobering. Less than two years earlier, Col. Hosman had told me, "You were so dumb you didn't know you couldn't do it," and once again, I realized he was right.

Mary had asked her parents if she could stay with them while I was in Asia and they had, of course, agreed. They had a large farmhouse north of Green Bay and a nice apartment above the main kitchen complete with a small kitchen and the use of a bathroom she would share with her younger siblings who were still at home. The bedroom area was spacious enough for a couch and a crib. *Holy crap,* I thought as she explained her new setting. We were pregnant, and I was about to leave her behind to have our first child by herself. I was much comforted knowing she would be living with her caring family, but it was still a sobering thought.

We pointed Cognito northeast along the same route we had used to arrive in Wichita six months earlier, and once again, we observed Thuds flying low over the Flint Hills. This time the perspective was much different. I was now a "fully qualified" (at least in my mind) Thud driver. Watching those jets head north toward Smoky Hill had taken on a whole new meaning, and we were both fully aware of the risks associated with being a Thud driver in the middle of a sustained air war.

En route to Wisconsin, we made a slight detour to visit my mother and George in Kentucky. I had not seen her since Bob's death and was not sure what to expect. She seemed pretty normal as she greeted us outside the house on the hill. I did not bring up Bob nor did she. She did say that Merry was doing OK, and that Merry and Susie were going to live together in Minneapolis when Wes departed for the war. In time, I had an opportunity to talk to my stepfather in private, and he painted a pretty gloomy picture of my mother's situation. I knew she had taken

Bob's death hard, but George painted a picture of a woman who was in danger of a complete breakdown. He said he was not sure they would be married much longer if she didn't improve. Certain my mother would not want me to bring up anything about Bob's death, I avoided the subject with her.

We did have one happy experience on this trip when we all jumped into the family Buick and drove to Nashville for a night at the Grand Ole Opry. The trip brought back many memories of similar excursions we had made in my youth when George would load us up in the car and head for Nashville and the Opry. I will always regret that I didn't appreciate what I was experiencing because I was too wrapped up in the rock and roll of the day, including Elvis. Following the Opry, we went to a pawn shop and I bought a five-string banjo. I thought I would learn to play it while I was in Thailand. I also bought a few instruction manuals but would later realize I would never be Earl Scruggs. Some things you can't learn from an instruction manual.

The morning we got up early to head to Wisconsin, Mary and I went to the kitchen expecting to find Mother and George there waiting to bid us goodbye. Only George was there.

"Where's Mom?" I asked.

"She's still in bed—and she told me she doesn't want to say goodbye to you."

"What should I do?"

George paused. "I think you should leave. With both you and Wes heading off to Vietnam, all she can think about is what could happen."

I nodded. I sensed that to push the issue and attempt to say goodbye to my mother would not be a smart thing to do.

Mary and I didn't talk much as we headed toward her father's home in Suring, Wisconsin. There were too many topics we didn't want to discuss. We both knew what the consequences of my deployment could mean if it ended with a bad result.

When we arrived in Suring the next day, her family greeted us with a matter-of-fact attitude. As we unloaded our few belongings from Cognito and headed up the back stairs of the Bartz home to Mary's new abode, we were excited to find a crib already in one corner of the room.

That winter in Wisconsin was very cold with a lot of snow. During the 20 days I spent there, the temperature never got above 32 degrees and was below 0 almost every night. Mary and I did clear the snow off the ice on a pond behind her house, and she practiced figure skating, and from time to time, I would play hockey with her brothers and some of her cousins. However, I did not care for the cold and the solitude of the Northwoods and soon became restless to begin the next step in my journey.

On Christmas Eve day, I helped Mary's dad, Bill, cut down the Christmas tree from out in the maple swamp. The family tradition was that Santa decorated the tree, so it was not trimmed until Christmas Eve. Christmas at the Bartz farm was a great experience for me as it was so festive with lots of trips to church, Mary's mom, Laura, playing the piano, and Bill leading the kids in every Christmas carol ever written. The three older girls were very good at harmony, and as a wannabe musician, I was impressed.

New Year's Eve was spent at Chute Pond at a cabin owned by Claudia's (Mary's sister) husband. Unlike several New Year's Eve celebrations prior to this one, I was not in a celebratory mood. I drank little and, for the most part, stayed close to Mary. We spent a lot of time staring at each other. On a few occasions, she would tell me the baby was kicking and I would put my hand on her slightly expanding stomach, wondering if I would get to see this fussy little kid of ours. What was there to say—it was what it was.

Early the next morning, Mary's oldest brother, Billy, drove us to the airport in Green Bay. It was cold, and the snowbanks by the side of the road were high. If there was a silver lining to any of this, at least the next day I would be out of this deep freeze. We parted at the airport with a somber embrace, and I was off. My maturity climbed a steep learning curve that day. I was happy Mary was in such a great situation, and I knew she would do well.

Years later, Mary would recall our last moments together as being more emotional:

My husband's SEA deployment began with Jungle School in the Philippines. He flew out of the smallish airport near my hometown right after New Year's Day 1969. We had celebrated New Year's Eve at the lake cabin of my sister's boyfriend's family with cousins and friends. I didn't drink, of course, being five months pregnant, and all was going well, laughing and dancing, until Percy Sledge's "When a Man Loves a Woman" was put on the turntable. Ed and I slow danced. That did it. I started seriously crying—in April I would give birth (my first time) without my husband by my side, and by the end of this new year, I might be without the baby's father. I guess we went into another room so as not to put a damper on the party. Ed tried to console me; although from what I could tell, he had no regrets about what was going to happen even though, years later, he admitted he really didn't think he would survive that year. Statistics don't lie.

January 2 rolled around. After all the Christmas celebrations, church services, meals, and family gatherings; the crazy, bloody farm pond games of hockey with my brothers and cousins; and the New Year celebrations, he was finally going to war.

My 18-year-old brother was selected, unsure by whom, to drive us the 60 miles to the airport that evening. Ed had packed a military duffle bag and carried his banjo, purchased in Nashville months earlier, with him. There was not much conversation during the drive or the wait at the airport. Ed tried to keep things a little lively, but the underlying tone for me and my brother was severely serious. My brother, of course, sensed all this but had no context for this situation—sending a family member off to war. Our large extended family was mostly from the farming community. I only had one uncle who served as a cook in WWII and two uncles who served a short stint in the Navy in the '50s. During the hour or two waiting for the flight to be called (no TSA back then), I was constantly on the verge of tears. Ed got his banjo out and started picking away like the novice he was, and I was thinking, Put that thing away and pay attention to me, your pregnant wife whom you're going to trade for a year of high excitement and danger!

Time to go, last hugs and kisses. He was fairly solemn but no tears graced his face. Off he went with a wave back. My brother and I walked off to the airport parking lot. I was bawling, so we didn't say a lot to each other on the way home.

Chapter 7
Making Dynamite

Charles Winship Dinan grew up in Wakefield, Massachusetts. From an early age, he demonstrated a great aptitude for mathematics and science. He used these skills to gain acceptance into the Massachusetts Institute of Technology (MIT). He graduated from MIT in 1928, but he always considered himself a member of the Class of '27. Just prior to finals his senior year at MIT, he was knocked out of a ring during a boxing match, and his injuries set him back a semester, causing him to graduate with the Class of '28. He was one of those rare students at MIT who commuted to class, every day riding the train into Boston from Wakefield. Somehow, he also found time to compete on the MIT track team. Upon graduation, he found that his degree in chemical engineering provided plenty of opportunities for employment and thus commenced a career in the field.

Alice Mary Drugan also grew up in Wakefield and was an exceptional student. She began her college career at Radcliffe but later transferred to Boston University, where she completed her degree in education and found a job in the Boston public school system as a teacher. She was brought up in the Irish-American community in Wakefield. However, like Charles, she did not follow the common practice of early marriage. Alice and Charles were married in 1941 when he was 33 and she was 31.

During the Second World War, Charles was busy assisting in the building of coke (a byproduct of coal) plants, which were vital to the steel industry. Consequently, he moved frequently as more coke plants were brought online. Charles and Alice began growing their family and, over a period of seven years, had four children: Mary in '42, David in '44, Charles Jr. in '45, and John in '48. As their family expanded, they decided to set down roots and ended up in Nutley, New Jersey, where Charles was completing a coke plant. In all respects, the Dinan family was the typical family of the 1950s and early '60s.

Mary, their only daughter, was born with a thyroid condition, which was, unfortunately, undiagnosed at birth. She developed a condition known as cretinism. This malady prevents the normal growth of the head and limbs and also impairs normal brain development. Not diagnosed until she was nearly a year old, it was too late to rectify the effects. Consequently, Mary would have impaired mental function all her life, and her hands and feet would never develop to full size. She attended schools for children with special needs, but she was never able to read and spoke only in short phrases, communicating mainly by hand signals. The prognosis for this condition also included an early death, with an average life span of 35 years. Mary would die when she was 37. The Dinan family was protective of Mary and provided her with a comfortable existence.

David Thomas III and Charles Jr. (Charlie) were only a year apart and did the things young brothers typically do: argue, fight, and wrestle, always blaming the other when being disciplined by their parents. David was smart and loved to read. Many nights, Charlie would go to bed and notice, as he drifted off to sleep, that David was under the covers reading a book with a flashlight. David also enjoyed music, playing the trumpet and becoming a pretty accomplished piano player. As the oldest son, David was a role model for the younger boys.

At a young age, Dave began to build model airplanes and got good at putting together the balsa wood–framed models that were covered with paper and dope. He would mount the high RPM engines on them and fly them around in circles on 50-foot (or more) tether wires. The goal was to always make them fly faster, and with his friends, he would participate in mock air battles using crepe paper tied to the back of the aircraft as the target of the other planes' propeller. Many hours of hard work could be destroyed in an instant during these encounters—and often were.

Seton Hall College was a 30-minute bus ride from the Dinan home, and David took that bus every day to attend Seton Hall Prep during his high school years. There he ran on the track team and participated in the Drum and Bugle Corps, but his real love was science. Dave possessed an outstanding ability to retain information. Consequently, he was a good test taker and didn't have to study much in high school. On one occasion, one of his high school teachers offered the class 20 dollars if anyone could memorize "The Rime of the Ancient Mariner" in a week. Dave won the prize easily. (This is a lengthy poem written in the late 1700s and the genesis of the phrase "wearing an albatross around your neck.")

With all his abilities and a good amount of test-taking skills, Dave was readily accepted at MIT. Charles was proud that his oldest son was following in his footsteps and would achieve a degree from his alma mater. Dave, enrolled in the physics department and AFROTC, was fired up to succeed at this next life hurdle. However, MIT was a whole new kettle of fish. Suddenly, Dave found himself surrounded by lots of folks who were just as smart as he was. Things didn't come as easily as they had at Seton Hall Prep, and the competition was far stiffer. Also, Dave was on his own for the first time and decided to spend a significant amount of time having fun.

During his first semester at MIT, he discovered Phi Kappa Theta (PKT), a Catholic fraternity with its own house near the campus. Before long, he had moved into the house and was now surrounded by a group of fun-loving "brothers."

Gene Montrone, one of Dave's PKT brothers, recalls that the brothers "lived together, worked together, played together, and prayed together. In those days, all events and parties ended up with a large group singing around the piano. As the lyric of 'Heart of my Hearts,' one of Dave's favorite songs, goes: 'We were rough and ready guys, but oh how we could harmonize.'"

Dave's social calendar soon overwhelmed his study calendar. The best way to sum it up was that Dave was having too much fun. Dave lasted three semesters at MIT before it was suggested he "take a semester off." Of course, this is not an uncommon occurrence among young folks making their first step out into the "real" world. However, Dave was not greeted with much sympathy when he came back to Nutley with his tail between his legs and his head hanging down. His father was furious and made sure Dave knew it. It had cost Charles about 25 percent of his income to send Dave to MIT, and this was not the result he expected from his oldest son.

Dave was allowed to move back into the family home and registered at Stevens Institute of Technology in neighboring Hoboken. Having learned a significant lesson, Dave turned things around and managed to complete his degree in physics with a total of four years in college. He graduated in the spring of 1966. Along with his degree, he pinned on 2Lt. bars, having successfully completed AFROTC. Having redeemed himself, for the most part, he got his airspeed up and set off for flight school at Craig AFB in Selma, Alabama.

Once again, Dave was out on his own, but this time, he was in an environment where his skills were suited to the challenge. He found the academic challenges much easier than the MIT or Stevens curriculum, and he was able to finish number one in his class in academics. His hand-eye coordination was also quite good, so his flying scores were high, and he managed to finish high in his overall class standing. His successes began to transform Dave into someone who, once again, wanted to have fun and wanted a bigger challenge.

Fun was back in! He found that being able to walk into a bar with an easy smile, no wedding ring, and an obvious air of self-confidence made it easy to make new friends. And when asked what he did, he would modestly proclaim, "Oh, I fly jets." Suddenly, a lot of new doors were opened.

One thing was missing, however—a really hot car. His brother Charlie had found a blue 1965 Corvette Stingray convertible back in Nutley. It came complete with an extra-large engine and cost only $2,500. Dave told Charlie to buy it, sight unseen, and drive it to Selma. Charlie was more than happy to participate in this road trip and drove straight through to Selma. Dave bought Charlie a plane ticket back to Nutley, and now he had his signature car, a blue Corvette convertible with plenty of room in the trunk for his guitar. Could it get any better than this?

By now, Dave's swagger had fully returned and probably been elevated to a new high. Now Dave had to decide what his next challenge would be. He was high enough in his class to get almost any assignment that came available. Like some of the others in his class, he had developed a strong desire to live on the edge. He knew he wanted to get a fighter. A few instructors at Selma wore The Patch, and the 10 o'clock news almost daily reported on the F-105 raids into North Vietnam. Being able to say, "Oh, I fly jets," gives you status, but saying, "Oh, I'm a Thud driver with 100 missions over North Vietnam," puts you in a completely different league. Testosterone is a wonderful thing!

Dave requested an F-105, and his efforts were rewarded with his first choice. He packed up his guitar in his blue Corvette and headed for McConnell AFB in the fall of 1967. What Dave didn't realize until he arrived at McConnell was that his class was part of the Air Force's first attempt to bolster its dwindling ability to fill F-105 cockpits with experienced fighter pilots. Hence, the decision was made to infuse their F-105 force with a complete class of inexperienced, recent UPT graduates. Dave's F-105 class would consist of 25 lieutenants and two captains. This was a bold experiment as the F-105 was probably the most difficult airplane in the Air Force inventory to fly. Hence, only UPT graduates who had finished at the top of their classes and were volunteers were selected.

The Air Force decided to compensate for the inexperience of these rookies by increasing the training program with an additional 28 hours of flying time. Thus, each student would receive 78 sorties (flights) for a total of 120 hours. (Putting this into perspective, an airline pilot must have 1,500 hours to command a commercial airliner, and their primary responsibility is taking off and landing safely—flying formation, inflight refueling, dropping bombs, fighting MiGs, or operating at higher speeds not required.)

Dave reported to McConnell in late November 1967 and was assigned to the 563rd TFS. The instructors they inherited were some of the most colorful characters ever to fly a fighter, all with 100 missions over North Vietnam. The instructors were unsure at first of what to do with so many novices but soon discovered these kids were eager to learn with good flying skills, and unlike some of the more experienced pilots they had trained, they would actually listen and do what they were told to do. The danger was asking them to do more than they were capable of. On the other hand, these seasoned veterans knew what these young studs would soon face and wanted to give them as much realistic training as they could. Hence, the training always operated on the fine edge between safety and the best effort to prepare them for survival.

Another issue that was on everyone's mind at McConnell in March of 1968 was President Johnson's announcement that American forces would no longer strike targets in the northern areas of North Vietnam. These were the highly defended targets around Hanoi and Haiphong. However, strikes in southern North Vietnam would continue in an attempt to interdict the supplies the North Vietnamese attempted to move to their troops and allies in the South. It was decided to continue to instruct these young students for the worst case—a return to full-on war against the North. Thus, the training would continue to include aerial warfare against MiGs, defensive maneuvering against SAMs, and tactics to avoid high-caliber anti-aircraft guns. Almost everyone assumed the Thuds would be going back to the North. (And they were right, but it wouldn't take place until 5 years later in December 1972.) Thus, these students would get the full-meal deal.

Dave performed well in his Thud checkout. He didn't finish at the top of his class, but he demonstrated a high level of competency along with most of his classmates. The serious nature of his new environment began to surface. In early March 1968, one of his classmates, 2nd Lt. Frank Perry III, was killed as he ejected from a Thud just before impact on a radar low level en route to Smoky Hill Range. It was a rather mysterious accident. Perry and the rest of his flight were operating at a low altitude and had just been ordered by the flight lead to make a radio frequency change. There was no response from Perry, and no one in the flight saw him impact the ground or spotted the wreckage. Assuming he had lost his radio, they called McConnell and told them to expect a "radio out" Thud to be returning to McConnell by himself using "radio out" procedures. They then continued with their mission. The accident was reported to McConnell by some local farmers who had observed the impact and found Perry's body close to the aircraft with the ejection seat nearby.

For most of the students, this was the first time they had been exposed to the loss of one of their fellow pilots. However, the instructors had been through this type of loss many times, and the students soon emulated their demonstrated behavior that this was the "cost of doing business." There was little time to ponder the dangers of the course they had chosen. In fact, as it was pointed out and demonstrated by their instructors, to concentrate excessively on the dangers of their situation made them more likely to become a "smoking hole."

Dave and his classmates also had the opportunity to observe how following your fighter pilot instincts can get you in a lot of trouble. Near the end of the course, the class deployed to George AFB in California to get a chance to hone their skills in a supersonic environment and actually shoot missiles and the gun in a simulated air-to-air exercise. This lesson took place on an air-to-air mission led by two of the most celebrated and colorful instructors in the entire Thud community: Maj. Robert V. "Boris" Baird and Lt. Col. James W. "Black Matt" Matthews. Baird had led the first strike mission over North Vietnam in 1965 and had been the first F-105 pilot shot down and rescued in the war. Black Matt was widely known for his bravery in combat and for his ability to tell fighter pilot stories for hours on end and never repeat himself. Both men were highly respected by Dave and his classmates.

The flight was a four-ship air combat maneuvering (ACM) mission with the objective of demonstrating and practicing defensive and offensive maneuvering against an airborne target. Each of the instructors had one of Dave's classmates on his wing, and once they were airborne, the instructors would first demonstrate the attack and defensive maneuvers and then let the students practice them. While the instructors demonstrated the maneuvers, the students were instructed to maintain "fighting wing," which requires two wingmen to stay in a 500-foot, 45-degree cone behind the instructor (translation: "keep the son of a bitch in sight and don't hit him"). Boris was the defender on the first setup with Matt making the attack. Matt made his attack from high above, coming in at supersonic speed, and Boris called for a "hard turn" to offset the attack.

Boris felt like Matt had not performed a very effective attack and decided to attempt to gain the offensive; in the process, he let his airspeed get too slow, and his aircraft snap rolled and fell into a spin. Unable to recover from the spin, Boris ejected and gave the jet back to the taxpayers. The student flying fighting wing behind Boris decided not to follow him through the spin and brought his jet back home. *Lesson to all students:* No matter how good you are and how much experience you have, there is always room to fuck up!

Not wanting to be outdone by Boris, Matt found a way to shift the attention to himself. A short month later, an article in the June 1, 1968, *Colorado Springs Gazette-Telegraph* ran the headline, "Sonic Boom in Fly-Over Damages Academy Buildings." Ray Herts reported:

> *Every good airplane deserves a good story. On 31 May 1968, a dedication ceremony took place at the United States Air Force Academy to honor graduates who had served in Vietnam. The ceremony included the entire cadet wing, the superintendent and commandant of cadets of the USAFA, a representative of Republic Aircraft, members of the press, among others. To conclude the ceremony, a flight of four F-105s were to fly over in formation at 1,000 feet and then fly over singly at 250 feet. The formation portion was flown as planned. But the flight leader, Lt. Col. James "Black Matt" Matthews, came back for the single-file pass and exceeded the speed of sound at less than 100 feet. The ensuing sonic boom broke hundreds of windows.*

As Dave's class prepared to depart for SEA in early June 1968, they graduated all but one of their original classmates and had some great stories to tell about some great American heroes. Dave's orders assigned him to the 469th TFS at Korat, Thailand. He sold his blue Corvette and headed off to war with a great air of confidence and sense of accomplishment. Preston T. Duke, one of Dave's classmates at McConnell, remembers him in an email to me:

> *I met Dave Dinan when our class reported for F-105 RTU training at McConnell AFB, Kans., late Fall 1967: 25 2LTs and 2 captains freshly graduated from UPT. Our enthusiasm and egos vastly overshadowed our lack of experience, but we had it on good account that the Air Force was about to hand us the keys to a single-seat fighter with carte blanche to fly fast, party hard, tell outrageous lies about our aerial exploits, and be somebody.*

Dave was living the dream: single-seat fighter pilot, blue Corvette Stingray convertible, and a handy guitar as a social icebreaker. I called him Dynamite Dave. My main social memory of Dave from RTU is of a late Saturday night foray, immediately prior to Easter Sunday, to The Stables, an off-base watering hole, where we engaged in light conversation and awesome dance moves with two local hooters who, upon later reflection, may have been slightly out of warranty. Suffice it to say that our Easter sunrise service did not fit the traditional mold.

We knew we were headed to SEA, sought that assignment and eagerly anticipated the challenge...

Chapter 8

Dave's War

During mid-June of 1968, there were still three squadrons of F-105s at Korat: the 34th TFS and the 469th TFS, which were both strike squadrons, and the 44th TFS, which was a Wild Weasel (SAM-suppression) squadron. Because the 44th was a highly specialized squadron that flew two-seat Thuds, they did not take on any of the 12 lieutenants who were parts of Dave's McConnell class that arrived at Korat in late June. The other 14 members of Dave's class went to Takhli to fuel the remaining Thud squadrons there.

The 100-mission program was still alive and well, so more bodies were needed to fuel the fight, but this large influx of "Thud babies" was testing some unexplored territory. Dave and his friends were divided up between the two strike squadrons, and Dave found himself rapidly incorporated into the 469th "Fighting Bulls." He was assigned his living quarters (we called our quarters "hooches") in the 469th area, and his roommate was his fellow classmate Bob Zukowski. Shortly after his arrival, Dave flew an area familiarization flight and, a few days later, flew his first combat mission into Route Package 1, the area of North Vietnam just north of the demilitarized zone (DMZ).

The instructors at McConnell passed on many stories about the heavily defended areas around Hanoi and Haiphong but had not talked about the kinds of defenses around the DMZ or in Laos. However, there had not been many air strikes against the northern targets in several months, and the Vietnamese had moved substantial amounts of weaponry out of the North into the South and along the Ho Chi Minh Trail. Defenses were especially heavy around the Laotian Communist capital of Sam Neua in northern Laos and the villages of Ban Ban and Tchepone in central and southern Laos. So, on his first combat mission, he rapidly became aware that the enemy defense efforts were alive and well.

Figure 4 - 1Lt. David Thomas Dinan III, 1968.
Photo courtesy of the National Museum of the
U.S. Air Force.

Then on his third mission, Dave got a chance to really grow his fangs. While flying as part of another mission in the DMZ, a Thud from Takhli had been shot down, and Dave's flight was called to the scene in an attempt to rescue the pilot who had ejected and was evading North Vietnamese ground forces. A search and air rescue (SAAR) mission is the effort that brings out the highest level of emotion (not to mention testosterone) in a fighter pilot. Saving a downed brother should be, and is, the highest motivator imaginable. This is especially true if the downed pilot is using his radio to describe his situation to the aircraft above. It is hard to describe the sense of urgency on the part of those listening to their desperate comrade.

In this case, Dave was called in to make multiple strafe passes on the enemy troops attempting to capture—or worse, capture and torture—the downed pilot. Dave's learning curve suddenly became very steep, and the realization of the risks involved became quite clear. Maturity and confidence grow rapidly when faced with this type of situation. Best of all, the downed pilot was rescued. For his demonstrated courage, Dave would be awarded the Distinguished Flying Cross (DFC) a few months later.

Part of the citation read as follows:

1Lt. David Thomas Dinan III is awarded the Distinguished Flying Cross for extraordinary achievement while participating in aerial flight as an F-105 Thunderchief pilot over North Vietnam on July 14, 1968. On that date, Lt. Dinan was a member of a flight diverted from a pre-planned mission to support the rescue of a fellow pilot downed in a fiercely defended area of North Vietnam. In a constant barrage of deadly anti-aircraft fire, Lt. Dinan, without thought of his own personal safety, made repeated passes in close proximity to the survivor, successfully silencing the fire and halting the advance of hostile ground forces attempting to capture the downed airman. The professional competence, aerial skill, and devotion to duty displayed by Lt. Dinan reflect great credit upon himself and the United States Air Force.

Not a bad day's work! That same day, a member of Dave's class who was flying with the 34th TFS was hit and attempted to recover his aircraft at Udorn RTAFB in northern Thailand. He managed to get the jet on the runway but, due to the loss of essential hydraulics, was unable to stop the aircraft and ran off the runway, destroying the jet. Luckily, the aircraft did not become a fireball and the pilot, Lt. Gary Confer, walked away from the wreckage. He caught a hop out of Udorn and was soon back at the bar at the "KABOOM" (Korat Officers Open Mess).

With both Gary's and Dave's stories of that day's events, an especially enlivened bunch gathered at the bar that evening. One of the best parts of experiences like these for the guys was the opportunity to retell and embellish their stories at the bar. With this large influx of lieutenants, all with new adventures to describe, getting to the bar after debrief was an unquestionable temptation. Because the flying schedule was spread out throughout the day, there was almost always a table of lieutenants telling loud, colorful stories with descriptive exclamations and arms flailing. Beer at the bar was quite cheap, which helped make the sessions even more outrageous.

In addition, there were a lot of new games to be learned. Fighter pilots were always looking for competitive pursuits to prove their manliness and make someone else buy them a drink. While sitting around the table as the drinks were running low, we played "three coins" or "dollar bill" games, both of which incorporated luck and a certain amount of skill to win.

Although several slot machines were in the bar, they were seldom used by the fighter pilots. Personal confrontation was a hell of a lot more fun. If money were to be wagered, the session would normally graduate to a dice game known as "4-5-6." The biggest problem with this game was that the initial low stakes at which it started always seemed to become quite high, and some rather drunken loser would end up writing an IOU to the winner. Leadership generally didn't like this game because it sometimes created animosity within the ranks, as large debts became common. In a few cases, leadership was called in to referee IOU indebtedness. They always ruled against the signer of the IOU, even if he lost his house, his truck, or his wife. If you made a bet and lost, dammit, you've got to pay!

More active games were played as well. Like at McConnell, there was a bell at the bar. Wearing your hat would cost you a round. Luckily, no one had to sweat getting a call from their wife at the bar. Phone calls to and from the States were expensive, and it was almost impossible to comprehend what was being said, so you could be confident that your wife wouldn't interrupt your drinking and storytelling.

One of the most popular games was "Dead Bug." Anytime the words "Dead Bug" were yelled out, everyone had to fall to the floor and immediately place his arms and legs toward the ceiling as if they were a dead bug. The last one to hit the floor would be obligated to buy the entire bar a round of drinks. The loser also inherited the "hammer" and was then the only person who could yell "Dead Bug" and got to determine who the new loser was. If the person with the "hammer" was around the bar, one soon found out that it was unwise to sit on a bar stool. Many a fighter pilot found himself knocked silly as he pushed himself backward from the bar, stool and all, with his arms wrapped around his head to cushion the impending impact. No comedy skit ever contrived is more humorous than watching a bunch of grown men play "Dead Bug." If you ever hear someone refer to an actual dead bug as a "deceased insect," you can be pretty certain he was a fighter pilot.

"Carrier Landings" was also a fun exercise. A portion of the bar room floor would be marked off with candles and covered with ice water. A start point was designated, and the players would take turns running as fast as they could and then belly flopping onto the "carrier deck," seeing how far they could slide. This game was played infrequently at Korat, but the boys at Takhli were really into it, mainly due to the better playground. At Takhli, you could take a long run through the dining room to pick up speed and then, after making a 90-degree turn to your left, enter the bar. There were three steps down to the bar, so the player could leap from the top of the stairs and become airborne for a long time before impacting the hard floor and cold liquid below. Also, the "deck" of the carrier was perfectly aligned with the back entrance to the bar, so if you did a good job, you could slide out the back door onto the concrete porch. Needless to say, a lot of bruised and sore ribs as well as skinned forearms resulted from this game, but injuries never became too painful until the alcohol wore off.

End-of-tour, 100-mission parties were normally the craziest, with pilots being thrown in the pool at Korat, which was right next to the bar entrance. Almost everyone would end up in the pool, and most went in under duress. Not having a pool within easy reach, the guys at Takhli threw each other over the bar. Normally, three or four guys would grab the victim and swing him back and forth, and on the count of three, attempt to launch him over the bar, many times with only limited success.

It is rumored that the wing commander at Takhli banned this practice after two victims sustained broken arms at the same party. Games that resulted in removing pilots from the flying schedule generally caused the "Wing Kings" to lose their sense of humor.

Couldn't these guys take a joke? Let's not forget that on the first supersonic flight by a human being, Chuck Yeager flew with broken ribs and hid it from his supervisors. This was after falling off a horse at Pancho's Villa, a local watering hole near Edwards AFB the night before his historic flight. This kind of behavior is among the finest traditions of the fighter pilot fraternity.

Dave and the other new McConnell lieutenants were also introduced to a contest between squadrons. On one occasion, Dave described a contest between the 34th and 469th where the challenge was to eat the most raw eggs. To qualify as a "counter," one had to eat a whole raw egg, shells and all, without gagging or spitting anything out. Breaking the egg was easy and swallowing its contents was OK if you didn't think about it, but the trick was to make sure you chewed the shells into very small pieces before you swallowed them. Failure to chew those eggshells into a fine paste gave you a fit when you got ready to expel them the next day. The older officers would normally limit their egg intake to one and then compel the lieutenants to keep eating more until they puked. As in all games of this type, the losers bought the drinks.

Within a short period, pilots lost track of what day it was as the War went on without interruption. Thud strike flights only attacked targets in daylight, but the early takeoffs began well before sunrise so as to be in the target area as the sun came up. This often meant being out of bed around 3 A.M. The high availability of lieutenants gave the older guys a chance to get a little more shut-eye. It soon became common practice that the earliest morning missions were mostly lieutenants, with a captain thrown in for adult supervision. So it was not uncommon to see a table of lieutenants at the bar as early as 9:30 in the morning, already done with work for the day, and then sometimes already "done-in" at the bar by noon.

Each day's missions began with a mass briefing for all the flights within a 2- to 3-hour window. Normally 20 or more pilots attended each of these briefings to get details on the latest threat intelligence in the target area as well as the best area for the possibility of escape from capture if you were shot down. Also, the latest weather information was updated. At one of these briefings early in July, something radically different occurred—the intelligence briefing was given by a female. As she walked out on the stage, the pilots turned their heads, looking at each other with questioning expressions. They all had the same thought: *What the hell is this all about?*

The attractive petite young brunette introduced herself as Lt. Valerie Galullo and proceeded to go through all the information the pilots were used to hearing at these briefs. However, their attention was more acute today as she used a pointer to describe defenses in the target areas and areas for the best chance of survival. Suffice it to say, their attention was not fully focused on the pointer. Only a handful of "round eyes," as Caucasian women were labeled by the pilots, worked at Korat, and they were all assigned to the hospital as nurses or as secretaries for the wing commander and his staff. This was the first woman to be involved with the operations of the Wing.

The lieutenants had no reason not to accept this woman in her nontraditional role, but the senior guys, especially the majors and above, were a little less benevolent. In their many years as pilots, they had never been exposed to a female in this role and perhaps felt like their own personal value was somewhat diminished by the presence of this inexperienced woman. The military culture, especially the old guys of the military, did not value women at that time. In addition to giving the mass briefing, Valerie would oftentimes conduct the intelligence debrief after each flight. She found that the young lieutenants were more than willing to accept her in this role, but she felt a good amount of pushback from the senior officers.

This acceptance extended beyond her intelligence duties. When she ate at the Officers Club or had a drink at the bar, it was always the lieutenants who asked her to join them. They soon bonded as friends, and she looked forward to her encounters with her young warrior friends. Within a few weeks, Valerie was joined by a second female intelligence officer, Maureen McCabe. Maureen was much taller than Valerie and thin. The two soon gained the monikers "Mutt and Jeff" from the aircrews when they were out of earshot.

Dave was frequently found at the lieutenants' table and was one of those who would urge Valerie to join them. Having both grown up in the Northeast and both being Catholics with similar interests, the two soon became close friends. For Valerie, her time at the lieutenants' table and with Dave was what she needed to escape the feeling of inadequacy created by many of the older fighter pilots.

Another break from her uncomfortable work situation was a relationship she and Dave and a few of the other lieutenants formed with Father Gene Gasparavic. Father Gene was the Catholic chaplain at Korat and was frequently found at the end of the runway before the Thuds took off on combat sorties wearing a vest with a cross on it and crossing himself as the aircraft finished being armed and pushed up its power to take the runway. Many pilots admitted to feeling a little uncomfortable taking off without one of the base's chaplains to send them off.

Father Gene was a delightful character who would express his wish that the Pope would soon allow Catholic priests to marry. "My candle is burning low" was his expression for this wish. One perk of being a chaplain was that you were issued a pickup truck and could drive it off base. On many occasions, Father Gene, Dave, Valerie, and other lieutenants would jump in his pickup and head to one of the many Thai restaurants in Korat City. Eating Thai cuisine, drinking a bottle of Singha beer, and listening to Thai music would always prompt someone to exclaim sarcastically, "War is hell."

Dave was beginning to feel competent in his role as a strike pilot and to feel comfortable assuming the risks associated with that role. However, he was also feeling disgruntled about the way the War was being conducted and the rules of engagement placed on him. In a letter to his brothers written less than a month after his arrival at Korat, he expressed his feelings:

I'm sitting in a little shed on the end of the runway with a pair of binoculars and three radios to keep me company while I sit here and watch (supervise) the flying operations. Mostly it's just boring as the dickens 'cause most of the time there isn't anyone landing or taking off. What I do is check the F-105s to make sure everything is normal before they take off and while they're landing. Thank God for small favors.

Things are really getting Mickey-Mouse around here, with a whole bunch of new regulations on what is proper behavior for officers and gentlemen on base—and what is ten times worse, they're telling us how to fly the airplanes in combat. Of course, I have a tendency to develop a short memory when this nonsense is going on. The way I figure it, if I want to go hand my ass out going after a target, that's my decision and no one else's.

Route Pack 1, the area just north of the DMZ, is not the piece of cake I thought it was. They have more guns up there than you can shake a stick at—and the little monkeys manning them have had 4 years practice shooting at jets. Some of them are awfully good. But we've been carrying CBUs—a real effective anti-personnel weapon—to make them keep their heads down. They don't like the CBUs at all—"Yankee Imperialist Pellet Bombs" is what Hanoi Hannah calls them. Anyhow, the HVN gunners don't know which aircraft has the CBUs, and is looking for gun sites, and which ones have bombs and are going after trucks and the like. So they can't risk shooting as much.

They have still managed to knock down three 105s since I've been here— they got the pilots out in two cases—in the other one the guy had to bail out over the middle of Doug Hoi, and there just wasn't any way to pull him out.

I've gotten 8 missions in over the North so far, and I've got another one this afternoon. So I guess I'll have 10 counters or maybe 11 by the time I've been here a month. If the war keeps up, I could have my 100 around the middle of March. On the other hand, if the war ends, I'll probably have to stay here an extra 6 months past the maximum of one year that they can keep me here in a combat zone.

… Well, John and Charles, that's about it for now. I've got to go get the good words on the target I'm going to strike this afternoon. Good luck to all of you, Mom, Dad and Mary also.

P.S. I'm in for a DFC—as of my third mission here. Will know in about 2 months if I'll get it.

The frustrations expressed by Dave in this letter reflected the talk around the lieutenants' table every day. In an effort to reduce aircraft and pilot losses, the pilots were told they would limit their exposure by dropping bombs and shooting their guns from higher altitudes and limit the number of passes they could make at each target. They knew they were being asked to do a difficult job with pretty tough restrictions on how they accomplished it. The argument was that it was just as dangerous to operate with all these restrictions and have to return to the same target time after time before it could be destroyed. Why not get the job done on the first attempt and go look for new targets?

Another sense of frustration was that some possible targets had been declared off-limits because of political considerations. Some targets were declared off-limits because of the possibility of foreigners, most notably Russian and Chinese, being in the target area. Some target restrictions were in place for humanitarian reasons, including hospitals and schools. The enemy soon realized that we would not attack those targets, and they became sanctuaries for the emplacement of guns and trucks that were moving cargo south to support the Viet Congress and the North Vietnamese Army (NVA). It was frustrating to be shot at by a gun in one of these restricted areas. Being shot at without the ability to retaliate is about as bad as it gets in the mind of a warrior. Every day as Valerie and Maureen gave their intelligence briefings, the part the pilots hated the most was the listing of restricted areas in the vicinity of their targets.

One morning Dave and several of his buddies were expressing their frustrations about the conduct of the War and decided it was time for some "civil" disobedience. The draft laws of the United States required that every male register for the draft when they turned 18 and were required to carry their draft card, which denoted their draft status. For years, the antiwar movement in the U.S. had used the burning of one's draft card as a symbol of objection to the War. After several beers and a loud pronouncement of their frustrations, someone suggested, "Let's go burn our fucking draft cards! What can they do, send us to Southeast Asia?"

In unison, they agreed to this symbolic gesture and marched out of the bar, trooping over to the hooch area and somehow getting a small fire going. As many of the older pilots looked on, laughing and shaking their heads at their younger brothers, they watched as the lieutenants dug into their wallets, found their draft cards, and threw them in the fire. Despite this symbolic act of defiance, they all knew they would continue to do exactly as they were told, but somehow this act made it easier for them to internalize their dilemma.

Early in his tour, Dave met one of the most interesting characters on the Korat scene, Roxie. Roxie was in her mid- to late-forties and a major in the Thai Air Force. No one knew what her official job title was, but to the pilots, she was the "VD Control Officer." Her primary job was to monitor the local establishments and attempt to keep our finely honed fighting force from spending all their time at the clinic getting penicillin shot in their ass.

Roxie was frequently in the company of the fighter pilots. She spent most of her time around the older guys but seemed to enjoy the good-natured style of the lieutenants a lot. The lieutenants enjoyed her a lot too. Like Father Gene, she had one outstanding quality—a pickup truck—and was not averse to hauling the young guys off base to Korat City to eat. She also took Dave and some of his friends on one of her "checky checks." They accompanied her into the bar and watched as she gained the immediate attention of all the girls. She then lined them up and had them lie on a table while she performed a close-order inspection of the merchandise. Following her inspection, she would annotate their inspection cards with a "pass" or "fail" symbol. The "fail" symbol was marked in red, and the receiver of this designation was not allowed to practice her trade until she was cured. She would then be reinspected and have her card updated with a "pass" mark.

This problem was especially prevalent among the younger enlisted guys. The crew chief would commonly talk about the "clap status" of the main forces as preflight checks were completed.

Pilots, despite their low aversion to risk, were infrequent visitors to the clinic. Dave and his friends had all heard the warnings at the McConnell Bar about the dangers of being shot down and taken prisoner with a case of the clap. You almost certainly would not be treated by the enemy. They also knew that if they reported to the clinic with this affliction, they would be placed on "Duty Not Involving Flying" (DNIF) status and would not fly again until cleared by the flight surgeon. About the most exciting activity between the Thai women and pilots were visits to the local massage parlors for a "scrub and a rub," as they were commonly called. On one occasion, Dave and his friends took Valerie and Maureen to get a "scrub and rub," later delighting in telling stories about the amazed looks on the Thai girls' faces as they were assigned to massage the much larger American women.

In mid-August, an incident occurred that made everyone aware that the Golden BB could strike anytime. "Golden BB" was an expression used to point out that it didn't matter how good you were or how hard you tried to avoid catastrophe, if your number was up, it was up. It is almost impossible to not become a fatalist if you are a fighter pilot. Capt. Nobe "Ray" Koontz of the 469th had just finished his 100 missions but agreed to ferry a Thud from the periodic inspection facility in Taiwan back to Korat. It was the sort of mission everyone enjoyed as it did not include getting shot at and gave the pilot some delightful relaxation in Taiwan.

However, through an unbelievable chain of events, Capt. Koontz ended up having to land early at Da Nang AFB in South Vietnam, an F-4 base. Da Nang had a midfield barrier, as did most F-4 bases, and because of some serious misunderstandings and bad weather, he attempted to engage the midfield barrier and tore the ass end of the aircraft off, losing all his hydraulics. He subsequently rolled off the end of the runway and the aircraft flipped. Capt. Koontz found himself inverted in a pond at the end of the runway and drowned.

The discussions at the lieutenants' table became more subdued as the word got back to Korat. Only a few days earlier, they had helped celebrate the completion of Ray's 100 missions, and now his body was being returned to the U.S. for burial. For a short time, at least, their youthful sense of invincibility came into question. However, the young have short memories, and soon the impact of this disaster was well behind them.

Dave also decided to grow a mustache in an attempt to turn it into one of those manly handlebar 'staches that had been worn by Robin Olds and a few other colorful characters of the era. Korat was, by far, the most "civilized" fighter base in Thailand, and grooming requirements were more strictly enforced than on the other "Wild West" bases. However, the exception was the loosening of restrictions on mustaches. Dave's beard was thick and black, and before long, he had a pretty presentable 'stache, which he started to train into those classic handlebars. At first, he caught hell from the rest of the lieutenants. "Are you sure you're old enough to grow a 'womb broom'?" Or, "Good idea, Dave, you could use a booger trap."

The base exchanges in Thailand all had ample supplies of mustache wax, and soon Dave was attempting to cultivate and train a trophy 'stache. However, like most pilots who attempted this style, he soon discovered that the damn thing made his face itch and took a lot of upkeep. It was also hard to manage use of the oxygen mask with this hairy appendage protruding from both sides of the mask. The only cool part was looking at yourself in the cockpit mirror. The presence of this large bushy 'stache somehow made one feel more suited to the role of "Yankee Air Pirate" as you prepared for battle. Eventually Dave tired of the handlebar and cut it back to the standard Air Force limits, the outer edge of the lips.

All squadrons were required to submit a quarterly report of their activities, and for the months of July through September 1968, the 469th Report contained some interesting facts and observations:

Continued bombing restrictions limited activity during this period to operations in Route Package One in North Vietnam and certain areas of Laos. Our main objective during the period as directed by 7th AF was the interdiction of vital lines of communications in Route Pack 1. Under the direction of Lt. Col. Victor R. Hollandsworth, the 469th maintained an outstanding record of superior mission accomplishment.

This reporting period was accented by the arrival of the first group of new Undergraduate Pilot Training graduates to fly with the 469th Tactical Fighter Squadron in well over a year. The majority of the group were 1st Lieutenants, and although they did not possess extensive flying hours or experience, the quality of their performance left little doubt as to their desire and abilities. They have proven themselves extremely capable as tactical fighter pilots.

The report also contained language that would wind down the 469th as an F-105 Squadron:

The 469th Tactical Fighter Squadron will become an F-4E squadron in November 1968. Present members of the squadron will join the 34th Tactical Fighter Squadron here at Korat [RTAFB, Thailand] and some will transfer to the 355th Tactical Fighter Wing at Takhli. This conversion will mark both the end and beginning of one of the most significant chapters in the history of aerial warfare. The record of the F-105s of the 469th Tactical Fighter Squadron stands by itself. We are confident the new members of the 469th will carry on the outstanding accomplishments of the Unit.

So, Dave became a victim of a sad truth: The Air Force was running out of F-105s and in the process of bringing in the new F-4Es to replace them. He soon received orders that assigned him to the 34th TFS as of November 1. The good news was that his good friend and roommate, Bob Zukowski, was also assigned to the 34th and he would not have to leave Korat. This meant he would still be around his many friends, including his best friend Valerie.

The level of activity began to slow down in August and September as Dave approached his 50th mission. Every pilot was allowed 5 days off each month for "R and R" (rest and relaxation). The bachelors referred to it more commonly as "I and I" (intoxication and Intercourse). Many married guys chose to not take their R and R in an attempt to accumulate more missions and head home. The bachelors typically headed down to Bangkok for other pursuits. A few married guys had brought their wives to Thailand, and they lived in Bangkok, so their monthly R and R's were more like conjugal visits. Bringing your wife to Thailand was discouraged, and bringing your wife to Korat City was highly discouraged. The last thing the Air Force wanted to do was notify a wife that she was a widow while she was living on the economy of a foreign nation.

Also included in your "contract" was a two-week R and R to the place of your choice. Most of the married men went back to the States or met their wives in Hawaii, while the bachelors typically went to Australia or some other garden spot in SEA. Dave decided to go to Okinawa, Japan, on his mid-tour R and R and wrote a letter to his parents while he was there:

I'm on Okinawa, at Kadena AFB, taking an R&R. I have to go back to Korat either tomorrow or the next day, but I've really enjoyed my stay here.

… As of now I have 43 counters under my belt—almost halfway through. Nothing exciting has happened for the last several times I've flown—it's been a real piece of cake.

… Glad to hear that Johnnie is O.K. That's the second concussion he's weathered, isn't it? I really don't know how that big lummox manages to get banged up so badly. Looking at him you'd think he could run into a locomotive and not be any the worse for wear.

… I guess Mary is overjoyed to be back in school. She can't keep going forever, can she? What happens when she runs out of years? Or maybe you just don't worry about it.

I don't need a thing. But if you feel ambitious some day when you're baking, a box of fat pills is always appreciated. However, if you do decide to send something, send it airmail and pack it as tightly as possible. Popcorn is a pretty good material to use. And get whatever you do send as airtight as possible.

… I guess that's all I've got right now. I've been writing to Charlie at Quantico, by the way, pretty regularly, so we're keeping in touch. Give my best to everyone.

One of the best days you could have at Korat was when you went to the Post Office and found you had a package from home. Cookies! Now the trick was to get them back to the hooch without the others seeing you and hide them. Otherwise, you were compelled to share them with your brothers, and they were gone in no time. The taste of home-baked cookies was magic! Also, as noted in this letter, Dave's younger brother, Charlie, had joined the Marine Corps and was going through initial training. The Dinan family was now getting a pretty heavy dose of the implications of this war. Later that month (October), Dave again wrote a letter to his parents:

I guess it's been a sizable time since I wrote. The big reason I didn't was that I've been busy as a beaver since I got back from R&R. We're short on pilots again, and carrying an increased rate of sorties—so I've flown every day but one since I've been back. Which makes it kind of rough. I'm always exhausted after I fly—probably half physical and half mental. Also hasn't been much happening here that's exciting or interesting.

... We've been saddled with so many restrictions the past couple of months (presumably in an effort to save airplanes) that it's almost impossible to do an adequate job. I'm sure more stuff gets through now than did a couple of months ago. And, of course, every truckload of stuff we don't blow up in NVN [North Vietnam] is a truckload the groundpounders are going to have to face sooner or later.

... There are really about four different wars going on—three in Laos that you never hear about, and South Vietnam. By the way, the rest of this stuff about Laos is supposedly classified—so if you tell anybody, say you read it in Time magazine. They've printed most of it.

Right now, we're flying more missions in Laos than we are in Vietnam. There is a full-scale war between the nationalist Lao and Pathet Lao up in northern Laos (where Dr. Dooley was). The Pathet Lao headquarters is at Sam Neua—and that is always an exciting mission. There are more guns there than anyplace in Pack 1—or at least they shoot more. The Laotian army desperately needs air support, however, and we provide it. The Air Force calls them "armed reconnaissance" missions, but they are out-and-out raids, and probably the most dangerous we fly.

... The second front is connected with SVN [South Vietnam]. The famous Ho Chi Minh trail. That's where most of the flying is now. It's a disappointing mission, usually. We bomb suspected tree parks and stuff like that, working with forward air controllers who are supposed to know something. I've never hit anything worthwhile there. The third war in Laos is down south near Cambodia. I don't know if American troops are mixed up in that or not—we rarely go down there.

... If this is confusing to you, it should be. I'm confused. The action, though, is around Sam Neua and in the southern panhandle of NVN. The way the pilots look at it is really pretty simple, though it sounds callous, I suppose. "If you're going to hang your butt out, it had better be for something worthwhile." In Pack 1 and Sam Neua, you can get a visual confirmation of the effects of your bombing. Fires, secondary explosions, and stuff like that. We do good work with the bombs.

We also carry rockets and 20mm cannon. The gun is probably the most devastating weapon we have, if it's used properly—namely up to ranges of about 6,000 feet. But, to get back to the original theme of his letter, we're forced by directives to use a __minimum__ range of 9,000 feet. And only two passes.

The bombs are good—but if that's the name of the game, a bomber should carry them. A fighter is built to fight—to make the war personal—and we can't do it. It is really discouraging to find a good target and then have to leave with ammo still in the airplane because somebody high up will get upset if you make three passes on a target.

I guess I've bent your ears enough about the war. How was Anne's wedding? I don't suppose Charlie could make it. Too bad. How's Johnnie doing? I suppose he's carrying a B average and worrying about flunking out. He's quite a guy.

I'm in fine shape—just griping more and enjoying it less. And I still can't spell. I've got 53 counters now, and 66 combat missions. A real old pro. By the way, I got my absentee ballot. I just might burn it in effigy. Not too much choice there.

I have to close now. It's now 9:30 p.m.—and I have to get up at 0330. I'll try to write more regularly in the future.

Around the same time, on October 27, 1st Lt. Robert "Bobby" Edmonds was shot down and assumed killed just north of the DMZ. His flight was encountering intense heavy anti-aircraft fire, and his aircraft was observed being hit. His canopy was jettisoned, but there was no chute and no beeper. Because there was no evidence of an ejection and the area of the crash was heavily occupied by enemy troops, no attempt was made to rescue him.

This accident was painful for the younger pilots. Bobby was one of the most frequent visitors to the lieutenants' table and one of the most popular. Both Dave and Valerie had developed a strong relationship with Bobby. Dave and his fellow pilots mourned their friend's loss and toasted him ceremoniously. However, they recognized the peril of dwelling on the dangers of their profession and knew thinking about the risks only took away from the concentration needed to do their job. Valerie, however, could not hide her emotions and carried Bobby's loss in her inner consciousness.

A few days later, President Johnson would announce the complete bombing halt of North Vietnam, and with that, all the counters that Dave had accumulated meant nothing. He was now committed to a full-year tour and would not be able to rotate home until July 1969. It was a bummer for all the pilots in Dave's situation, and on top of that, he was now initiated into a new squadron, the 34th "Rams." The lieutenants' table was the scene of several "attitude checks." When the discussion became too depressing, someone at the table would shout, "Let's have an attitude check!"

To which the rest of the table would respond, "This place sucks!"

The initiator of the attitude check would then respond, "No, no, let's have a positive attitude check."

"This place *positively* sucks!!"

On November 5, Richard Nixon was elected President of the United States, and as the pilots listened to the rhetoric of the new guy, it appeared doubtful that anything would change much after he was sworn in. Moving to the 34th TFS turned out to be not that big a deal. Several of his friends went with him, and the leadership of the 34th was impressive. The move was seamless.

As Dave and others departed the 469th, they were reminded of its history with the F-105. In its 4 years of flying the F-105, the 469th TFS paid a high price. Pilots assigned or attached to the Squadron lost 58 planes to combat and accidents—enough to equip the Squadron more than three times over.

Twenty-one Squadron pilots died in combat, six more in accidents. Eight pilots survived accidental crashes, 10 were rescued from enemy territory, and 13 more became POWs, one of whom died in captivity.

A few weeks later, on November 17, Dave was scheduled on an early afternoon flight to interdict the routes leading out of North Vietnam headed south to the Trail. His load was two 2,000-pound bombs, which was a load not carried often but was a great load because the drag on the aircraft was very low and "two granders" were big enough to do a lot of damage. Dave's call sign was Gator 2, and after joining on lead after takeoff, he proceeded north to join up with a tanker to top off prior to heading for the target. About 100 miles north of Korat, Dave noticed his navigation signal from Korat had broken lock, and he looked in the cockpit to change frequencies. While doing this, he glanced at his oil pressure gauge. *Oh, shit!*

His oil pressure gauge read "0" and the Oil Pressure Low light on the caution panel was illuminated as was the Master Caution light. He undoubtedly had not observed the Master Caution light come on because of the angle of the sun entering the cockpit over his right shoulder.

"Gator 2 has zero oil pressure!" was his immediate call to his lead, Gator 1.

After a short delay, "Roger that 2, better head for Udorn. I will call them and coordinate. Gator 3, chase 2 and let him know what's happening."

Dave acknowledged the instructions and made a slight turn to the left to head toward Udorn, some 80 miles in front of him. Gator 3 suggested that Dave use afterburner to burn down some of his fuel for the upcoming landing. Dave did this and partially extended his speed brakes to allow #3 to keep up without using afterburner. Shortly thereafter, "Two, you have sparks and some fire coming out of your engine!"

"Roger that," Dave acknowledged and deselected afterburner. Dave's engine RPM was now beginning to decrease, and the exhaust temperature was becoming quite hot, and he knew it was unlikely he would be able to make it to Udorn. He also knew that his only chance to make it included getting rid of his bombs and fuel tank. There was a discussion about proceeding to the Udorn ordnance jettison site, but it was determined it was too far off his track to Udorn, and Dave elected to select the Panic button, allowing his bombs and fuel tank to fall toward the rather barren Thai countryside. Since the bombs were not armed, it was unlikely they would detonate—and they did not.

Feeling the aircraft jump and become more nimble with the relief of this burden, Dave, for a moment, thought he might have a chance to recover at Udorn, but as he watched his altitude continue to decrease, he realized it was not to be.

A couple thousand feet above the ground, he raised his ejection handles, assumed the position, and pulled the triggers to initiate the process. Soon he was hanging in space with a fully blossomed parachute above him. On the way down, he observed the jet crash about a mile from his position and burst into flames. His landing was in a soft rice paddy field, and a few minutes later, a helicopter from Udorn was at the scene to pick him up.

Arriving at Udorn, he was given a quick examination by the Base Hospital and released with only a few minor cuts and bruises. He was able to catch a late-afternoon shuttle back to Korat and slept in his own bed that night. Just another day at the office. The leadership of the 34th listened to his recounting of the incident and saw no reason to take any action, and a few days later, he was back on the schedule.

Following Dave's ejection, good fortune shined on Korat, and they did not suffer any more losses for three months. Takhli, on the other hand, was not so lucky, losing several aircraft before the end of the year. One loss was on Christmas Eve, and the pilot was not recovered. Dave and the rest of the aircrews continued to fly through the holidays without interruption. Christmas and New Year's Day were regular days on the tour.

There was a grand piano in the main bar at the Officers Club, and from time to time, the pilots would get out their Squadron Song Books and sing a collection of rather raunchy, traditional Air Force songs. They varied from PG to XXX rated, but the guys loved to sing, and Dave oftentimes used his piano skills to keep everyone on key. This time of the year, they would set aside the song books and sing a few Christmas carols. Another song they almost always sung when they got together was not religious but had a bit of a religious ring in the refrain:

"Oh hallelujah, hallelujah, throw a nickel on the grass, save a fighter pilot's ass. Hallelujah, hallelujah, throw a nickel on the grass, and you'll be saved."

On January 20, Dave almost suffered another catastrophe. While releasing a bomb, it improperly came off and hit one of his wing fuel tanks. The tail fin from the fuel tank broke off and hit the right rear stabilizer, causing severe damage. Luckily, Dave was able to recover the aircraft despite the damage. As was customary when you did damage to an aircraft, you bought the crew chief a case of beer, and Dave was more than happy to oblige.

On January 26, Dave celebrated his 25th birthday with Father Gene, Valerie, and a few others. He had every reason to celebrate. He had dodged a few bullets and was now a seasoned combat fighter pilot. He had completed over half of his tour and felt confident about making it to the end. Three new replacement lieutenants had just shown up in the Squadron. Danny Seals, Joe Widhelm, and Ed Sykes were part of the second all-UPT class that came out of McConnell and were already joining in at the lieutenants' table, soaking up the stories as told by Dave and the other veterans. The new guys seemed like a bunch of dumbasses, but they might be able to figure it out. Life was good!

Note: Lt. Dave Dinan's letters were compiled by Homecoming II Project 15 March 1991 and updated by the P.O.W. NETWORK 1998.

Chapter 9

Valerie's War

Being brought up in a middle-class family in Waterbury, Connecticut, in the 1950s and early '60s was a pretty good gig. Korea was on a lot of people's minds and the Cold War was intensifying, but life in the U.S. and Waterbury was peaceful. The Galullo family was no exception. Dad Pasquale and mom Elsie and their two daughters Patricia and Valerie resembled the mold of one of the most popular TV series of that time, *The Adventures of Ozzie and Harriet*. The sitcom ran from 1952 to 1966 and was canceled when the country entered the turbulence of the late '60s. The story of a middle-class, suburban family, which centered on their dealing with everyday family issues, paralleled the time in which Valerie was raised.

Pasquale and Elsie were introduced to each other by their parents, and each was the youngest of eight children. Consequently, when Valerie was growing up, there was an ample supply of cousins to play with. Pasquale was a talented mechanical engineer. He designed and built the home Valerie grew up in. Like most white-collar families of the time, Elsie was a homemaker with fabulous cooking and baking skills. One expectation stressed to the young daughters was the need to earn a college degree in case their future traditional family lost the breadwinner and they needed to support themselves.

Valerie recounted some of the good times had with her cousins:

When we went to the lake with our cousins as kids, we went to Lake Quassapaug, where my cousin was a lifeguard. We swam every day. Our uncles also had cottages at Sandy Beach. We used to swim over to Sandy Beach and sneak in. We also had a real bark canoe that always leaked. We would take it across the lake, turn it upside down, empty out the water, and paddle back. We had a huge wooden raft that we could dive off in the deep water and huge truck inner tubes for floating. We always had patches and rubber cement to repair leaks.

We had a great hammock that was always being repaired as it had a split where your butt would fit. Since we had to wait a half-hour after lunch before going back into the water, we would fight over who got the hammock. Too many bodies in the hammock would cause it to rip again. We would stay all day until dusk. My cousin would always stand at the edge of the dock in his dry clothes. He would get dizzy and fall in and have to go home in wet clothes. My sister, Patricia, and the older cousins didn't want the younger ones around. However, we learned to swim on our own at a very young age, and they couldn't get rid of us when they took the raft into deeper water. We were all strong swimmers and divers.

The Italian neighborhood that Valerie grew up in was close, and before she was five, she was introduced to sewing by a neighbor. She was intrigued by the handiwork of her neighbor and soon began to emulate her, using leftover materials the neighbor gave her. She was soon making clothes for her dolls, which impressed her young friends. Recognizing their daughter's skill and enthusiasm for sewing, her parents bought her a sewing machine when she was in the sixth grade, and soon she was making herself clothes. As she progressed, it was evident that one of her greatest strengths was the ability to develop a skill and then remain focused enough to master it. It was a quality that gave her a good deal of self-satisfaction, but other kids, recognizing she was better at something than they were, became little buttheads and discredited her as being "different."

Valerie also demonstrated great focus in her studies. Later, Valerie recalled her school years:

I attended grammar school at Barnard Elementary. It was a K-8 school. In 6th grade, I was in 7th and 8th grade English. I graduated valedictorian. My friend Charlie Donato never forgave me for beating him out. He was salutatorian, and his wife couldn't wait to meet me at our 25th high school reunion. All she heard about me was that I graduated #1.

I attended Croft High School in Waterbury and was an honor student all 4 years. I graduated around #7 in a class of 435 students. During my high school years, I was in Drama Club. I made all the costumes for A Connecticut Yankee in King Arthur's Court and the musical Oklahoma! I completed 4 years of sewing in 3 years and learned to do tailoring too. I wore designer clothes in high school that I had made. I was voted Best Dressed. I also was a member of The National Honor Society and received honors in Spanish at graduation. I belonged to every school club and usually held an office. I never learned to type as sewing was my elective choice.

By all accounts, this was a great start for a young woman getting ready for college. Since she had been so good at Spanish and was interested in a career where she could use her language skills, Valerie elected to attend the University of St. Joseph in West Hartford and entered the Latin American Studies program. At the time, she considered the foreign service area as her best career option. By her own admission, her college years were not that much fun. Perhaps that was due to being a perfectionist, which takes time and attention. Instead of developing new relationships and having fun, she concentrated on her studies. Valerie remembered:

My college years were not happy ones. The girls couldn't stand that I had such a beautiful wardrobe, all made by me. They were very catty, meow! I was a loner and studied like crazy. St. Joseph's was traditionally an all-girl liberal arts college. We were all pure female virgins! We also had our share of frat parties at Trinity College, Wesleyan University, and a few others. I really wasn't into frat boys. I found them very immature and they couldn't hold their liquor! We had the first male students admitted while I was there. They were day students and one was in my U.S. History class. It was great to hear someone else talk in class besides myself. Most of the girls were quiet in class, however, I was not.

I always got the highest grades. We had to draw a map of Europe and put in all the countries from memory on a blank piece of white paper. My map was perfect as I have a photographic memory and loved drawing and art. I did make some costumes for a play, but never to the extent I did in high school. College was very hard. I had to get good grades by studying every spare moment, was not very involved in any clubs on campus. Having religion shoved down my throat made me give up being a Catholic and I investigated many religions. Nothing ever stuck. I really have no fond college memories.

As she prepared to depart college and start a career, she considered exploring the world beyond Connecticut. She had also become curious about the fact and fiction of the biggest news story on her TV every evening—the Vietnam War. By the time of her graduation in 1967, the war was in continual expansion, and Americans were taking sides on the issue. The issue of the draft was also up for contention. Young men were required to register for the draft when they turned 18 and then were at the mercy of their local draft boards as to whether they were selected involuntarily for military service. Several exemption categories included medical exemptions, marriage, and student status, but the draft was at least as unpopular as the War itself.

In 1967, President Johnson signed an Executive Order that allowed the military services to recruit enlisted women above the 2% level previously set as the ceiling. More women were recruited to fill noncombat billets in order to release more men for combat duty. In addition, women were beginning to be sent to combat zones in noncombat roles.

Valerie remembered:

My main reason for choosing the Air Force was to get out of Waterbury. I knew I would have no problem getting in, as the war was on and language majors were in demand. I enlisted and was sworn in at New Haven, Connecticut, on 7-5-1967. I also wanted to know what was going on with the war in Vietnam. Military Intelligence would guarantee that I would be sent to the war zone. They needed some females with brains besides nurses, and we were the first group of women to be sent there.

We also had to be "lookers" and I was very svelt and looked like a model. The recruiters had height and weight restrictions for females besides your overall appearance. Since I only missed 1 item on their test, I could pick just about any job classification. I was first assigned to Lackland AFB in Texas after graduation and ran the inspections on the barracks and wrote some Operations manuals. The married guys over me liked to harass me, but they got it back from me, as I was very sharp-witted and funny! They couldn't break me. I attended Military Intelligence School at Lowry AFB in Denver, Colorado, a few months later for 9 months. It was a combined Air Force, Navy, and Marines school. My friend Linda and I had a contest on who could bring home the best-looking guys. The best-looking were always duds in the personality department. I cooked a lot of dinners in Denver for former second lieutenants as I was the only female in the class from all the branches. It was the best time in my life.

As she approached graduation from Intel School, Valerie received her orders and, as she wished, was assigned to the combat zone with one nice provision: she would be going to Thailand and not Vietnam. She was ordered to the 388th TFW at Korat RTAFB. She had joined the Air Force to get out of Waterbury and travel, and this was about as far as she could go. She was excited about her assignment but somewhat tentative. What lay in store for her in "This Man's Air Force"?

If Valerie was tentative, it turns out, so was the 388th. A few female officers served in the Wing, but they were behind the scenes, all assigned to the hospital. A few civilian women worked in the Headquarters section, but Valerie would be the first woman who would be part of the combat team and work with aircrews on a daily basis. Valerie would learn that the 388th raised objections to Air Force personnel about her being assigned to Korat, but they were overruled, and within a short time of her arrival, a second female officer, Lt. Maureen McCabe, would be added to the Intelligence Section.

So in mid-summer 1968, Valerie set off to Southeast Asia for an adventure that would change her life forever. She described her trip to Korat after arriving in Bangkok and her arrival to the base:

I took a small transport plane, probably a C-47, from Don Muang Airport, Bangkok, into the jungle to Korat Royal Thai Air Force Base. By auto, it would be a 4-hour drive on a dirty and filthy road. I stepped off the plane onto pierced and tall planking. It had large sheets of metal with round holes placed over the ground. My black pumps, shined to look like patent leather, had small heels that stuck in the holes. I had to walk on my toes. I was a Reserve Officer on active duty attached to the 388th active Fighter Wing from Ogden, Utah. No one ever came to meet me or welcome me to their wing. It stayed this way for my entire tour.

I breathed in the air and have never forgotten the disgusting smell of jet fuel and human feces that almost choked me as I gasped for clean air. The extreme heat was like a wall of fire and was suffocating.

Valerie did not have a great first impression of Korat! It would have been even worse if she had ridden that bus from Bangkok. The bus was hot and the ride was dusty and included one potty break at a little Thai café. The facility was unisex without a door and featured a squatting potty. Having just finished Jungle School on my trip to Korat, I considered the whole experience an upgrade but can imagine Valerie would not have shared my view. Also, the air at Korat was not as hot and humid as what we had left in the Philippines, and the smell of jet fuel, to me, was heavenly.

Also, I am certain that new pilots were greeted in a friendlier fashion as there was always a need for more bodies to fill the flying schedule (even if they were dumbass lieutenants). In her description of life in Thailand, she always refers to the young pilots as 2Lts. However, all the flying lieutenants at Korat had made 1Lt. a few months before their arrival, so they were clearly a pretty mature bunch (in their own minds— a view not necessarily shared by the senior officers).

Valerie continued, "I found out later that the wing commander of the 388th had tried to use his rank to get my orders canceled. He was unsuccessful, and I was the first of a handful of line officers, who were not nurses, who were to be assigned to SEA. Since I was the 'first female' to brief pilots in the war zone, the *Pacific Stars and Stripes* wanted to interview me, and I declined. I was there to do a job and did not need any publicity as I was definitely not wanted at the 388th."

It seems that Valerie's first impression of Korat was a lasting one. She described some of her misgivings with the 388th:

I had nothing but disgust for the 388th and my time at Korat. I never fit in and all my 2Lt. friends were being killed off. Even the nurses and civilian women on the base never socialized or traveled in our circles. Everyone had their own cliques. Once, the head nurse checked up on the females in the Intel Compound. It was to make sure we were wearing the correct uniforms. What an ass! She was a coarse, big-mouthed female. The female civilians never even bothered with us at the club. They were too busy trying to catch a major or colonel; however, most were already married.

Our wing commander tried many times to get the females to come to his trailer to use his bathtub, a novelty for sure. I am also sure he was successful at getting laid. However, I was not a bathtub kind of gal, but he tried, in vain, numerous times to get me in his bathtub. He even showed it to me at a party he hosted in his trailer once.

On the other side of the base, at the C-130 Headquarters, their female Intelligence officer was treated with respect and was very welcomed and accepted by her comrades. She was treated so well, she returned later for another tour. I had good reasons to want to forget the past. The officers at the 388th never accepted me. I was their equal or better intellectually, but big egos flying fast planes could not wrap their heads around a smart female briefing them before they got in their F-105s.

Valerie referred to the C-130s at Korat, but she was undoubtedly talking about the C-121s that were co-located with the Fighter Wing. The C-121 was a military version of the Lockheed Super Constellation and was used for intelligence gathering and airborne control. The two units were like oil and water, and I don't recall ever meeting or talking to a single C-121 crewmember during my entire tenure at Korat.

I have little doubt the female intelligence officer for the C-121 units was treated more civilly than Valerie. I suspect that much of the pushback by Thud drivers came from the senior officers in the "Wild Weasel" Squadron (the pilots of the 44th TFS). They were much older, on average, than the strike pilots and, I am certain, capable of showing their displeasure with females being involved in their manly endeavors. A few of these guys had fought in World War II and Korea, and this presented a dynamic change.

Valerie was assigned some other duties from time to time.

I also investigated fraud at the Enlisted Men's Club, discovered how the locals were stealing. It gave me another job, but nothing was ever done about the stealing. We had so much and the locals had nothing. They would stuff toilet paper in their clothes upon leaving the base. I even had all my underwear stolen from the outdoor clothesline. What Oriental had an ass as big as my American one?!

I had other jobs to occupy my off hours. One was meeting all the visiting dignitaries visiting the base wanting to know how the war was going. The USAF didn't mind dragging out their females to impress the reporters, etc., who were visiting.

I also remember briefing a crew from an SR-71 that was on base. They thought I was not for real; I had to convince them that I was their debriefing officer. The SR-71 was a beauty to behold. They bought me a drink at the club later.

When I would go to the Officers mess, the 388th always sat on the right as I entered. No one even greeted me... I would have to look for an empty seat at a table and sit with my new strange friends. I was totally isolated. I also had to find my way to the mess hall.

I met my 2nd Lt. friends at the O'Club. They invited me to sit at their table for drinks. I only remember Bobby Edmonds and Skip Holm from our group. Bobby Edmonds was the first in our group to be shot down. Skip Holm had a dad who was an officer at SAC [Strategic Air Command]. There was always talk of Skip being returned to the States before his tour was up. There were five of us who always hung out together at the club. We drank too much, as there was nothing to do, sang dirty songs about Barnacle Bill the Sailor, and told really bad jokes.

Singing dirty songs was a favorite pastime of the fighter pilot fraternity and it certainly didn't hurt that Dave possessed some pretty good piano playing skills.

Valerie could not remember the names of the other lieutenants at her table, but Ron Hoffmeyer was likely there frequently as he was a good friend of Dave's. Others might have been Ron Stafford or Bob Zukowski, neither of whom would survive the War. Another likely candidate would be Marshall Tilley, also one of Dave's classmates at McConnell.

Over time, Dave and Valerie developed a mutual attraction. Valerie wrote, "I have no memory of how David and I became an item. I remember he had dark hair, hazel eyes, and the biggest dimples ever on his cheeks. He was the smartest and most talented young man I had ever met. He wasn't bothered that I had a brain too. We were probably close friends for 4 to 5 months before we had serious feelings for Catch-22."

"Catch-22" was a term coined by Joseph Heller in 1961 in his popular book of the same name. Catch-22 describes a situation in which one becomes trapped by some bureaucratic rule or military regulation from which he cannot escape. For instance, it is assumed that anyone who would volunteer to fly an F-105 into combat was insane. However, if he attempted to get out of his combat duties by claiming he was insane, he would be demonstrating that he was sane and, therefore, must continue to fly combat missions.

The Catch-22 that Valerie refers to was the cultural code of the time, especially in the military. If you were an "item," you should be married. However, if you were married in the military with a military spouse, the military was reluctant to station you at the same base with your spouse when it came time for reassignment. Also at that time, if you became pregnant, you were subject to immediate discharge from the military.

Valerie also recalled going off base for meals with the group. They probably used the two-baht bus on most occasions, but they also likely often jumped in the back of Father Gene's pickup for the short ride into Korat City.

We would also go off base to eat together at Veena's Hideaway. Rumor was that some officer set up Veena and her restaurant. We never got sick eating off base. David always brought a fork for me as I never mastered eating with chopsticks.

David was a hand holder. He was very generous and sweet natured. He always said he wanted a large family; me, not so sure. I always called him David and not Dave. He bought me a beautiful tiger's eye ring set in white gold for xmas. Years later, I lost the stone. He also gave me a hand-carved jewelry box with a red velvet lining on Valentine's Day. It soon was smashed and unrepairable, too much moving on my part.

We had a job to do in the Intelligence Compound and never dwelled or talked about our lost friends that next day. We partied the night before, and by dusk the following day, they were gone and forgotten. I had never been faced with so much death, and it became impersonal with my temporary new family of 2nd Lts. I only cried once as I was returning from the post office. A group of F-105s were returning to base and I was overcome with sorrow and loud sobs.

Chapter 10

My War

As I watched Green Bay disappear out the plane's window, several thoughts went through my mind. Less than two years ago, I was a student at Madison and enjoying life. Since then, Mary and I had evolved into a family in the making, and the Air Force had entrusted me with the responsibility of operating the most sophisticated war machine on the planet. What's more, they were now asking me to employ that machine in combat. For lots of reasons, I thought, *Don't fuck it up!*

Later that day, in the comfortable climate of northern California, I arrived at Travis AFB where, much to my delight, I found all 15 of my classmates from McConnell ready to depart for the Philippines on the same flight. We began to focus on the future and the upcoming adventure. "Let's go kick some Commie ass!" was the general theme of our conversations.

We were loaded on an older Boeing 707 aircraft operated by some lowest-bid contractor and were off. We stopped for gas somewhere in the middle of the Pacific and eventually ended up at Clark AFB in the Philippines. The cold and snow of Wisconsin was soon forgotten, and now keeping cool was our highest priority. Clark was a garden spot, and the Officers Club was really nice. They had a great pool and bar, and we soon found ourselves acting like fighter pilots again. The bachelors found a way to get downtown to Angela's City on the first night and came back with stories of the Wild West, Filipino style. We had two days to wait until the next Snake School class convened, so we drank too much and lazed around the pool. Some of the guys played golf at the Clark course and said it was fantastic.

Of the 16 in our class, only three of us were bound for Korat and the 34th TFS: me, Danny Seals, and Joe Widhelm. The others were headed to Takhli to augment the three Thud squadrons there. Snake School was a three-day course and ran continuously, starting a new session every three days. Two of my classmates and I showed up for class about 5 minutes late on the next start date, and as we entered the School Building, the instructor at the door informed us that the class was full and we should show up in three days to start with the next session. "Well, crap," I exclaimed, but it was pointed out to me by one of my tardy partners that this was no big deal. "Just three more days off the tour."

However, since it had been several weeks since I last touched an airplane, I was concerned about the erosion of my newly honed skills. But, it wasn't over yet. That winter, a particularly bad strain of the flu (Hong Kong flu) was going around the world, and on my second day of extended absence from Snake School, I found myself as sick as I can ever remember being. I went to the local flight surgeon, who gave me some stuff, but it was not a cure. I spent the next five days in my bathroom thinking I might die and alternatively being afraid I wouldn't. Finally, I came around and showed up for the fourth possible start date since my arrival at Clark. By now, all my classmates had departed for Thailand, and I suspected I would be in a heap of trouble when I got to Korat.

Snake School was actually a fun program except for the one night we spent in the jungle and got to know the animals you never saw in the Tarzan movies—rats. When the sun went down, they were everywhere. We were warned to make sure we washed up well after eating or suffer the chance of a rat bite and the series of rabies shots to follow. The jungle area we were in was beautiful, and we all went skinny-dipping in a nice jungle pool. Another highlight was a training extraction from the jungle by a Jolly Green Giant (a rescue helicopter). They dropped a cable with a harness on it, you secured yourself in the collar, and you were then pulled out of the jungle by the Jolly Green. Really neat training that I hoped to never need!

I was really concerned about showing up late to my Squadron at Korat and knew my commander would be upset about my tardiness. My classmates, Joe and Danny, had checked in several days earlier. So following Snake School, I hightailed it for Korat, spending a night in Bangkok en route. The next day I took a taxi to Dong Muang Airport, the

U.S. military base in Bangkok, and caught a hop to Korat on the C-47 rotator that made the rounds of all the air bases in Thailand.

I arrived mid-morning at Korat, and Base Operations gave me directions to the Wing Headquarters. I made the short trip over and signed in. I expected someone to ask the obvious question, "Where have you been, Lieutenant?" but it didn't happen. They checked my orders and went through my travel tube and, satisfied, told me to report to the 34th for further instructions. The 34th Operations Building is just inside Fort Apache, the secure and guarded complex within the base where each of the three squadrons had their headquarters and where all intelligence briefing took place. I saw the 34th Ram patch as well as a banner over the door that said, "The Men In Black." Inside near the dispatch desk, I discovered Danny Seals chatting with some other pilots. We shook hands and I asked him how it was going.

"Pretty good, Ed. I have four missions, and the flying here is great." I explained my tardiness to which he just shrugged, pointed toward the Squadron commander's office, and told me I should probably let him know I was here. I was about to get an ass chewing but might as well get it over with. I knocked on the doorframe and a tall, thin, 40-something man with dark hair looked up and smiled. "What's up?" he asked. I noticed the nameplate on the desk: "LTC Harvey Prosser."

"Col. Prosser, Lt. Sykes reporting," I somewhat nervously stated, expecting him to ask me to close the door so he could really give me a beating.

"Glad to have you, Sykes. We heard you were on your way. We can always use some good help around here." I realized then that what I had heard at the bar several times was really true. You could get credit for a full SEA tour and never show up for duty as long as you had your pay tube and stayed in the pipeline. The pipeline was that period of time between leaving the States and showing up for your SEA assignment. You were given a pay tube that contained your financial records so you could be paid while en route.

"Sykes, I'm not going to put you in the hooches yet. We're going to deploy to Takhli for about a month so they can do some repairs on the runway here, so I'm going to have you stay at the BOQ [bachelor officers' quarters] until we return from deployment."

"Yes, sir," I replied.

"I will introduce you to the ops officer, T. C. Glass, in a few minutes. Your flight commander will be Maj. Holly. He's flying today, but you'll meet him soon enough, and we'll schedule you for a local area checkout with him in the next day or two. Right now, the most important thing to do is get you a party suit. Take the two-baht bus downtown this afternoon and get fitted at Tommy's Tailor. We have an end-of-tour party tomorrow night, and you need to have your party suit for that."

Wow! I thought. I show up in a war zone, and the most important thing I need to do is get a party suit—shit hot!

Col. Prosser then introduced me to Lt. Col. Ike Glass, and I knew at once that he was a classy guy.

So instead of getting my ass chewed, I was given a friendly welcome and was feeling comfortable in my new surroundings even though I had only been here 20 minutes.

I met Danny a bit later, and he told me where to catch the bus and how to find Tommy's. I first proceeded to Billeting and got some keys to my quarters, a nice air-conditioned room behind the Officers Club. I did have to share a bathroom but didn't care. *Not bad digs.* I then walked over to the main drag and, on the way, met a furry brown dog who I recognized at once. It was Roscoe, the dog that had been featured in the movie *There Is a Way*, which I had watched about a year earlier while in flight school. It was the first time I had met a movie star, and I was impressed at once by his humility and friendliness. Note to self: see if the BX sells dog biscuits.

I caught the two-baht bus for downtown Korat City. Two baht, at that time, was about 10 cents USD. I followed Danny's directions and soon found myself at Tommy's Tailor. I introduced myself to a smiling young Thai man who I found out was Tommy and told him I needed a party suit for the 34th TFS. Within seconds, he had the tape measure that had been around his neck measuring every proportion of my body and writing it all down on a sheet of paper.

"Lieutenant, your suit will be ready in the morning."

"How much does it cost?" I asked.

"400 baht," he replied. A quick calculation told me that was only $20 in "real" money.

"What are you going to wear on your feet?" he asked. Good question. I told him I would probably wear my combat boots. He then explained that I would look much better in fancy boots and urged me to visit the boot shop down the street for suggestions. So I went to the bootery and again was greeted by a friendly young Thai man who showed me samples of many boot styles. The American-style cowboy boots were my immediate choice. He quickly measured my feet in detail, and I ordered a pair of black cowboy boots with "ED" imprinted on the outside of each boot. I had seen this done by some of the locals in Lubbock and had always thought it would be neat to have my own personalized cowboy boots. Now I was going to get them for the small sum of about $35. I was told they would be ready by noon the next day.

I returned to Korat on the two-baht bus and after a much-needed nap took a short walk to the Officers Club bar. Entering the bar, I spied several lieutenants sitting near the stage to the right side of the entryway. I had seen a few of them at the 34th earlier that day, so I approached the table and was greeted by the standard, "Hey, it's the FNG [fuck'n new guy]," and other "fresh meat" comments. They all turned out to be members of the 34th. Several of them—Dave Dinan, Bob Zukowski, Ron Stafford, and Marshall Tilley—were all members of the first all-UPT class from McConnell. Neither Joe nor Danny were there from my class. Although this group had all been at Korat for only a little over 6 months, all of them had more than 70 missions, with a large number accumulated before the bombing halt.

They didn't hang around long as they were all part of an early morning gaggle of all lieutenants. They explained that the old guys didn't like to fly the early morning goes, so they sent the lieutenants. The lieutenants found a real advantage to flying the early goes because they felt the NVA gunners didn't want to give away their positions early in the day and then get bombs on their heads all day long. Listening to them discuss some of their experiences and knowledge, I realized I had a lot to learn and the learning curve would be steep. *Guess I really am an FNG.*

The next morning, I wandered down to the Squadron and met Maj. Manfred C. Holly, my flight commander. Manfred was a cool guy. He was laid-back, and I could tell that if you didn't screw up, he wouldn't bother you much. He was not a highly experienced fighter pilot guy, having grown up in Air Defense Command and spent most of his career as an interceptor pilot—F-102s and F-106s. He had only recently learned to hurl his pink body at the ground and participate in real dogfights. While at the Squadron, I met the guy who was flying his EOT (end-of-tour) mission that day, and he offered me his bike. Almost everyone rode a bicycle around the base, and as guys left, they gave their bikes to the new guys.

That afternoon, I went downtown and picked up my party suit and boots. The suit fit like a glove and looked real sharp. I loved the black suit with my name and wings embroidered on the chest with Snoopy doing a dance below the wings. It was set off by a black Thud patch, a Squadron patch, and a patch labeled "Yankee Air Pirate" with a skull and crossbones on it. The boots were great and fit well. A little over 24 hours in SEA, and the only things I had accomplished was the gift of a bicycle and the acquisition of some really nice party apparel. This war gig ain't so bad.

I got back to the base in time to see the EOT party begin with the pilot being hosed down as he left his aircraft and then towed around base on a trailer accompanied by a parade of elephants and fire trucks and all kinds of noise. We ended up at the O'Club and, after a couple of beers, retired to our rooms to put on our party suits. The party was held in the stag bar, which was a basement underneath the KABOOM. I joined the party in my new party suit and boots without having flown a single mission. Joe and Danny were both there, so we spent some time swapping lies. Eventually the party transitioned to a sing-along. Someone handed me the 34th Song Book, and I began to learn some of the dirtiest and most outlandish songs I had ever heard.

As the party was about to break up, I witnessed something I had never seen before and will probably never see again. Two pilots at the bar suddenly began to yell at each other, one of them threw a punch, and the fight was on. I recognized the guy who threw the punch as Dave Dinan, but I didn't know the other guy. People quickly jumped in to break up the scuffle and coerced the two to shake hands, but it was clear that Dave Dinan was not afraid to mix it up when someone pissed him off.

At the party, Maj. Holly came up and let me know that we would fly an area familiarization flight the next day. We would brief around 10 A.M., provided there were aircraft available once the combat commitments were met. "Great!" I stated, and then the recurring thought returned. I had not flown in nearly two months and was not sure I had retained as much as was needed to operate a Thud.

The next day, January 26, 1969, I put my fears to the test. As it turned out, I had nothing to worry about. Manfred briefed me on all the U.S. bases in Thailand and where the tanker orbits were. He also briefed me on some explosive ordnance jettison ranges. Once I had gotten in my jet and fired it up, I realized that my skills were intact, and my comfort level rose swiftly. We took off and flew some basic formations and then headed to one of the tanker orbits where Manfred begged for a couple thousand pounds of gas and got it. I'd also retained my refueling skills and felt a true joy being back in the air.

Manfred gave me a quick look-see of Udorn on Thailand's northern border and Ubon on the east. These two bases would be our primary recovery destinations in case of battle damage. He also showed me Nakhon Phanom (NKP or "Naked Fanny," as it was commonly referred to), which was the site of most of the special operations units and was a last-ditch emergency recovery base if you couldn't make it to Udorn or Ubon. The runway there was short and consisted of metal panels rather than concrete. It was extremely short for a Thud, but it had been used in dire situations. Most of the Jolly Green Giant rescue helicopters were based at Naked Fanny.

After an uneventful landing, we returned to the Squadron and, right away, checked the schedule for the next day's flights. About halfway down the schedule, I saw it—Holly and Sykes were scheduled for an 11 A.M. takeoff on a combat mission. Manfred looked at me and grinned. "Looks like you get your cherry tomorrow."

"Yes, sir," I replied.

"The mass briefing is at eight. Meet me here at the Squadron just before that, and we can go over together."

"Yes, sir."

We then debriefed the day's flight, but there wasn't much to talk about. Manfred seemed to be happy with my rekindled skills, so we broke for the day. I had a few whiskeys at the Club that evening and found many of the lieutenants gone. It was Dave Dinan's 25th birthday, and several guys had gone downtown to celebrate. Both Joe and Danny and a few others were there, and I told them that tomorrow would be my first combat mission. Although a rite of passage experienced by all, it had been uneventful in most cases. One of the buddies explained that easy targets were normally chosen for a rookie's first ride, and the hard targets came later. The conversation then revolved around some of those better targets. Sleep came easily that night as I was reassured by that day's flight and an expectation that my first strike sortie would be a ho-hum event.

I showed up early at the Squadron and met Manfred there and, along with several others, made our way to Richter Hall for the mass briefing. In the hall, I was surprised to see two female lieutenants on the briefing stage getting ready for the briefing. I couldn't recall seeing any females in the Air Force other than nurses in the clinics since I had entered the Air Force, and I guess the last place I expected to see them was in a combat zone. The mass briefing consisted of pilots from all three of the fighter squadrons at Korat who were preparing to fly the midday missions. There were probably 40 pilots in the room, half of them from the newly arrived F-4Es who had taken over the 469th. As I watched the F-4 pilots file in "holding hands," I was glad I flew a single-seat jet.

The first item of the day was the weather forecast for the target area. The weather briefing for all of SEA was pretty much the same every day depending on the season. We could almost always count on scattered thunderstorm buildups on the refueling tracks and target area, most of the typical early-morning fog burnt off, and pretty good visibility in the target area. The weather was not a challenge. Also, the winds up to 19,000 feet were light, so corrections for winds on bombing deliveries were not much of a factor. Those difficult Kansas winds were history—thank God!

The second briefer was the taller of the two women, and she talked about the threats in the target areas. Her briefing consisted of reported active gun sites from the observations of previous flights as well as intelligence reports from higher headquarters. She also talked about the possibilities of SA-2 missile sites in the area. With the shutdown of air operations over North Vietnam, the Communists had begun to move missiles along the trails in Laos and Cambodia in an attempt to threaten air operations in that area. MiG threats were also discussed, but for the most part, MiG operations at this time were limited to momentary dashes toward incoming flights in an attempt to get them to retreat or jettison their ordnance.

Next, the shorter woman talked about "safe areas"—those areas where you were most likely to encounter friendly forces in the event of an ejection. She also noted that the situation on the ground was fluid and that her information was not 100 percent dependable.

During the threat briefing, it was beginning to soak in that today I would probably get shot at for the first time, and the question weighing heavy on my mind was, *What in the hell am I doing here?* I had wanted to be a fighter pilot for as long as I could remember, but I was finally coming face-to-face with the consequences of my quest. I was no longer listening to the briefing but was being consumed by this question, *What the hell am I doing here?* As I looked around the room, I noticed the other pilots didn't seem to have the same concern on their faces. It was just me. I then looked above the briefing stage where there was a large portrait of Karl Richter and below it an inscription in large letters: "The Mission of the Air Force is to Fly and Fight and Don't You Ever Forget It!"

I had always thought the mission of the Air Force was to fly and had never given serious thought to the "and Fight" part. Why had it taken me this long to figure it out? *I guess it really doesn't matter now because you put yourself in this situation, and you better deal with it, dumbass.* So, in a matter of a few seconds, I had discovered and then solved a big philosophical dilemma and was now prepared to move on. Over the next several months as I moved through my tour in SEA, I would often ask myself, *What the hell am I doing here?* and that inscription under Karl Richter would come back to answer me.

After leaving the mass briefing, we went to the Frag Shop where we picked up fragged targets for the day. Manfred casually looked them over and wrote them down on his mission card. The fragged targets came from somewhere on high but, as I would find out later, had little to do with which targets you actually struck. The war was simply too dynamic and, after all, "Flexibility is the Key to Air Power." We then went to a briefing room, and Manfred gave me a thorough briefing on how we would conduct business, stressing radio discipline. Only three calls are allowed: "Two," "Bingo," and "Lead, you're on fire." He also spent some time talking about how to conduct business in the target area, constantly moving to avoid getting hit, and the importance of always having a plan of action in case you do get hit.

We next went to Life Support to pick up our flight gear: helmet, g-suit survival vest, and the heavy and uncomfortable parachute. For me, the most sobering part of this visit was the removal of all your personal stuff from your pockets and your Velcro-attached patches and placing them in a small locker secured by a padlock. When you left Life Support, the only items you should have of a personal nature were your dog tags, your military ID, and the key to your locker. This was always a reminder to me of the bad things that could happen to you—the worst being captured and becoming a POW, which could mean torture or worse. The last thing you always did before you left the room was check your survival radios in your vest. Even more than the .45-caliber pistol and 50 rounds of ammunition you carried, these two radios were the most essential part of any survival effort if you got hammered.

A few aircrews rode the van out to the jet, but there was little talk. As I got off the van at my aircraft, I heard Manfred say, "See you on 1, cowboy," referring to the channel one check-in frequency. Greeting the crew chief, I let him know this was my first combat mission.

"Hope you kick some Commie ass, sir," he said with a smile and proceeded with the walk-around. I had only carried live bombs once before, at George, so I spent a little time arming wires attached to the bombs. When those wires were pulled away from the bombs upon release, they mechanically allowed a propeller in the front of each fuse to spin, arming the fuse and allowing it to explode on contact with the target.

"I filled up your water bottle with ice water just a few minutes before you got here, sir."

"Thanks, Chief," I replied as he strapped me into his jet. Everything through the start checks went perfectly, and exactly 10 minutes after starting, I heard Manfred call, "Chevy, check."

I responded, "Two," and we were off.

Figure 5 - Picture of author in front of an F-105 at Korat in Thailand. Photo courtesy of the 388 TFW.

The takeoff was less exciting in the Thud on a hot day with a full load of bombs. Thirty seconds after watching Manfred begin his takeoff roll, I released my brakes and engaged afterburner and I was off on my life's greatest adventure, up to then. Feeling the afterburner engage and stabilize, I engaged water injection and felt the gentle nudge of its additional boost. Even with all this thrust, the extra weight and heat soon made you realize that if you had a problem and didn't stop early, you were going to have an even bigger problem. By the time you got airborne, almost all of Korat's 10,000 feet of runway were behind you.

Getting airborne with gear and flaps up, I turned off the water and, at 350 knots, disengaged the afterburner. Spotting Manfred up ahead in a gentle turn to the north, I joined up. Once he had rolled out and headed toward the tanker orbits, he gave me a hand signal to check over his aircraft, which I did. Of primary concern was the general condition of the aircraft with special emphasis on those little propellers on the front of the bomb fuses. If one was spinning, the bomb was armed and best be jettisoned in the local jettison area. I gave Manfred a thumbs-up, and he gave me the lead and checked me over. All was well so he retook the lead and continued north.

I was experiencing a good deal of thirst, so I sucked on the plastic tube attached to my water bottle, but I realized that wasn't going to get the cotton out of my mouth. By the time we reached the tanker, I was out of water. The day before, refueling had been easy, but today was a little tougher for some reason—possibly because of the heavy load strapped to my fuselage or because I didn't get as much response from my engine because of the weight or because the fuel level was getting low—but most likely because I was nervous.

We got our gas and headed further north. Manfred contacted a forward air controller (FAC) and found out where the target area was. I was not paying a lot of attention until I heard the FAC say something about numerous reports of anti-aircraft fire in the vicinity. I thought the first mission was supposed to be a milk run. As we approached the target area, Manfred called, "Chevy, green 'em up for one pass, take spacing."

"Two."

I reduced power and fell back a mile or so behind lead and set up my switches for a ripple release of all my bombs on one pass, thank God. I also went through the combat area preparation Manfred had described in the briefing. "Go to 100% oxygen, turn off your air conditioning, and make sure you select afterburner before you roll in on the target." All of these steps were to maximize your chances of survival if you took a hit—the first two to reduce and endure the large amount of smoke that could enter the cockpit, and the last one to ensure afterburner would be available in case you lost hydraulic pressure. Without air conditioning, I suddenly began to sweat even more than I already was. *Oh crap, so this is what I signed up for.*

I now spotted the FAC under Manfred's nose and heard him describe some type of target under the trees that covered the entire area and then call, "Raven is in for a mark." The little aircraft rolled toward some indistinguishable point on the ground and soon released a rocket; a large puff of white smoke coming up through the trees followed. "Put your bombs 100 meters north of my smoke lead," I heard the FAC exclaim and shortly after, "Chevy, Lead is in," as I watched Manfred begin a hard turn toward the smoke.

Suddenly, a huge barrage of exploding flack went off right behind Manfred's jet. I pressed my microphone button and exclaimed in what I'm sure was a high-pitched voice, "Lead, they're shooting!"

After what seemed like a long delay, the calm response came, "Roger."

The hidden message was clear: "What the hell did you expect them to do?"

His response was somewhat reassuring, but I began to have some doubts about what I was about to do. *This is crazy, I can't do this.* This thought was background noise as I offset my roll-in, heading past Manfred, and rolled in on the target.

"Put your bombs 50 meters south of Lead's Two," I heard the FAC state.

"Roger, Two is in."

I could now plainly see Lead's bombs impact and see what I thought might be a secondary explosion to the left of his string of bombs. As I put my pipper below the target and began to work it toward the target, I caught something unusual out of the corner of my eye. Something that looked like an orange golf ball was fixed in space in the middle of my windscreen. I focused my vision on it in time to see it begin to move and rapidly pass over the top of my canopy. I knew at once it was a tracer and I was drawing fire.

This is stupid as hell crossed my mind as I continued to track my pipper toward the target. The pipper finally reached where I wanted it and the release conditions looked good, so I pushed the pickle button and remembered to hold it as I felt the bombs ripple off the jet and the aircraft breathe a sigh of relief as its burden was released. I immediately began a hard right turn followed by a series of jinks to avoid being hit, and as the nose came above the horizon, I easily found Manfred in a left turn well out ahead of me.

"Chevy, join it up," I heard him exclaim, and I began to pursue him at well over 500 knots of airspeed, feeling an instant state of relief.

The next sense of accomplishment came when I heard, "Nice bombs, Two," as the FAC confirmed that we had gotten some nice secondary explosions from the target area.

He didn't ask us to strafe the target, and Manfred didn't volunteer. I joined Manfred, turned on my air conditioner, and came off 100% oxygen. Then we cruised back to the tanker anchors after checking each other over for any sign of battle damage or hung ordnance. No problems.

The exit refueling was much easier than the first one. Cruising back toward Korat, I knew I had just been shot at, but it had missed. I had done something that, on its face, seemed pretty stupid. But I had passed the test. It was a test that would be repeated over 100 times in the year to follow. With each passing mission, my comfort level would increase, and after a time, I would be able to say "Roger" in the same calm way that Manfred had while under intense fire.

Opening my canopy and feeling the hot but fresh air of Korat was refreshing. Unstrapping and taking the jet off my back was even more satisfying. As I climbed into the crew van, the driver handed me an iced-down, wet hand towel, which I wrapped around my neck. It felt so good. Manfred was already in the van, and he greeted me with a wide grin on his face.

"Did those mean old Commies shoot at you?" he asked mockingly.

My only response was a laugh and a "Yes, sir."

Hanging up our gear, we proceeded to intelligence debrief and were greeted by the shorter of the two women who had presented the mass briefing. She was carrying a clipboard with a checklist on it. Manfred greeted her with, "How's your day going, Valerie?" I checked out her name tag as he began to run through her checklist: "Valerie Galullo." Manfred and I described the AAA we had observed, and he gave her a rundown of the battle damage assessment (BDA) that had been given to him by the FAC. Valerie was pleasant enough, but it wasn't obvious whether she liked her job. However, it was a nice touch to have a few round-eye women around, and she seemed to be as competent as any man might have been in the role. I would discover later that my view was shared by most of the lieutenants, but many of the older pilots didn't like women in the combat zone.

The following day, I again flew a two-ship with Manfred, becoming more comfortable with the battle operations. I also started familiarizing myself with the landscape of northern Laos. This time in a less hostile environment, we ended up making several bomb passes and strafing some trucks along a road. I did see some muzzle flashes coming from the area along the road, but Manfred pointed out that it was probably small arms fire as the area we were in did not have much larger gun activity reported. However, he also pointed out that regardless of what you thought might be the threat, you should always assume the worst.

God, I'm a seasoned combat veteran, I thought. Not so fast, GI! I would later figure out I was not.

After two missions, it was clear that this was not the war I had signed up for. All the stories about the large strike packages against sites like the Dommer Bridge and the Thai Nguyen railyards, which I had heard the instructors describe and embellish at the bar at McConnell, were replaced by mundane operations by comparison. However, as evidenced by the not-so-infrequent losses in Laos, the risk was still high. And since this was a "secret war," these losses were no longer being reported by the news media. Not that it mattered to us. The one realization that struck me the hardest was that, unless we decided to go back North, I would never wear The Patch.

The next day, I was scheduled as checkout at mobile. It was a safety practice to send some dumbass lieutenant out to a little shack next to the runway and advise pilots when they had a problem or were screwing up. You would use binoculars to check that each plane had its gear down before landing and make sure each Thud got water injection on takeoff (the yellow flame turned orange as the water was introduced). For the most part, it was a boring way to spend half a day, and most guys took a book or girly magazine to read or wrote letters home. Occasionally, you were able to shoot the flare pistol to check its operation or you would find a little excitement by encountering a cobra while driving out to the shack—especially after heavy rainstorms.

That same day, Col. Prosser told me I had been assigned "Balls 93" as my personal aircraft. Balls 93 referred to the tail number on the jet, 093, and he let me know I was expected to get to know the crew chief and be familiar with the flying status of the aircraft. He also told me that anytime I was on the schedule and 093 was available, I would be scheduled to fly "my" jet.

Upon finding that the 093 was not flying then, I jumped on my bicycle and rode out to the flight line, where it was being loaded with bombs. I introduced myself to the crew chief. His name was Marty, and he seemed too young to have so much responsibility.

"So, how's our jet doing today, Marty?"

Marty obviously took a lot of pride in our jet, and as he led me around the big machine, I noticed his name was stenciled on the right side of the canopy. We continued the walk-around with Marty letting me know about some of the recent malfunctions and damage the jet had suffered. Coming around to the left side of the aircraft, I focused on the side of the canopy: "LT ED SYKES." Goose bumps! If this was a dream, I never wanted to wake up.

The next day, Dave Dinan, Bob Zukowski, and I flew over to Takhli to deliver three airplanes to their avionics shops to receive an update to the bombing system. The two of them had decided to play a joke on me, and about halfway to Takhli, Dave accelerated to around 500 kts and pulled the aircraft nose up to about 50 degrees high and called over the radio, "Dead Bug!" *What the hell?* I thought as I watched them roll over on their backs. I followed, and the three of us were inverted. I next saw the landing gear come up (down) on their aircraft. Of course, I did the same thing, but my gear was last to come out, so Dave said, "You buy the drinks, 3." It was a fun way to fuck with the new guy. After landing, we went to the Takhli Club for lunch and I bought them a drink, and we yucked it up before catching the C-47 back to Korat.

As Col. Prosser had explained when I arrived on base, we were about to deploy to Takhli so they could repair runways at Korat. Most of us flew missions out of Korat, but some shuttled over in the C-47. The move happened rapidly, and soon we were all working out of temporary quarters. I had already been told that Korat was the "show place" base in Thailand, and our monthlong experience at Takhli proved it. Takhli had a much different personality than Korat. It was like a Wild West show with more of a crazy fighter pilot environment than the sometimes subdued nature of Korat. *What fun!* It was also great to see most of my McConnell classmates again. Our whole class was still intact; no one had had to eject or gotten shot down yet.

Three significant events took place while we were at Takhli. First, the long-expected Tet Offensive kicked off in South Vietnam. The NVA along with the Viet Cong began to send more troops into the South in a massive attempt to capture more territory in the South. With the beginning of Tet, the amount of goods moving down the trail through Laos as well as armaments to protect the flow increased substantially. Consequently, our emphasis focused on the trails through Laos in an attempt to stem the flow of goods.

Second, the 34th suffered its first combat loss since my arrival. 1st Lt. Bob Zukowski was leading a two-ship with Maj. Sheldon Cooper on a bombing mission along the Trail when, without any distress calls or attempt to eject, he never pulled out of a bombing pass. Sheldon observed him hit the ground, and there was no rescue attempt given the circumstances. Bob, on his 128th mission, was really one of the superstars of the young guys. His death gave us all a wake-up call. I did not know Bob well, but he was a real class act in every respect. His roommate and close friend, Dave Dinan, did not show much sign of distress or mourning following this loss, but his attitude seemed to harden as he became less fun-loving.

A few days after Bob's loss, Col. Prosser got Dave and me together to inform us we would become roommates when we returned to Korat. We both replied, "Yes, sir," but there was not much comment or reaction beyond that. I was happy I would be moving into the hootch area with the rest of the Squadron, but the circumstances of my move were not the best, and I wasn't sure what was going through Dave's mind. *Doesn't matter, war is hell,* I thought. I made a few attempts to get closer to Dave before departing Takhli. However, he never went out of his way to be friendly or approach me in search of an expanded relationship, which I didn't take personally, considering recent events.

The third major event during the Takhli excursion was the announcement that the 34th would be converted to F-4Es in May. Once again, the loss of F-105s was resulting in the reduction of Thud squadrons in SEA.

This announcement probably meant that all the lieutenants who were now in the 34th who had come over with Dave would be going home in May before the turnover. With only a few months remaining on their tours, most of them started talking about their next assignment. All expected to go to Training Command and become either T-37 or T-38 instructors. The discussion was primarily about which of the eight bases they might be assigned and which airplane they would get. All but a few wanted a T-38, and most had Williams AFB in Phoenix as their first choice, but most recognized they would not get that. They also realized that, unless things changed rapidly, none of them would ever wear The Patch, although many of them came close.

Upon our return to Korat, I got my stuff out of storage and hustled over to the 34th hootch area to where I knew Dave roomed. The door was open, so I dragged my stuff in and received a cordial welcome from my new roommate. I couldn't sense any ill will. In fact, he shared his experiences and gave lessons, assuming a sort of "mother hen" approach to our relationship. He talked in detail about his earlier ejection and about being hit along with some of the more exciting experiences he had had in the lower route packages of North Vietnam.

Prior to the bombing halt, Dave had 81 counters and was outspoken about being cheated out of his opportunity to wear The Patch, which required 100 counters and allowed an early return to the States. He offered me a lot of advice concerning tactics and aircraft operations, but the one piece of advice he gave me that I will never forget: "Never strafe a gun site. Only dumbasses do that." Strafing gun sites had led to the demise of a number of Thud drivers, and I had heard that advice many times since the McConnell bar, but somehow, Dave's emphasis made it sound like the 11th Commandment.

Shortly after our return to Korat, I came in one morning to find the initials "FCIF" written next to my name on the scheduling board. The Flight Crew Information File was an opportunity for the line jocks to become aware of the latest policies, regulations, or comments about our conduct that needed to be curtailed or modified. A card file contained a card with your name on it, and you were required to initial the card, signifying you had read the latest FCIF guidance before you could fly. Normally, you got a flavor of the memo as you watched others' reactions after reading the guidance. Comments like, "Who the hell does that asshole think he is!" or "What a bunch of bullshit!" were common reactions.

However, as people read today's FCIF, few comments were made to give away its content. As I read it, I could see why. It described the situation in South Vietnam as being very tense, with our ground forces being pushed hard. In early March, the Viet Cong and their NVA allies had kicked off the 1969 Tet Offensive, and conditions on the ground were dire with a high number of U.S. casualties. The importance of stopping the flow of goods to the South at all costs was stressed, and there was an inference that pilots and aircraft were expendable in this effort. As I left the Squadron to go to the mass briefing, I noticed a sign on the door, "Mandatory Meeting All 34th Pilots at 1700."

No one really talked about this guidance, but everyone seemed to be trying to sort it out for themselves. The meeting that afternoon started off differently than most. Colonels Prosser and Glass counted noses to make sure everyone was there. I also noted that none of the enlisted personnel were around. When satisfied that the right folks were in attendance, Col. Glass locked the door.

Col. Prosser began by talking about the intent of the FCIF, emphasizing the need to do whatever we could to make sure that our troops on the ground were supported to the maximum extent possible. But then he stated clearly that "pilots and aircraft are not expendable!" As he gazed around the room, he seemed to affix his gaze on each lieutenant in the group, making sure we understood that we were not to take any stupid risks to satisfy what we might think the intent of the message was. As he finished, my mind drifted to the loss of my brother-in-law, Bob Carmody, and wondering what I would have done if I thought I could have made a difference in his fate.

"Any questions, gentlemen?" There were none so Col. Glass unlocked the door.

Chapter 11

St. Patrick's Day

On March 17, 1969, Capt. Evan Jones—"Jonesy," as he was known by all except the most distant friends—woke up in his hootch by himself. His roommate of a few months had just shipped back to the States, and he had his own personal kingdom. Jonesy had only been at Korat for a few months, and a new adventure awaited today as he worked into the routine of his new environment. Today was to be his first shot at leading a flight to northern Laos. He had led a few missions in southern Laos where the trails all merge and flown into South Vietnam. Jonesy had flown over 100 missions as an F-100 Misty FAC 9 months earlier and knew the area well. However, this was his first effort as a flight lead in North Laos. Checking the schedule the day before, he had found his name at the top of a two-ship, call sign Simmer, with Lt. Dave Dinan as number 2 in his flight, and they would probably end up in "Barrel Roll" (Northern Laos).

Jonesy was different from many of his fighter pilot comrades. Quiet and seldom raising his voice, he didn't cuss and rarely drank too much. However, his appearance and demeanor bely the toughness and resolve within. Growing up in upstate New York, he was convinced he wanted to fly fighters after watching an F-4U buzz him one day on the shores of Lake Ontario. After getting his commission through AFROTC, he attended flight school in Del Rio, Texas, and was assigned to fly F-100s in North Africa at Wheelus AFB in Libya, where he accumulated a lot of fighter time pulling around darts, participating in good training exercises, and getting frequent cross-country flights to Europe on weekends. (Many of the other pilots were married and elected to stay at home on weekends, but being a bachelor, he would build his flying time on weekend trips to the many bases in Europe where the U.S. operated at that time.) In 1967, he was assigned to Vietnam as an F-100 strike pilot.

Shortly after arriving in Vietnam, he did some damage to his squadron commander's airplane and decided to get off base. He found a notice that they were looking for Misty FACs, so he volunteered. Misty FACs were a group of F-100 pilots that would go out looking for targets of opportunity and then call in other fighters to kill the targets. The Mistys only flew two-seat F-100s with the back-seater keeping busy with the housekeeping tasks, such as maps, radios, observation, and anything else he could do to help the undertaking. Although they were designated as forward air controllers, they also carried ordnance and were more properly called hunter killers. Jonesy flew 100 missions as a Misty and ended up getting shot down on one of his last missions along the Ho Chi Minh Trail in 1968. He and his back-seater were rescued as they were able to get out of the target area and into Laos before bailing out of the jet. After his Misty tour, Jonesy decided to volunteer to go back to SEA as an F-105 pilot. Once he had completed a quick checkout in the Thud, he was assigned to the 34th TFS.

Jonesy did not know Dave well but had found him to be a highly seasoned veteran of the War who seemed to know his way around Laos and had a good grip on the tactics that worked well in the field. They had agreed to meet for breakfast at the KABOOM at 0800 hours and do a little informal mission prep. Dave had been there many times, but Barrel Roll was new territory for Jonesy.

That same morning was "just another day of the tour" for me and Dave. The afternoon mission pilots wandered toward the KABOOM for breakfast in no particular order. Dave Dinan and I made the short walk from our hootch to the Club with the standard small talk, mostly about what the targets for the day might be. Dave also asked if I knew Jonesy.

"Just met him at the bar the other night, and he seems like a sharp guy. He said he had just completed a tour in South Vietnam in F-100s as a Misty FAC."

Dave nodded without comment. "Guess we don't need to worry about wearing green today," Dave remarked.

I had to think about it for a moment and then realized he was referring to the fact that it was St. Patrick's Day and our green goatskins met the

171

requirement. Laughing, I nodded acknowledgment. "Not sure there's anyone around here that I want pinching me."

The hootch girls were sitting on the lawn in a circle eating from their stacked lunch buckets. "*Sawadee, kup,*" they gleefully offered as we walked by. "*Sawadee,*" we returned. They were on their second layer of rice and what looked like chicken of their multilayered lunch buckets. When they smiled at us, some of the older ladies revealed the red beetle nut stain on their teeth. One of the girls had brought a plate of fresh pineapple chunks and placed it in the hootch refrigerator earlier that day. Dave handed her a five-baht note as payment, and both bowed slightly.

After entering the Club dining room, we seated ourselves across from each other at the 34th's reserved table. It was not really reserved, but only the 34th Rams sat there. A few others on the afternoon brief were seated nearby, and we exchanged some standard fighter pilot banter. Popcorn and Lek, the two Thai waitresses who always waited on our table, came by and greeted us. We asked them how their *Teelocks* (boyfriends) were, and they replied with the standard response, "No Teelock, no Teelock," smiling and giggling. We gave them our order. A few minutes later, Jonesy entered the dining room and came over and sat next to Dave.

Over the standard sausage and eggs, they talked about the mission in general. "You know this is my first flight lead to Barrel Roll, and you've been there many times—keep me out of trouble," Jonesy said.

"Not much to it," reassured Dave.

Following breakfast, we went to Ft. Apache for the mass gaggle briefing for the early afternoon flights. Weather was forecast to be standard for this time of year with multiple layers of thin clouds and a number of early afternoon cumulus buildups. Valerie and Maureen briefed the threats of the day and the most likely safe areas if you were forced to eject.

Following the mass gaggle brief, Dave and Jonesy went to flight planning and filled out lineup cards and reviewed the numbers. They proceeded to a briefing room where Jonesy ran the briefing guide items and discussed the target area with Dave. Dave knew this area like the back of his hand and gave Jonesy some tips on what to look for. As they departed the briefing room to make their step time, Dave reassured Jonesy, "I won't let you fuck it up." Jonesy grinned with a nervous "Thanks."

On the standard van ride to their jets, the driver proceeded down the back row of revetments and yelled out "104" (the tail number of the Thud in front of the van). "That's me," said Dave, and he departed the van. Dave was greeted by the crew chief and served up some standard banter. "How drunk did you get last night? Over the hangover yet?"

"Yes, sir, feeling pretty good."

"How's the jet been flying?"

"Not bad, sir, have had a couple of air conditioning gripes and had a Doppler write-up last flight, and the avionics guys came by and tweaked it."

Dave checked the forms and did his walk-around with the crew chief following close behind. The jet looked good and clean. As he checked the arming wires on the centerline bombs, he noticed a member of the weapons loading crew had signed the bombs and put a few "greetings" on them. Also, one of the two bombs with fuse extenders had a flower necklace wrapped around it. Dave could smell the heavenly odor of jasmine as he inspected the arming wire. Maintenance troops commonly wrote a greeting (e.g., "Ticket to Hell, Uncle Ho") on the bombs, and flowers were not unusual. The flowers never made it through takeoff, but it's the thought that counts.

Satisfied his jet was ready, Dave ascended the ladder and slowly strapped in with the help of his crew chief. Since the ground checks had gone well, he hit the start button at the briefed start time. He could see the smoke from the cartridge drifting over his left shoulder and the jet sprang to life. Watching the engine gauges, he was comfortable that this start was normal. He occupied himself with running his flight control checks and getting the radar online. Shortly after completing his checks, he heard Jonesy on the radio, "Simmer, check."

"Two," Dave crisply replied.

As Jonesy got taxi instructions, Dave waited for Jonesy to taxi in front of him and then advanced the throttle, then his jet slowly moved out of the revetments in trail with Jonesy. Two other Thuds waited in the arming area, and Jonesy pulled up parallel with them and waited for the arming crews to arrive. As they finished the first two birds and those birds took the runway, the arming crew approached the planes. Jonesy and Dave put their hands in view to ensure they didn't do anything to possibly injure one of the arming crew, who inspected the aircraft for leaks and inspected the pressure gauges to make sure all systems were working as they should. Dave noticed Father Gene approaching the flight and waved a friendly greeting. Father Gene would stand by the aircraft and once the pins were all pulled and the pilot got a thumbs-up, he would cross himself with a wide-open hand. Dave was especially glad to see his good friend Father Gene; not sure why, but Father Gene giving his blessing reassured him today.

Jonesy asked for takeoff clearance. After hearing, "Simmer Flight, cleared for takeoff," Jonesy gave the armament chief the signal to remove the chocks and gave Dave a signal to lower the canopies, then they put the lids down in unison. Simmer took the runway, ran up the engines, and once Dave gave a head nod signaling he was ready to go, Jonesy released brakes and lit his burner. Thirty seconds later, Dave engaged the afterburner as he released his brakes, and once he was sure his gauges were good, he selected water. Once he felt the afterburner light, followed by a slight nudge as the water kicked in, he focused on the runway. His 2000 and 3000 speeds were close to forecast, so he was off on his 118th mission.

Dave rejoined after takeoff and did the standard safety and bomb checks before moving out to a comfortable route formation for the trip to the tanker. They were assigned Lemon anchor today, and as they approached, their tanker headed straight at them. The tanker was directed to start his turn to Simmer's heading. The tanker called "turning," and shortly after that, Dave called, "Simmer 2 has a tally, 10:30 just above the clouds." A moment later, the response was, "Simmer 1, tally." The rejoin went smoothly as did the refueling, and Simmer was off to the north, crossing the Mekong River and into Laos within 10 minutes. Shortly after crossing the river, Simmer was assigned a FAC frequency, and Jonesy checked in with Zorro 32. "Zorro 32, Simmer, flight of two is with you, say position."

"Roger, Simmer, I'm just north of Ban Ban on Route 7. Give me a call when you are 5 minutes out. What you got today?"

"Roger Zorro, we have six 750-pounders each and 2000 rounds of 20 mike-mike. What have you got for us?"

"Roger, Simmer. Not much in the way of trucks today but would like you to cut a road interdiction point just north of Ban Ban. Give me a call when you are in the area and I'll give you a mark."

"Roger that."

En route to Ban Ban, Simmer encountered a few thunderstorms that required Jonesy to use afterburner to top them. Jonesy was sure Dave would give him some grief over having to use up this valuable gas en route. As they approached Ban Ban, Jonesy gave Zorro a call, "Simmer is 10 west of Ban Ban."

"Roger, Simmer, look for me headed north about 10 clicks north of the village." Another 30 seconds passed.

"Two has the FAC, left 11 o'clock just over that ridge line."

"Tally, Zorro, rock your wings for me." The little O-1 could be seen rocking his wings just above the crest of a karst that was masking him from the target area on the other side of the ridge.

Jonesy was somewhat relieved that his first mission to Barrel Roll looked like it was going to be a milk run. He told Dave to plan on dropping four bombs each on the first pass to help make the road cut more effective. Dave acknowledged the call.

"Ready for your mark, Zorro."

"Roger that, Simmer. I'm going to pop over this ridge line and mark your target."

Zorro could be seen rolling over the ridge and, shortly after pointing his craft toward the northeast, shooting a rocket. In a few seconds, a white puff of smoke emanated from the jungle canopy.

"Drop your bombs in that area 100 meters west of my smoke."

By now, Jonesy was east of Zorro and had a clear view of the target area.

Jonesy rolled in and aimed directly for the smoke and pickled twice to release four bombs. Dave rolled in behind him and was coming down the slide when he heard Zorro say, "Just a couple hundred meters north of Lead's bombs, 2."

"Roger that," replied Dave. However, in making the correction, he only hit the pickle button once, releasing only two bombs.

This circling of the target continued for another pass with no sign of secondary explosions. Dave again only released two bombs.

Thinking both aircraft were out of bombs, Jonesy called for a rejoin intent on departing a rather boring operation. Zorro, too, seemed bored and let Simmer know he could take credit for a road cut in their intelligence debrief.

"Simmer 2 still has two bombs remaining," came the call from Dave.

Jonesy initiated a call. "Zorro 32, we have two bombs left. Got a good target?"

"Roger that, Simmer, there is a ZPU site just north of your previous target that has not been active all day, but I suspect he will come up later today. Want me to mark him for you?"

ZPU anti-aircraft guns were small-caliber, rapid-fire weapons that typically become ineffective above 4,500 feet. Jonesy asked the FAC to mark the target, and Simmer 1 would high-angle strafe the site while 2 delivered his last two bombs.

"I've got enough gas for one strafe pass," Jonesy declared and set up for a high-angle strafe pass. He rolled in on the marked ZPU site and fired a long burst from a steep dive angle and pulled off well above ZPU range. He saw no muzzle flashes and assumed the site was inactive.

Dave called in behind Jonesy and had decided to press a little closer. As he was rolling in, he saw something: continuous muzzle flashes coming from a site on the side of a hill to the left of his target. He continued his pass, dropping his bombs on the suspected ZPU site, and as he pulled off, he looked to his left and clearly saw the gun emplacement. It was in a grove of trees with a large camo netting tent over the top to hide its presence.

"Simmer 1 is bingo, let's join it up," declared Jonesy with a clear note of frustration in his voice.

"Got time for one more pass? I've got the target," Simmer 2 replied.

"Roger that," replied Jonesy as he prepared to roll in on the same target, moving his aim point a little further toward the south. Meanwhile, Dave was setting up for his pass from a lower altitude and was trying to reacquire the position of the gun he had located.

"Lead is off left, join it up," was the call Dave heard as he saw Jonesy complete his pass and begin a pull-off above him. As he rolled in toward the gun emplacement he had spotted, he did not have the site visually. However, he knew where it was, so he lit the burner and pulled in to the target area. "Two's in, last pass," he called as he squinted in the area of the target. The muzzle flashes he had seen on the previous pass reappeared, and he brought his pipper toward the target. He now realized he was much lower than he wanted to be but decided to continue the pass. He began firing before he got into range and held the burst until he was within 1,500 feet of the muzzle flashes that continued to appear.

Knowing he was too close and too low, Dave made a hard right turn and started to climb out of the target area. Suddenly, he felt a hard concussion, and his aircraft bounced violently followed by a second lurch. He knew at once he'd been hit, and since he was already headed to the southwest and Thailand, he rolled out of his turn and began a climb. Smoke began to rapidly fill the cockpit, and he could feel his eyes burn.

"Two is hit, smoke in the cockpit," was to be his last radio call. He continued to climb. As he stared through the thick smoke in the cockpit, he could see the caution panel was lit up with several warnings, including the fire and overheat lights. He heard the ATM shutting down and realized the engine had also failed and was winding down. "Shit!" He could see the horizon through the thick smoke and zoomed the aircraft straight ahead until he could feel the slight burble indicating he was getting close to stall. He assumed the ejection position, raised the handles, and without hesitation, he pulled the triggers.

This wasn't Dave's first rodeo, so he knew what to expect. The canopy disappeared in an instant and the noise became deafening. The smoke was gone in an instant as he felt the powerful push of the rocket behind his seat as it accelerated him up the rail and clear of the cockpit. He remembered the sensation of being propelled out of the cockpit and, almost in slow motion, watching the cockpit disappear below. Soon after, the lap belt fired, freeing him from the seat, and the butt slapper pushed him clear of the seat. He felt gravity grabbing him and pulling him to the earth, but the tug of the parachute dramatically slowed his acceleration toward the earth.

Suddenly, the world was quiet. Dave assessed his situation and realized his ejection had gone picture-perfect. His chute was fully blossomed above him and his seat kit was dangling well below him. When he heard the engine of a prop aircraft, he looked toward the sound and spotted an A-1E headed in his direction. He waved excitedly at the "Sandy" (A-1E) and was relieved to know he had been spotted. Knowing he was in sight of friendlies, he reached for the survival radio on his right shoulder and turned off the emergency beeper. He estimated he was about 4,000 feet above the menacing karsts and jungle below and assessed where he might land. He could not sense that his chute was drifting at all, so he surveyed the area right below him and spotted what appeared to be a small grassy field in the jungle on the side of a karst to his left. Making it to that field would avoid a tree landing, and the surrounding jungle would give him a good place to hide.

Dave hurriedly pulled his survival knife from its g-suit pocket and used the special blade on the knife to cut the clearly identified red risers on his chute, causing the chute to drift forward. Reaching up with both arms, he pulled on the left risers, and his chute gracefully turned to the left and toward the clearing. Dave now felt like he was 2,000 feet in the air and could begin to make out the contours of the earth below him. The field he was aiming for was steep, and he was faced with two bad prospects: landing on a steep hill or ending up in a high perch in one of the trees below.

Suddenly, the earth was now rushing up at him. Realizing that he was going to land in a tree on the top of the clearing, he raised his arms to cover his face as he entered the tree. At first, the limbs were soft as he engaged with the outer foliage of the tree, but then he felt the pain of the impact of his legs against the larger limbs. Momentarily, the chute caught a limb near the top of the tree, and Dave found himself suspended in space high above the grassy field below. He was about to prepare his nylon letdown strap when the chute lost its grip on the limbs above and he fell toward the oncoming jungle floor.

This is going to hurt!

Dave impacted the ground and became unconscious immediately. Hitting the steep grade, his body rolled down the hill, wrapping itself in his parachute as he picked up speed. A tree stood by itself in the middle of the field, and his body crashed into the tree with such great force it became wedged between the tree and the steep face of the karst.

When Jonesy had heard Dave's call, he immediately turned back toward the target to find his stricken wingman. As he rolled out to the southwest, he called, "Position, 2?" Nothing. "Where are you at, 2?" Still no response. Jonesy then went over to "Crown," the agency that handled SAAR efforts, and reported Dave's suspected loss. He then heard the undeniable sound of an ejection beeper and knew Dave had ejected. Within a few seconds, the beeper went silent. A call then came over Guard frequency directing him to switch to a SAAR frequency. Jonesy did this at once and soon heard a Sandy and other aircraft talking excitedly about the position of the pilot. Jonesy spotted the aircraft up ahead and headed that way, spotting Dave in his parachute just before he disappeared into the jungle.

Jonesy now checked his fuel and realized that if he did not get fuel soon, he would join Dave on the side of a karst. "Sandy, I'm going to have to get to a tanker. I will get some gas and come back and give you a hand." He pointed his jet toward the tanker anchors and went over to tanker control frequency.

"Simmer 1, give me a vector to the closest tanker you got and ask him to come this direction as far as he can." Tanker control gave him a vector close to the heading he was on, and Jonesy began a best rate of climb toward the tanker. As he approached the river, he heard the control start the tanker into a turn and soon saw the big bird on the horizon in the middle of his turn. Jonesy went straight to the tanker boom and started refueling at once, requesting a full top-off in anticipation of returning to the ejection site.

"Control has asked us to keep you with us and await further instructions," the tanker relayed to Jonesy. "Roger that." They drove south and began an orbit in the normal anchor area. Ten minutes passed. "Any word yet?" Jonesy asked. "Nothing yet, sir."

On this clear day, flying on the wing of the tanker awaiting instructions was no challenge at all, and Jonesy's mind began to wander. Less than a year earlier, he had been a Misty FAC operating along the lower Ho Chi Minh Trail and had been hammered on his last pass on a target. He had headed his stricken F-100 toward Laos and had ejected when he was informed by another Misty that his aircraft was on fire. He had managed to climb to a high altitude, so his ejection included a long freefall before his chute opened at 14,000 feet. He had landed in a large tree and managed to lower himself to the ground and was rescued in less than 30 minutes. Surely Dave would experience a similar fate. However, he was upset with himself for not being able to assist Dave as his stricken jet breathed its last and Dave was forced to abandon ship.

Another 15 minutes passed when another call came that Jonesy would never forget.

"Control just informed us that Simmer 2 is dead, and you are cleared to return home."

"Roger that," Jonesy replied in disbelief. Jonesy had seen a lot during his combat flying experiences, but this event was, by far, the worst. He departed tanker frequency, checked in with the controlling agency, and headed back to Korat with only half a flight. Losing a wingman was bad enough, but not being able to assist in any type of rescue hurt even more. The disbelief would haunt him for a lifetime.

***

My mission that afternoon had gone without incident. We had been working with an F-4 fast FAC in southern Laos, attempting to take out some of the river fords that handled logistics support going into South Vietnam. As the crew chief climbed the ladder to bring my helmet bag, I noticed right away that he did not have the usual happy smile on his face.

"How's the jet, sir?" he asked, and before I could give him a response, he announced, "104 just got shot down." That got my attention. I hadn't heard any beepers nor had I been aware of any SAAR effort, so this was unexpected news.

"Who was the pilot?" I asked.

"Lt. Dinan, sir."

"Did they pick him up?" I asked.

"I don't know, sir."

Climbing into the crew van, the normal sense of relief one has after a mission was absent. Normally, you were tired and relieved to be back, with a cold towel and cold water waiting for you. Today, I wrapped the cold towel around my neck and filled a paper cup with some cold water, but my senses were not the same. Now as our Thai driver took me and my flight mate to Ft. Apache, we discussed what might have happened. Neither of us had heard a beeper or any mention of a SAAR effort. Getting off the van, I downloaded my chute and gear and gathered my belongings from my locker before heading to debrief, where I was sure I would get more information. As I exited Life Support and headed for debrief, I met Bob Howard, another Squadron pilot.

"What's the word on Dave?" I asked.

"Dave's dead," he said. "A PJ went down to get him out and found his body on the side of a karst. Not sure what happened." (A PJ is a pararescue specialist tasked with rescuing downed military personnel.)

Your mind begins to race when hearing such news and fill with questions. I didn't ask any of them. At debrief, we gave our reports to Maureen.

"What do we know about Dave?"

Maureen looked distressed. "The Command Center got a call from NKP, and they said the PJ found his body wrapped in his parachute on

181

the side of a hill. It was wedged between a tree and the karst. They were getting shot at and the PJ was having trouble pulling the body loose, and they left it on the side of the hill."

I headed back to the Squadron and met Jonesy on the porch that surrounds Ft. Apache. He had a rather hollow look on his face and shook his head as we approached.

"What happened?" I asked.

"Not sure. He asked for another pass after I called bingo, and I let him do it. He made one call saying he was hit. He was behind me, and by the time I turned around, he was gone. I never saw him eject but did see his chute enter the jungle. I had to get gas, and a Sandy told me he had seen him in the chute and Dave had waved at him on the descent. I held onto the tanker for about 30 minutes, and they told me to come home and said the pilot was dead."

We simply stared at each other for a few seconds, then Jonesy left for debrief. As I entered the Squadron, the dispatcher saw me and said, "Lt. Sykes, Col. Prosser wants to talk to you—just a second." He went back to Col. Prosser's office, and shortly thereafter, Col. Prosser emerged and motioned me to join him at the dispatch counter.

"You've heard about Dave?" he questioned.

"Yes, sir."

"We're almost certain he is dead, but we need a little more confirmation," he stated matter-of-factly. "I expect we will have positive confirmation shortly. I am appointing you the Summary Courts Officer for Dave."

"Yes, sir," I stated while thinking, *What the hell does that mean?*

Col. Prosser handed me a couple pages that contained some type of regulation.

"This will explain it, Ed. I will let the Wing know you're doing this, and they will give you an assist. You will be responsible for gathering up all of Dave's things and preparing them for shipment back to his family, and you will also need to write a letter of condolence to the family. This regulation will outline your duties. I am taking you off the schedule for the next two days so you can get this done. Any questions?"

"No, sir."

"Good. Go to Wing Headquarters in the morning, and they should have instructions for you once Dave has been officially declared dead."

I grabbed the document and decided not to join in on the conversations going on in the Squadron. I returned to the hootch and read the regulation.

The instructions were pretty simple. Gather all of Dave's belongings and put them in a container that would be provided by Transportation. They would pick it up when I let them know it was ready. All his effects were to be returned to close family. Also, a letter was to be attached informing them who I am and what my duties were and explaining any details that I felt comfortable with. I surveyed our hootch and his closet and realized that gathering his belongings would be easy. However, the letter would be tougher.

Enough of this. I walked over to the bar and joined several of my fellow lieutenants, and of course, the topic was Dave and the circumstances of his ejection. Jonesy was not there, but by now, the rumors were coming together, and it was concluded that Dave had been strafing a gun site and got in way too close. After several Scotch and waters, I joined my compadres in a toast to Dave. Had it not been for all the Scotch I drank, I'm sure I wouldn't have slept very well that night. The haunting feeling of returning to the hootch and not hearing Dave rustling around and talking about the day's events was troubling.

The next morning, I went to Wing Headquarters where they were waiting for me. Transportation would be by later that morning to bring the containers needed to package Dave's stuff, and when I had the letter prepared, I should bring it by the Wing for handling. Back at the hootch, I was going through Dave's stuff and laying it on his bed when there was a knock at the door.

Guessing it was Transportation, I went to open the door. Instead, it was Valerie, dressed in civies and obviously distressed. I could tell she had been crying.

"Ed, I would like to have a few of Dave's things."

Not expecting this, I stared at her without responding.

"Dave and I were dating," she explained, "and there are a few personal items that we shared that I would like to have."

I could never remember Dave discussing Valerie, and I had no idea they were dating.

"I'm sorry, Valerie, but the regulation is very clear that only the family is entitled to Dave's personal effects," I finally managed to blurt out, after quickly thinking through the Summary Courts Regulation.

Seeing my resistance, she said, "These things wouldn't mean anything to his family. They were between Dave and me."

"I'm sorry, Valerie, the Regulation doesn't allow me to do that."

She sniffled and, choking back tears, replied, "Would it make any difference if I told you we were going to be married Saturday?"

Faced with this dilemma, I once again fell back on the Regulation and again explained it to her.

Now crying openly, Valerie pleaded, "Please go talk to Father Gene. He was going to marry us this weekend."

Sensing my out, I promised her I would check with Father Gene right after the boxes were delivered, and she seemed satisfied with that response. After a long soulful look, she turned and departed the hootch. Within a half hour, Transportation delivered the containers and gave me a number to call when the boxes were ready for shipment. I then proceeded to the Chapel and was surprised to find Father Gene there.

"Father Gene, glad I caught you. Thought you might be out on last chance."

"No," he replied, "the Baptist has that duty this morning. Are you here to ask about Dave?"

"Yes," I replied, realizing that Valerie had already been by. "Valerie says they were going to be married this weekend?"

"That's right, Ed. They've been taking instruction from me for the past few weeks, and we were going to do a very private ceremony Saturday morning."

I did not explain my dilemma to Father Gene, but we continued talking about Dave for several minutes. I knew that Dave and Father Gene had developed a close relationship during their period at Korat, and it was obvious Gene was affected by the loss. Walking back to the hootch, I realized that my decision was a moral as well as legal one and decided it was probably best to go with the requirements of the regulation. I spent much of the rest of the day trying to figure out what to say in my letter to Dave's parents and packing some of his bigger effects, mostly clothes and a few paintings and some Thai souvenirs. I took some time out to run around the track while I thought about my letter, Valerie, and the loss of my friend. The initial shock of Dave's loss was wearing off, and I was now facing the final reality of his departure.

Once again, I went to the bar that evening and spent some time with my friends, but I suspect I was not good company. Once again, I drank a few too many Scotch and waters but figured it would help me better face the night. I was surprised Valerie had not returned that afternoon. Perhaps she had decided not to bother me with her request. I was certain I did not want to seek her out. The next morning, I packed Dave's belongings into the containers Transportation had left with a couple empty containers to spare. A knock at the door. Valerie.

She was even more distressed today and not even trying to hold back her tears. "Did you talk to Father Gene?" she asked.

I nodded. "Yes, and he confirmed what you had told me."

"So, can I have a few things that were personal between the two of us?"

"Valerie, I'm really sorry, but the Regulation is very clear about—"

"Ed, would it make any difference if I told you I'm pregnant?"

Without hesitation, I replied, "What would you like, Valerie?"

After some hesitation, she asked about a little teddy bear, a necklace, and a bracelet.

I had packed the teddy bear but knew where it was. The bracelet and necklace were still in the nightstand next to Dave's bed. She gathered up these items and began to sob. She glanced at me briefly. "Thanks, Ed." She turned toward the door and was gone. I stood there trying to digest what had just happened. I knew I would never forget this moment, but I also decided to not tell anyone about it—not Col. Prosser or any of my fellow officers. After a few minutes of deliberation, I also decided not to include this incident in the letter I was about to send to Dave's parents. I felt they had enough to grieve over. Perhaps they already knew, or maybe Valerie had a way to contact them.

I finished the letter to Dave's parents, telling them how much the loss of our brother Dave was felt by us all and how he was a brave warrior. As I wrote the letter, I remembered my mother and sister's grief at the loss of Bob and knew nothing I could say would significantly ease the pain. I then went to Wing Headquarters and gave them the letter before going to the Squadron to let Col. Prosser know I had completed my Summary Courts duties.

"Thanks, Ed," he replied. "I'll let the scheduler know he can put you back on the schedule.

"One other thing, Ed. I would like you to move in with Jonesy in the next day or two," Col. Prosser explained.

"Yes, sir," I answered. Not sure if I liked this change, it was certainly low on my list of concerns at this moment. I liked Jonesy and suspected it would work out fine. *Hope Jonesy is comfortable with it.*

As I left the Squadron, I decided to walk around Ft. Apache and see if there was any sign of Valerie in the Intelligence area. She was nowhere to be found. As I departed the Intel area, I encountered Col. Nelson, the wing commander. "Hi, Ed, how you doing?" I explained that I had just finished my Summary Courts duties and that I had been Dave's roommate.

Col. Nelson studied me before making a statement I will never forget: "Dave's not dead." He hesitated, observing me intently and noticing the questioning look on my face. "He just went somewhere else," he finished.

In an instant, my view of the world changed. *Dave's not dead, he just went somewhere else.* I transitioned into that somewhat callous reality that most warriors find when they experience the loss of a brother. You must block it from your mind and move on as if your brother just went somewhere else. I would lose several more friends over the years, but I would never again experience the sense of loss I had with Dave. They just went somewhere else.

Chapter 12

Picking Up the Pieces

During my time in Thailand, a spoof recording written by Lt. Col. Joe Kent went around of a war correspondent and an Air Force F-4C pilot at Cam Ranh Bay. Each answer given by the Captain is further refined by the Air Force Public Affairs Officer (PAO), who is also part of the interview. As we listened to it, we identified with the Captain at once, and in many ways, it indicates the amount of skepticism in the minds of the fighter pilots.

WHAT THE CAPTAIN MEANS

Correspondent: Captain, what is your opinion of the F-4C Phantom?

Captain: It's so fuckin' maneuverable you can fly up your own ass with it.

PAO: What the Captain means is that he has found the F-4C highly maneuverable at all altitudes, and he considers it an excellent aircraft for all missions assigned.

Correspondent: I suppose, Captain, that you've flown a certain number of missions in North Vietnam. What do you think of the SAMs used by the North Vietnamese?

Captain: Why, those bastards couldn't hit a bull in the ass with a bass fiddle. We fake the shit out of them. There's no sweat.

PAO: What the Captain means is that the surface-to-air missiles around Hanoi pose a serious problem to our air operations and that the pilots have a healthy respect for them.

Correspondent: I suppose, Captain, that you've flown missions to the South. What kind of ordnance do you use, and what kind of targets do you hit?

Captain: Well, I'll tell you, mostly we aim at kicking the shit out of Vietnamese villages, and my favorite ordnance is napalm. Man, that stuff just sucks the air out of their friggin' lungs and makes a sonovabitchin' fire.

PAO: What the Captain means is that air strikes in South Vietnam are often against Viet Cong structures, and all operations are always under the positive control of forward air controllers, or FACs. The ordnance employed is conventional 500- and 750-pound bombs and 20-mm cannon fire.

Correspondent: I suppose you spent an R&R in Hong Kong. What were your impressions of the Oriental girls?

Captain: Yeah, I went to Hong Kong. As for those Oriental broads, well, I don't care which way the runway runs, east or west, north or south—a piece of ass is a piece of ass.

PAO: What the Captain means is that he found the delicately featured Oriental girls fascinating, and he was very impressed with their fine manners and thinks their naivete is most charming.

Correspondent: Tell me, Captain, have you flown any missions other than over North and South Vietnam?

Captain: You bet your sweet ass I've flown other missions—missions other than in North and South. We get fragged [scheduled] nearly every day for uh…those fuckers over there throw everything at you but the friggin' kitchen sink. Even the goddamn kids got slingshots.

PAO: What the Captain means is that he has occasionally been scheduled to fly missions in the extreme western DMZ, and he has a healthy respect for the flak in that area.

Correspondent: I understand that no one in the 12th tactical fighter wing has got a MiG yet. What seems to be the problem?

Captain: Why, you screwhead, if you knew anything about what you're talking about—the problem is MiGs. If we got fragged by those peckerheads at Seventh for those missions in MiG valley, you can bet your ass we'd get some of those mothers. Those glory hounds at Ubon get all those frags [missions], while we settle for fightin' the friggin' war. Those mothers at Ubon are sitting on their fat asses killing MiGs, and we get stuck with bombing the goddamned cabbage patches.

PAO: What the Captain means is that each element of the Seventh Air Force is responsible for doing their assigned job in the air war. Some units are assigned the job of neutralizing enemy air strength by hunting out MiGs, and other elements are assigned bombing missions and interdiction of enemy supply routes.

Correspondent: Of all the targets you've hit in Vietnam, which one was the most satisfying?

Captain: Well, shit, it was getting fragged for that suspected VC vegetable garden. I dropped napalm in the middle of the fuckin' rutabaga and cabbage, and my wingman splashed it real good with six of those 750-pound mothers and spread the fire all the way to the friggin' beets and carrots.

PAO: What the Captain means is that the great variety of tactical targets available throughout Vietnam makes the F-4C the perfect aircraft to provide flexible response.

Correspondent: What do you consider the most difficult target you've struck in North Vietnam?

Captain: The friggin' bridges. I must have dropped 40 tons of bombs on those swayin' bamboo mothers, and I ain't hit one of the bastards yet.

PAO: What the Captain means is that interdicting bridges along enemy supply routes is very important and a quite difficult target. The best way to accomplish this task is to crater the approaches to the bridges.

Correspondent: I noticed, in touring the base, that you have aluminum matting on the taxiways. Would you care to comment on its effectiveness and usefulness in Vietnam?

Captain: You're fuckin' right I'd like to make a comment. Most of us pilots are well hung, but shit, you don't know what hung is until you get hung up on one of those friggin' bumps on the goddamn stuff.

PAO: What the Captain means is that the aluminum matting is quite satisfactory as a temporary expedient but requires some finesse in taxiing and braking the aircraft.

Correspondent: Did you have an opportunity to meet your wife on leave in Honolulu, and did you enjoy the visit with her?

Captain: Yeah, I met my wife in Honolulu, but I forgot to check the calendar, so the whole five days were friggin' well combat-proof—a completely dry run.

PAO: What the Captain means is that it was wonderful to get together with his wife and learn firsthand about the family and how things were at home.

Correspondent: Thank you for your time, Captain.

Captain: Screw you--why don't you bastards print the real story instead of all that crap?

PAO: What the Captain means is that he enjoyed the opportunity to discuss his tour with you.

Correspondent: One final question. Could you reduce your impression of the war into a simple phrase or statement, Captain?

Captain: You bet your ass I can. It's a fucked-up war.

PAO: What the Captain means is…it's a FUCKED-UP WAR."

***

In December 1969, I had completed my tour. When the 34th TFS had turned over to the F-4s in May '69, I was assigned to the 44th TFS also operating out of Korat. The losses of F-105s continued, and in the fall, all the Thud operations were consolidated at Takhli. I shipped home in December '69 with my body still intact, now a seasoned veteran of aerial combat without a strong focus on what my future might hold. After a rather unfriendly exchange with some civilians at the San Francisco airport, I was greeted upon my arrival in Green Bay by my beautiful young bride and a new 8-month-old son, Bartz, who wasn't sure he wanted to share his mommy with this new man in her life.

I had flown 118 combat missions in the War and had been lucky enough to avoid being shot down. However, on two occasions, I had come close to ejecting over enemy territory. I had also watched more friends get "stuffed" and leave their jets, and in some cases, their bodies were strewn throughout the jungles of Southeast Asia. My good friend Jim DeVoss, with whom I had shared Paul Sheehy as an instructor at McConnell a year earlier, survived a high-speed ejection but suffered many broken bones. The helicopter crew that picked up his mangled body was asked by Control about the condition of the pilot to which the copter pilot replied, "The pilot would like a bravo, echo, echo, romeo" (beer). Jim was airlifted back to the States with the expectation that he would never walk again.

While in Thailand, I was not happy with the conduct of this war. I wanted to be reassigned to fighters upon my return to the States but knew that was highly unlikely as few fighter units remained in the States; most were in SEA. So, after a few nights at the bar and assessing my future with this man's Air Force, I decided to let them know I had had enough. In the fall, I had gone to Headquarters at Korat and put in my Date of Separation (DOS) papers. By doing this, I was letting the Air Force know that when my obligation was up in the spring of 1973, I wanted to bid them sayonara.

Near the end of my tour, I had received my orders for my next assignment. I had requested a return to the F-105 in the States but knew that was unlikely for a young lieutenant. I ended up being assigned back to Reese AFB as a T-37 instructor and figured I could tolerate that for the 3 years that remained of my stay in Uncle Sam's Air Force. Mary and I settled in at Reese, and I soon found that I did not care much for being an instructor in the "Hummer." So I found a way to avoid some of the tough duties as an IP by becoming an academic instructor. Teaching T-37 systems was much more to my liking, and my reduced flying duties gave me the opportunity to enjoy new things. I took up golf, and we bought a brand-new house in Lubbock and soon had a second child, our daughter Jennifer. Life was good, but I was not happy that my outlook for flying fighters again was fading.

I also knew I would probably never get a chance to wear The Patch. At Reese, instructors could wear any patch they liked on their left arm, and that was where my Patch would have ended up. As I had departed Thailand, I had found a patch in a tailor shop that some other frustrated Thud driver had probably designed. It was an oval patch with the peace symbol and the inscription "Participant Southeast Asian Wargames" around the outside.

Figure 6 - Southeast Asian War Games Participant patch. Photo by author.

It was my own personal protest against the War and, to a greater extent, my dissatisfaction with never getting to wear The Patch. I wore it during my entire time at Reese and fully expected some senior officer to ask me to remove it, but none ever did.

However, after two years of this pleasant existence, I became restless for another shot at a fighter cockpit. In early 1972, the Air Force put out a call for volunteers with SEA fighter experience to complete a short 6-week checkout in the F-4 and return to SEA. Mary knew I was unhappy with my role as an instructor and didn't raise too many objections when I discussed my volunteering for another bite of the apple. I wasn't sure I wanted much to do with the F-4, but with the unexplainable itch to return to combat flying, I applied.

Then in the spring of 1972, the Air Force announced a program called "Palace Chase" that would allow active-duty pilots to get an immediate release from the Air Force and go to work for an Air National Guard (ANG) unit. The Air Force was also pulling many of the remaining Thuds out of SEA and placing them in ANG units, so I immediately applied with the Kansas Air National Guard (KANG) back at McConnell, which had just had their F-100s replaced with F-105s and was designated as the schoolhouse for all the new Thud units in the ANG and Air Force Reserve. Wow! I called the KANG, and they seemed interested. The Air Force was good enough to let me borrow a T-37 to fly to McConnell, so on a beautiful day in early May, I flew to McConnell and, after a short interview, was hired on the spot. They offered me a full-time instructor's job, and when they showed me the pay scale for a GS-13 instructor, I gasped. Going back to the Thud with a $10,000 pay hike—oh happy day! When asked when I wanted to start, I declared, "June 15," which was my 29th birthday.

The flight back to Reese was one of exceeding joy. As Mary listened intently, I explained our new situation, and we agreed being back in the Thud with a nice raise and little threat of getting shot at again was cause for celebration. "Call the babysitter and let's party!"

A few days later, I got a call from Personnel letting me know they were getting ready to issue orders for my upcoming F-4 assignment back to SEA. Crap, I had forgotten all about that. Maybe I could deal with the lack of the thrill provided by flying a fighter in combat. I explained my new situation, and after a bit of hassling, they allowed that they had other volunteers who could fill my slot. *Just dodged that bullet!*

Mary and I and our growing family headed back to Kansas with a new outlook on life. I was soon back in the most wonderful office ever, the cockpit of a Thud. I fell in love with the Guard from the beginning. The atmosphere was more relaxed than active duty, and I soon discovered that I was dealing with a group of professionals in all areas. Everyone was highly experienced and could be counted on to do the job right. I was delighted to find that my old Thud instructor from active duty, Paul Sheehy, was also an instructor in the KANG. Paul gave me a tune-up ride in the Thud, and it felt good to reacquaint myself with this wonderful machine and the man who had taught me to fly it.

In December 1972, President Nixon became frustrated with the slow peace negotiations with the North Vietnamese in Paris and ordered a full-on assault of North Vietnam by the Air Force. As Mary and I watched the nightly news, it was a sight to behold. The Air Force threw the full weight of their muscle into the effort, and after 11 days and nights of unending attacks, the North Vietnamese sued for peace. The cost to the Air Force was high, with several fighters and B-52s shot down and many crew members killed or captured, but the North Vietnamese got the message. Watching this unfold before our eyes, Mary and I realized that I had literally dodged a bullet by escaping that F-4 assignment. I had mixed feelings as I watched, somehow wishing I could have finally participated in an effort where the objective was to WIN!

However, America's politicians were not prepared to protect the peace that had cost the lives of 58,000 Americans. On April 30, 1975, as we celebrated my son's sixth birthday, most Americans watched on their television sets as it all came crashing down. Saigon was being evacuated as the Viet Cong and NVA troops approached to occupy the city. The scenes of people clinging to the supports of overloaded helicopters in a desperate attempt to be evacuated to the awaiting boats off the coast were not easy to watch. Thoughts of what might happen to many of those who were unable to make the journey were also on my mind. Some Americans were rejoicing, others were in despair, and many, like me, were simply numb. *What the hell was that all about?*

***

As the South Vietnamese collapsed and became the slaves of their Communist northern brothers, I reflected on how my life had been affected by the War. I had benefited personally by my participation in the conflict and had now settled into a steady diet of my lifelong goal—a fighter pilot with prospects of being a fighter pilot for the rest of my career. Combat had also given me a sense of calm when dealing with extreme stress. Without a doubt, I was one of the lucky ones.

My mother and my sister Merry had not been dealt as good a hand. Following Bob's death, Merry was assisted by Bob's aunt and uncle, who lived near Los Angeles, in making the arrangements for Bob's funeral and preparing her departure from California. Merry decided to have Bob buried, as Bob had previously requested, at a cemetery near one of Bob's favorite aunts near Patterson, New Jersey. Mother objected and insisted that Bob be buried at Arlington. As Bob's widow, Merry had the final say in this decision, and Patterson it was. I was unable to attend the funeral as I had already used up my credits with the Air Force by making the trip to California to join Merry after Bob's death. Merry and Mother, along with Sue and Wes, attended the funeral along with a lot of military brass and Bob's Olympic boxing teammates, including "Smokin' Joe" Frazier, who would soon become the heavyweight champion of the world.

Following the funeral, Merry and mother drove back to Kentucky together, which was stressful for Merry. Mother was upset about Bob's burial site and not about to let it drop. As soon as they returned to Kentucky, Merry and her two small children got the hell out of Dodge. She headed to Minneapolis, Minnesota, where Merry could be away from Mother and close to a favorite aunt and uncle. Our depressed and angry mother was now living in a house with good-natured George, and we all sensed the impending disaster.

Wes had finished flight school in early 1969, and his assignment was the C-7 Caribou, a small two-engine trash hauler widely used in Vietnam to move everything from soldiers, ammunition, to Vietnamese peasants with their dogs and pigs, and even toilet paper. After a short checkout in the "Bou," he also headed for SEA and was stationed at Cam Ranh Bay in South Vietnam. Sue and her young daughter had lived with Mother while Wes was checking out in the Caribou at Seward AFB outside of Nashville. Mother was beside herself with fear during this period with a son flying fighters in SEA and her surviving son-in-law about to head that way. She was becoming a bit unstable. When Wes departed for SEA, Sue decided to join Merry in Minneapolis until Wes returned.

Following this, my mother and George soon parted ways, and mother lived a nomadic lifestyle that eventually found her settled in a community for Gold Star Mothers in southern California. Although she really wasn't a Gold Star Mother, as she had not lost a son in military action, they gave her entry to the community because of Bob Carmody. She lived out most of her life in this comfortable environment. I would visit her from time to time and discuss her favorite topics—politics and baseball. She would tell me that her biggest regret was that she would probably not live long enough to see the Cubs win a World Series. We never talked about Bob. Following a fall on a bus in 1998, she moved back to northern Wisconsin where she died in 2000. The turmoil Vietnam brought to her life always made me consider her a victim of the War.

My sisters eventually settled in Arkansas. Merry never remarried, and Sue and Wes went on to have three children during a 20-year Air Force career. Merry was clearly a victim of the War, but Wes, like me, came out of the experience a stronger person.

So that should be the end of my Vietnam story. And it was, until 2009, when I found myself compelled to take on one more War-related mission. Afterburner now!

Part 2

The Stream Where
the American Fell

Chapter 13

My Country Is Better Than This

For those who have been shot at, even if those shots missed, it is impossible to erase the memory of the events that shaped those situations. Over the years, my mind often drifts back to the days when I willingly took off with my magnificent steed loaded to the gills with some of man's most dangerous killing tools knowing there were others a few ridges over who were intent on terminating my life. The brothers who fought beside me make up a large portion of those memories.

One of those brothers who often showed up in my recollections was Dave. I could see him sitting on the side of his bed in our hootch saying, "Never, ever strafe a gun site! Only dumbasses do something that stupid." I never did, especially after Dave's demise. But even more indelible is the memory of Valerie as she sobbed and collected a few trinkets before leaving for an unknown journey. What had become of her? What about Dave's family? I had never received a response to the letter I sent them as part of my Summary Court duties. Surely his body had been recovered and returned to the U.S. for burial.

The date that always brought Dave's memory back was March 17, St. Patrick's Day. In Kansas, this is the day when avid gardeners plant their potatoes. And every year, I would plant my potatoes and peas and think about Dave and the anniversary of his death. Each year, the memory became more haunting as I realized I had been spared another year and Dave had been gone another year. St. Patrick's Day 2009 was especially significant as, on that beautiful Kansas day, I planted my potatoes and peas and thought about the fact that I was still enjoying life while Dave had been dead for 40 years.

During that 40 years, Mary and I had raised four children and now had 10 grandchildren. I had had a career that accomplished my life's dream of being a fighter pilot and, later, started my own business and made it succeed. By now, I had spent 36 years on my small Kansas farm and was happily raising a small menagerie of livestock—cattle, hogs, chickens—and a great dog and cats. *Dave, you have missed so much.* And what about Valerie?

In 2009, my oldest son, Bartz, his wife, and six children were living in Washington D.C. He was attending a language school to learn Russian in preparation for an assignment to the U.S. Embassy in Moscow. Bartz, an Air Force lieutenant colonel, was an Air Force Academy graduate who had spent most of his career flying F-16s and was preparing to close out his career on the military staff at the Embassy (in my opinion, a crappy way for a fighter pilot to close out a career). But 2009 was a good year in baseball, and I was able to take advantage of Bartz and his house in Washington D.C. to attend numerous baseball games at Nationals Park. I had been at Nationals Park the year before when they opened it up and found it a delightful place to watch baseball. Consequently, I attended games in May, June, and July of that year, sometimes taking my grandchildren to watch the Nationals play. I also visited Arlington National Cemetery at least twice during that summer to visit some old friends.

On one occasion, I walked across the bridge to visit the Vietnam Veterans Memorial Wall. I had visited on several occasions over the last 36 years. On each visit to the Wall, I would check the book that annotated where you could find the names of the veterans who had died, and each time I found 1st Lt. David T Dinan III with his status marked "KIA-BNR" (Killed in Action-Body Not Recovered). It had been 40 years since David had died, and I had always wondered whether his body had been returned to the U.S. for proper burial. As I viewed Dave's name on the Wall, I noted a small cross next to his name that denoted that his body had not been recovered.

I recall thinking that of all the bad things that could happen to you as a fighter pilot, this might be the worst. To die fighting your country's war in some faraway jungle and have that same country leave your body to rot on the jungle floor to be preyed upon by all the possible predators the jungle conceals is about as bad as it gets, I mused. What was especially chilling to me was that on the day Dave died, my country had known exactly where that body was and still had not returned to recover it.

A revelation suddenly occurred to me—I was embarrassed by my country! *My country is better than this!* In that moment, I decided to find out why my country had abandoned my friend despite its claim that we tried to leave no man behind. It was also then that I decided, on behalf of my country, I would return my friend to the nation of his birth.

Over the next several years, as I turned rocks over in search of Dave's remains, people would occasionally ask me what motivated me to continue the search, and I would talk about Dave's family and our friendship and, in a few cases, Valerie, but I would never state what was really my driving force: My country, on this issue, is an embarrassment, and I want to fix it.

So, where to start? I was clueless! I took some time reflecting on undertaking such an effort and decided the first thing I should do is attempt to locate any of Dave's remaining family and make sure they were comfortable with me doing a little digging. I remembered that Dave was from New Jersey, and one night, as I was having trouble getting to sleep, my mind threw out a name—"Nutley." It's amazing how much the human brain has stored in its hard drive and how it can sometimes spit out a detail without any reasonable explanation. I got out of bed and wrote "Nutley" down on a notepad in my office (because the brain can also withdraw its information from one's consciousness as easily as it spits it out) before going back to bed.

The next morning, I found "Nutley" printed on my notepad and went straight to Google to see if it was a real place. Bingo, it's a town in New Jersey. I then Googled "Nutley New Jersey Dinan" but got no results. I then did a people search for the whole state of New Jersey and got no satisfactory results. Looking back, I suspect if I had used one of my grandchildren to conduct the search, I would have been more successful.

The last half of 2009 and the first part of 2010 brought my research on Dave to a slow crawl as I labored to complete an attached garage on my house in Kansas. I did almost all the work myself, so I found little time to research Dave's case, but I did stew about it as I worked. I developed a plan for my next step. Following the War, my old Squadron, the 34th TFS, had been part of an Air Force relocation to Hill AFB in Utah. By now, the unit was flying F-16 Falcons and was part of the same fighter wing we had flown with in Thailand at Korat RTAFB, the 388th TFW.

Over the years, I have often flown jets into Hill AFB on navigation "training missions," which was code for ski trips, to take advantage of Utah's great mountain landscape. On several of these trips, I would stop by the 34th TFS building and visit with the members of my old unit. I remembered that the 34th kept an extensive history collection on the Squadron, which I had often gone through to recall all our accomplishments from 1969. The scrapbooks and logs contained lots of pictures and collected articles and books related to that era, and it was fun to reminisce. Perhaps this collection of goodies might contain some clues concerning any efforts to recover Dave's remains. So in July 2010, I scheduled a trip to Utah for late September in hopes of gathering information on Dave and, if I was lucky, finding out something about his remaining family.

In early September, I received an email from Harry Padden, a former Thud driver who was looking for support for a project to restore an F-105 to flyable condition. He asked me to write a letter to my congressional representative asking for support of a bill to give a retired F-105 to a restoration group. "Can do, easy," I replied along with this request:

> Harry, I was in the 34th for the first months of 1969 until they switched out to F-4s. My roommate was Dave Dinan, who was shot down and killed in March. I don't think his body was ever recovered and I am trying to determine where he ejected—somewhere in N. Laos. You seem to know more about the 34th than anyone else and I would like to call you and get some information concerning the 34th and their historical records. Any chance that would work for you?

A few days later, Harry replied that he didn't know Dave because he'd left Korat in February 1968. Then he suggested asking Ken Mays, who coordinates the 34th reunion events, and Howard Plunkett, the Squadron's historian. Harry included contact information for both men, so that's where I began.

Ken replied to my email the next day. Although he also did not know Dave, he put me on the mailing list for the reunions, which I hadn't known about. The next reunion would be in Colorado Springs, May 2012. I love a good party, so I wouldn't miss this one!

I also received an email from Howard Plunkett that day:

Hi Ed,

The 34 Fighter Squadron, an F-16 outfit at Hill AFB, was inactivated this past July. I've had no contact with them recently. DPMO [Defense Prisoner of War/Missing Personnel Office] is the agency charged with recovering remains in SEA and elsewhere. You might try writing to them about any efforts in recovering Dave's remains. In their records, his ID# is D379 and his case file has Reference No (REFNO) 1408.

There is some data about Dave on their site. They have him coded "BB-KIA," body not recovered. This status is different from "MIA." Based on the PJ sighting his dismembered body, as described in my record of his shoot-down, he was a known KIA. ... Howard

Howard's email was good news and bad news: bad news was that the 34th had been deactivated; good news was the contact information for the folks who might be able to facilitate the recovery. I called Howard and, in a short chat, discovered he had an extensive history of almost everyone who flew the Thud in combat, and he offered to help me in any way he could with my search. He also gave me a good contact number for Guy Benz at DPMO who could assist me with getting more information on the government's search for Dave's remains. I decided not to call DPMO until I had more information concerning Dave's family and obtained their permission to do some digging.

When I got that email from Howard on September 9, I was scheduled to go to Hill AFB later that month. I decided to go anyway as I suspected there would be several folks in the area who might know about the Squadron history records and might let me take a look. Up until now, I had conducted my limited amount of research out of my home office in Kansas and was getting a little anxious to set out on some exploring, even if it didn't offer a lot of promise.

After landing at Salt Lake City, I took the drive north to Ogden and Hill AFB. Along this route, one can view the Great Salt Lake and the expansive salt flats north of the lake, with a line of mountains to the west.

I grew excited as I viewed this scene, remembering all the times I had taken off from Hill on some exercise in a Thud, Phantom, or Falcon and headed west across these flats toward the vast fighter playground offered by the Hill ranges. The training afforded by this and other vast range complexes, especially the ranges northwest of Las Vegas, provide American and Allied pilots the closest simulation of war possible without actual ordnance being employed.

One of the positive outcomes of Vietnam was the assessment that the Air Force did not provide enough realistic training opportunities to prepare fighter pilots for combat against a competent, well-armed enemy. Hence, the Air Force developed the "Flag" exercises to provide this realism. I flew a Thud in my first Red Flag out of Nellis AFB (Las Vegas) in the early '70s and was amazed at how fast the tools I had learned in actual combat returned to me. Even more amazing, however, was that feeling of total awareness that comes over you—an awareness I had only experienced in combat, knowing you were a target and the penalty for fucking up might be death, or worse.

I guess when one of my grandchildren ask me, "What did you do in the War, Grandpa?" I can say, "I demonstrated that I was ill prepared and incompetent and that led to great improvements in our Air Force."

These musings crossed my mind as I viewed the vast spaces north of Salt Lake and realized how vital the use of this space for training ranges is to the defense of our country.

Upon arrival at Hill, I was given directions to the 34th TFS building, which I eventually located. It was in a different place than I remembered from previous visits, but I noticed the 34th logo on a sign in the front. It was early afternoon, but the building was locked up with no sign of life. A message on the door had a contact number for anyone with questions. I called that number, which turned out to be the 388th TFW Command Post.

When I explained what I was looking for, I was told there was a civilian named Greg, who worked at the Fire Department, who was a history buff on the 388th, and he might be able to answer my questions. Although

Greg was out to lunch when I called the Fire Department, I told them I would wander over and try to meet with Greg after he returned.

I got to the Fire Department about the time Greg returned from lunch. He was a pleasant guy who truly liked to talk history. Although the 34th had been deactivated a few months earlier, it was programmed to be reactivated in a few years as one of the Air Force's first operational F-35 squadrons. When I asked about the 34th's history files, he was aware of the files but was pretty sure they had been boxed up and sent to the Air Force Museum at Wright-Patterson AFB for safekeeping. They would be returned to the Squadron when it was reactivated.

So, my trip to Utah was yielding little. However, being on a fighter base and driving through the Utah desert, observing some of the fighter "playgrounds" that make up the Hill ranges, got me fired up for pursuing my challenge. Knowing the 34th might become an F-35 squadron was also exciting. Not much left to do but go to the bar. I hung around the Hill Officers Club for a few hours and had a few glasses of Scotch, hoping a few fighter pilots might show up. None did. It appears the bar is not the center of camaraderie and esprit that it was back in the day.

I returned from Utah with nothing of substance to show for my excursion. However, I was invigorated with the idea of investigating this mystery. So, what next? I continued to look in vain for the Dinan family. It occurred to me that I should try to find Jonesy. Because he was flying with Dave on that St. Patrick's Day, he might have some answers to my questions. If he had a good memory of the site where Dave died, he might be able to return to Laos and positively identify the site. He could also have some information about the Dinan family.

Jonesy and I had developed a close friendship when I moved in with him after Dave's death. Despite being roommates and talking about many things, I don't think we ever discussed Dave or the circumstances of his death. After all, in our fighter pilot world of semi-reality, Dave was not dead—he just "went somewhere else." And I am sure I never told him about Valerie. In reality, in a different way, she had "gone somewhere else" too.

After the War, Jonesy and I ran into each other from time to time as we had parallel careers, both becoming commanders of fighter units. He attended some events at the Kansas Air National Guard, but following my retirement in 1992, we lost touch. I had been to a few F-105 Combat Reunions over the years but had never encountered him at any of those. I recalled that Baine Lyle, a member of the 34th while we were both members, had told me he saw Jonesy from time to time and thought he lived in the Austin, Texas, area. I sent Baine an email and he responded. He didn't know what happened to Jonesy but knew someone who might and would check it out.

After a few more passed messages with Baine's contact, I got an email from Jonesy, updating me on his life and providing me with his phone number. I called Jonesy with great hopes of getting some helpful information. We spent several minutes reminiscing and catching up about our families and the other standard stuff, and then I told him what I was up to. He also had been frustrated by the circumstances of Dave's shoot-down over the years, mostly because he had done so little to effect the outcome. He had not seen Dave get hit or eject, and after he was told Dave had ejected, he was unable to stay in the area because he was low on gas and had to head to a tanker. It was on the tanker that he learned of Dave's demise and he did not return to the area. Bottom line was that he had no information concerning Dave's final location. He also stated that he had never communicated with any of the Dinan family and had no idea where they might be. We promised to keep in touch, and I would keep him updated on my progress.

Well, it was great talking with Jonesy, but that sucked! Now what?

As 2010 came to a close, I was beginning to think I was a pretty piss-poor investigator but vowed to do better in 2011.

Over the holidays, my youngest son, Ezra, brought his fiancé, Jill Rockoff, to visit us in Kansas. She was a lovely, fun-loving young lady who had played hockey for Dartmouth. She and Ezra were settled in Boston with good jobs and were going to marry the following August. We talked about the wedding, and Mary and I agreed we would host the rehearsal dinner. We also decided to make a trip to Boston in February to do some planning and pick out a site for the dinner.

So, what does this have to do with Dave Dinan? It turns out—a lot.

**

Mary and I packed our bags and made the trip to Boston. When we arrived, Jill and Ezra drove us to Hingham, Massachusetts, where we had been invited to stay with her parents, Mark and Beth Rockoff. Their house was from the early 1800s and perfectly restored and maintained. It probably looked much like it did in the 1800s minus the stables and outdoor bathroom. Mark's a doctor who makes a daily commute to Boston Children's Hospital where he works. Beth, a semiretired homemaker, demonstrated her outstanding cooking and baking skills while we were there.

We spent an entire Saturday making hotel and restaurant reservations for the upcoming wedding, and Mark provided for us a great seafood dinner that evening. During dinner, he mentioned that on Sunday mornings he always walked down the street to the Atlantic Bagel and Coffee Company to pick up a week's supply of bagels. I asked him to let me tag along.

"I leave the house at 7:30—are you up for that?"

"Don't leave without me!" I replied.

Early the next morning, we made the short hike to the Bagel Company, Mark placed his order, and we sat down with some coffee waiting for the order to be filled.

"Have you lived here all your life, Mark?"

"No, my family is from New Jersey originally."

Doesn't hurt to ask. "Are you familiar with Nutley?"

"Sure, it's a short drive from where my father owned a tailor shop in Elizabeth."

I went on to ask him if he had ever known any Dinans while he was growing up. "Don't remember anyone by that name, but I'll ask my father if he knew the family." I mentioned my search for Dave's remains, and he was interested as I explained the circumstances of Dave's death and my unsuccessful quest to find the Dinan family. When Mark's order was ready, we took the bagels and headed back to their house.

Mark's curiosity about Dave had clearly been aroused because when we got back, he took me to his computer and entered, "David Dinan USAF 1969." There were several hits, and as we went through them, I recognized them as items I had seen before. However, he selected one I had not seen; as I read it, I exclaimed, "I didn't know he graduated from MIT."

The cobwebs in my brain suddenly broke loose, and I remembered Dave telling me back in the day that he had started out at MIT but had finished somewhere else with a degree in physics. I also remembered discussing science issues with him as we had similar backgrounds.

Mark then volunteered, "I'm an MIT graduate. I'll bet I started at MIT about the time Dave graduated. Next week, I'll contact the Alumni Association and see if they have any information on Dave's family."

What a coincidence. We traded email addresses, and as we left Boston, Mary and I were delighted with Ezra and Jill's upcoming union, and I was hopeful that I may have caught a break.

Upon my return to Kansas, I sent Mark an email to make sure I had his email address right. That same day, I got an email back from Mark indicating that he thought Dave had attended MIT before he started there. He planned to track down any information he could from the alumni office. However, when he had asked his dad about the Dinan family, he couldn't recall one from Elizabeth, New Jersey.

Mark, bless his soul, got right on it. I was pretty sure Dave had not graduated from MIT but was not 100 percent sure. After a little research through his MIT files, Mark found the following article in the MIT alumni magazine:

Air Force pilot Lt. David Dinan III '65 was killed in the line of duty in March of 1969 in Laos, after he was forced to bail out of an F-105 jet that had been hit by ground fire. His name was the last to be added to the War Memorial in Lobby 10. The Class of 1982 has been the sponsor of the Vietnam Memorial plaque in Building 10 since its inception and paid for the engraving. MITMAA [MIT Military Alumni Association] is working with MIT staff to ensure that the Memorial is updated to include all conflicts since 1969. If you know of an MIT alum who was killed in the line of duty, please contact Fran Marrone.

Fran Marrone was an administrative assistant for the Dean of Students and had written this article for the alumni magazine. Mark sent an email to Fran:

Dear Fran – I am an MIT grad (Class of '69) and am interested in trying to learn more about another MIT grad, David Dinan, Class of '65. As you may recall from the article below, he was killed in the Vietnam conflict in 1969. Proving once again what a small world we live in, I discovered this past weekend that my daughter's future father-in-law was a pilot with Dave at that time; he is still trying to help recover Dave's remains. Is there any information you can provide me about Dave or anyone at MIT who could help me gather information about him? Specifically, I understand he was from NJ (like me), but do you have a home address for him (or any additional information)? Many thanks and I look forward to hearing from you. With best regards, Mark A. Rockoff, MD (Children's Hospital–Boston)

Within a few days, Mark received an email back from Fran:

This is a bittersweet surprise that you would contact me concerning David Dinan's memorial at MIT. Yes, this is very much a coincidence that your daughter's future father-in-law was a pilot with David. Perhaps he (the future father-in-law) is in touch with or will now, because of this email, be in touch with Charles Dinan Jr., David's brother. Back in November, we were fortunate to receive email correspondence from Charles, who wrote thanking MIT for adding David's name to the Lobby 10 War Memorial. I do not have a home address, but I do have an email address for Charles (whom I have cc'd here). I have also copied the chair and the officers of our ROTC committee so they are aware that you would like to contact the Dinan family. … Please do not hesitate to contact me if I, or members of our ROTC at MIT, may be of assistance. Very best regards, Fran

She also included an email she had received from Charlie Dinan who had seen the same article in the alumni magazine:

Fran, I happened to come across an article in the MIT Alumni/ae Association Newsletter that mentioned that my brother's name was added to the MIT Lobby 10 War Memorial. As the article correctly stated, David was killed when he ejected from his F-105 after being hit by ground fire over Laos. He was twenty-five years old and was within about six weeks of completing his tour when he died. If my parents were still alive, they would have been proud to have had him included in the War Memorial. My dad (MIT '27) always held Tech in the highest regard. The memorial is an indication of why that was. When I return to Cambridge, I'll be sure to visit the memorial. Best regards, Charles Dinan Jr.

In turn, Mark wrote an email to Fran in which he copied Charles Dinan, asking for permission to forward Charles's email address to me and mentioning my desire to help recover Dave's remains.

Shortly after Mark's email to Fran, he received an email from Charlie:

Mark, I would be honored to correspond or speak with your daughter's future father-in-law. I maintained a short correspondence with David's commanding officer shortly after he was killed, but I haven't had contact with any other of his associated pilots. Incidentally, prior to opening my email, I was reflecting that David went down on St. Patrick's Day, 1969, almost forty-two years ago. The Air Force has been in touch with my parents, when they were alive, and me regarding the recovery of David's remains for at least the past twenty years. Unfortunately, it doesn't appear to be possible in spite of the more than significant resources expended to locate them.

While I have been residing in the Pittsburgh, PA, area for the past thirty or so years, we grew up in Nutley, NJ, and my younger brother, John, still resides there.

Charlie

Mark sent Charlie another email and included me in the address block. Up until this point, I had no idea Mark was having these conversations, and when I read the string of emails, I was in disbelief. In the course of a little over a week, I had information that would allow me to seek approval of my quest from Dave's family.

Mark's email to Charlie mentioned my career in the Air Force as well as Bartz's role serving as a military attaché for the Air Force, both of which could be of use in recovering Dave's remains. Grateful to Mark for his assistance, I emailed him back, thanking him for using his MIT connection to locate Dave's brother. Not only did it make my job easier, it once again proved the small world theory.

It was too good to be true! A mystery was solved. I sent an email to Charlie asking for a phone number and a good time to call. I received a reply later that day:

Ed, thanks for the e-mail. It's amazing how some things fall into place. As I stated in my email to Mark, the government has been corresponding with us for the past three decades or so regarding the recovery of David's remains. Recovery does not look good; according to the search documents, it appears that David's remains were never buried and nature took its course. The last correspondence I received was a few months ago and it referred to a lead, an interview with a village elder, but no new information. I'd be happy to talk with you regarding the information I have received.

This is the second major coincidence regarding David. After I left the Marine Corps in 1971, I was talked into joining the Army Reserve. One of my fellow officers was prior Air Force and was apparently a forward controller in a C-130. He relayed a St. Patrick's Day incident where he went into some detail about the shoot-down and recovery efforts. At the time, he was not aware that I had a brother who was a pilot. It is a small world.

I'm currently retired but have a small tax consulting and preparation business. I'm tied up this weekend but have quite a bit of free time during weekdays and evenings. Mornings are usually best. Unfortunately, my commitments are often unpredictable at this time of year. So far, Tuesday morning is clear. If I don't pick up, leave a message and I'll get back to you.

Mark stated that your son is in the Air Force. So is mine. David is an RC-135 pilot and had been stationed at Offutt AFB but is now flying out of Greenville, Texas, at the L-3 Communications facility. My daughter was also in the Air Force and had been stationed at Hanscom AFB in Massachusetts. They couldn't talk her into going to flight school.

Best regards, Charlie

Outstanding! Knowing Charlie and his family were military types, I was fairly confident I would be able to obtain their permission to continue the search.

I sent another email to Charlie:

Charlie, great to hear from you. I will give you a call Tuesday morning—say between 8–9 AM Central. I am anxious to hear your information.

I was flying on another mission the day Dave was shot down and did not participate in the recovery effort.

Dave was flying in a two-ship with Evan Jones (Jonesy). Jonesy became my new roommate after David's death, and we discussed the incident many times. I am currently in contact with him.

I am also retired on a small farm in KS and raise cattle and hogs and do a lot of gardening. My oldest son is currently stationed at the U.S. Embassy in Moscow. Most of his career he flew F-16s. None of my other three children joined the military.

211

I look forward to talking to you. By the way, I am a big Green Bay fan—I'm assuming you like the Steelers?

Cheers, Ed Sykes (wgfp)

Colonel USAF (Ret)

I'm sorry if I'm boring the readers with these back-and-forth emails, but to me, it was no less than a miracle of the first order. This extensive exchange of information over such a short period of time could have never been accomplished back when I was an electrical engineering student at the University of Wisconsin. As I celebrated my good fortune with a nice Scotch on the rocks, I gave a silent toast to my electrical engineering brothers and sisters who had helped bring this amazing capability to life over a period of only a few decades.

Chapter 14

Put Me in, Coach

It was now early March 2011. It had taken me nearly 18 months to find the Dinan family. After an email exchange with Charlie, we had decided on a time for a call. So, what do I talk about with Charlie? I immediately decided not to discuss Valerie. If the family knew about her, my hope was that Charlie would introduce the subject. Based on his emails, it appeared that the family had lost hope for the recovery of Dave's remains. Obviously, I did not want to initiate an effort to recover remains if the family had any objections.

After some introductory chit-chat, I could tell that Charlie was a conversationalist and was completely at ease discussing the death of his older brother. I explained, without a lot of details, what my relationship with Dave was all about: friend, roommate, and fellow warrior. We both spent some time discussing our careers and families. Charlie's younger brother John was a dentist who had served in the US Army and had a son on active duty who was also training to become a dentist. Charlie told me about his own son, who was an RC-135 pilot, and about a daughter, who had also spent time as an Air Force officer. He also talked about his four years in the Marine Corps and later with the Army Reserve. I obviously wouldn't have to give anyone in the Dinan family a course in Military 101.

Charlie went on to discuss the situation with Dave. The conversation is best summed up with an email Charlie sent me at a later date:

After my mother died and I assumed the next of kin position, I received one or two "updates" per year, but they appeared to be irrelevant to the search for David's remains, and then in 2008, I received one that essentially wrote off the search, indicating David's remains would have been carried away by animals or washed away, and the additional information in the report was extremely confusing, contradictory or irrelevant. In retrospect, 2008 was thirty-nine years after David was shot down; I should have expected the contradictory information and confusion.

"Would you or your family mind if I did a little digging into this situation?" I queried.

He said they would have no objection to me attempting a recovery but felt it was probably too late to find any remains. However, he volunteered that he had saved all the correspondence he had received from the Defense Department and would be happy to share that with me. He would send me a package containing everything he had so I could update myself on what had been accomplished so far by the Department of Defense (DOD). Finally, he said he would ask the Air Force for the latest files they might have on David.

We then spent a good deal of time talking about our Vietnam experience. Charlie had entered the Marines in 1968, and although he never actually touched Vietnamese soil, he had been stationed on a Marine landing craft with the prospect of invading North Vietnam on a mission to free several POWs. The operation was canceled prior to the invasion taking place.

We talked for over an hour then promised to renew the conversation in a few weeks after I had reviewed his package. As the conversation ended, I had a sense of ease concerning working with Charlie. Despite his doubts about my ability to resolve the situation, he seemed more than happy to assist in any way he could. There had been no mention of Valerie, and I decided to put that topic off for a later date.

Having found the Dinan brothers and been alerted to the presence of DPMO by Howard Plunkett, I finally felt like I was making some progress. While I was awaiting the arrival of Charlie's package, I experienced another fortuitous encounter.

For years, I had made a habit of doing a no-notice visit to my old F-105 buddy, "Frosty" Sheridan. Frosty lived about 3 miles from me, and I frequently passed his house on my way home. I would always call a few minutes before stopping and ask him if he had a cold Heineken. He always said "no," but I knew that always meant "of course." This day in early March 2011 was no different. When I pulled into his driveway, he met me at the door. "What the hell do you want?" was his normal greeting, and "Go fuck yourself, you old fart!" was my standard retort, after which we would continue insulting each other around the kitchen island as he opened a couple Heinekens.

Frosty is one of my heroes for many reasons. First, he has The Patch. He flew 100 missions as a Wild Weasel out of Korat in 1968 and departed just before I arrived in 1969. He was assigned to McConnell as an instructor in 1969. Shortly after his arrival at McConnell, he was diagnosed with multiple sclerosis (MS) and subsequently grounded and eventually discharged from active duty on disability retirement.

I first observed Frosty entering the McConnell Officers Club in 1973 while eating lunch. He was struggling into the dining room on walking crutches. I asked one of my tablemates who he was, and he replied, "That's Frosty Sheridan, 100-mission Thud driver. He has a bad case of MS and is not expected to live for more than a year." So, the fact that we were drinking brews together in 2011 might be your first clue that Frosty is one tough old coot. Besides MS, Frosty had survived major heart problems and pancreatic cancer. On top of all this, he had managed to build a respectable résumé, including project manager positions in the production of the B-1 and B-2. The only flaw that he will carry for life is that he spent a tour in Strategic Air Command (SAC) as a B-52 pilot, which I remind him of anytime his ego gets out of hand. "I'd rather have my sister working in a whorehouse than fly a bomber," was the type of comment you might hear a fighter pilot make.

As we swapped stories (mostly lies), I told him of my good fortune of finding the Dinan brothers. About that time, Frosty's wife, Patti, wandered in and listened intently as I told Frosty about finding the Dinans and how they had given me permission to begin a search for Dave's remains. I also mentioned that I was preparing to contact DPMO and inquire about the status of their search. Patti then asked, "Have you talked to the League?" I probably replied with something dumb like, "American or National?" She then told me about the National League of POW/MIA Families.

Patti and Frosty met in a bar near George AFB in 1957. It was 11 in the morning, and she was "recovering" from a party the night before, and he was there because that's what fighter pilots do. At the time, he was flying F-100s in Germany and was home on leave visiting old friends at George. She was drinking brandy alexanders, and he was drinking whiskey sours. That must be a great combination because after 20 minutes of sizing each other up, he proposed and she accepted. Not sure 20 minutes is a world record because I've heard guys use "Will you marry me?" as an introductory pickup line, but in this case, it actually worked, and two years later, they married.

Fast-forward to 1969 at McConnell AFB when he'd been diagnosed with MS and is facing a disability retirement, and they have two small daughters and a third on the way. Because there were so many families in the McConnell area who were affected by the War, everyone knew several men who had been declared as POWs or MIAs. Patti was no exception, and upon seeing the metal POW/MIA bracelets (each bearing the name of someone in those categories) worn by others at the base, she bought one of her own and then bought 25 to sell to help Voices in Vital America (VIVA), the organization that was distributing the bracelets to support the families. Frosty wondered what his bride was up to, but within a few months, she had sold 10,000 and was gaining a good amount of notoriety.

McConnell AFB put her on their Speakers list, and she soon became a spokesperson representing VIVA for POW/MIA issues. In 1971, she attended her first National League of POW/MIA Families Annual meeting in Washington D.C. She wasn't allowed to join at the time because she wasn't an affected family member, but she soon became a spokesperson supporting the group. The League is an organization of family (and now associate) members who are active at pushing DOD, Congress, successive administrations, foreign governments, and anyone else who will listen to expend effort and resources in accounting as fully as possible for those missing in the Vietnam War, and now in other wars and conflicts.

"You have to talk to Ann Mills-Griffiths and get her advice on how to proceed. Ann is the executive director of the League and has been forever," Patti advised. "She'll know what you need to do to get started finding Dave's remains. She knows more about this subject than anyone else on the planet." I thanked Patti for her suggestion, and she wrote down the information I needed on a scrap of paper and gave it to me.

I then drank more of Frosty's Heinekens, promising to bring him more beer the next time I came. "You lying sack, you never bring over any beer." To which I accused him of being too old to remember all the beer I had brought to restock his fridge. I went home, and before taking a "combat nap," I got on my computer and easily found the League website. It was instructive, and I discovered that the next League annual meeting was only a few months away in Washington, D.C. I put together my plan. I would attend the League meeting and invite the Dinan brothers to meet me there. I would also set up meetings with DPMO and Ann Mills-Griffiths and solve the Dave Dinan situation in a matter of months. Heineken has an outstanding quality. It makes me think that anything is easy—oh, and that I'm pretty much invincible.

A few days later, Charlie's package arrived, and I anxiously opened it, expecting to find extensive investigations concerning Dave's case. The first thing I figured out was that the Case had an identification number, 1408. The next thing I figured out was that the action reports were difficult to read. The reports used date/time groups with all times in Greenwich mean time (GMT), or "Zulu," as it was referred to by the military. Plenty of acronyms were used that I didn't have a clue what they meant. Also, the geographical coordinates were sometimes given in the standard latitude/longitude format but were more often given in GEOREF coordinates, which consisted of both letters and numbers. On top of that, many positions were based on a heading and a distance from a TACAN station, which were used for navigation in SEA. Further explanation of these systems is not required, but suffice it to say, it was not always easy figuring out the locations described in these reports.

Another tough part of the process was dealing with how jurisdictional areas were laid out in Laos. Provinces were comparable to our states and districts like our counties. Towns and villages, however, were difficult because several had different spellings and some actually had two common names. I was able to sort out, after close scrutiny, that the most likely place of Dave's demise was in Xieng Khouang Province and within the Khoun District. Many villages were in the area, but none were clearly identified as being closest to Dave's site. I was certain that Dave had been hit while operating around Ban Ban, but as I consulted the internet, there was little mention of Ban Ban even existing. The reports cited many villages that had been visited by teams, but none of them triggered any memories for me.

As I read through the several individual reports of investigations related to #1408, I concluded that most of the investigations mentioned had to do with other cases that were carried out in the area where Dave was lost. Many of the reports referred to F-4 losses, and the F-105 losses were aimed at recoveries of remains that had different shoot-down dates then Dave's. The Air Force had clearly suffered a lot of losses in this area over the years given the importance of the logistic routes there.

However, I did locate one report that was specifically aimed at investigating Dave's site. It had been carried out by a team from "Stony Beach" (whoever they are) in 1992. They had investigated the Case in the village of Xan Noy in 1992 and had been told by the villagers that an American had been killed a few miles from their village, and some of the villagers might be able to locate the site. It was decided to visit the site the next day, but when weather prevented the helicopter from flying the next day, the search was called off. After reading these reports, I concluded that in the 42 years following Dave's death, no one had ever attempted to actually visit the site of my friend's death. Once again, I thought, *My country is better than this.*

I decided to call Charlie and ask him if he had any more information about attempts to visit the site. I also decided to ask him about Valerie. After leaving messages back and forth over a few days, I finally reached him. Other than the reports he had sent me, he had received little else about Dave's case. I expressed my disappointment with the DOD efforts, and he agreed but also seemed to indicate that the family had pretty much accepted that the episode was over.

I then decided it was time to discuss the missing link. "Charlie, did you ever hear anything about a woman named Valerie?"

After a long hesitation, he answered, "Yes, I did."

I suspect he would never have brought up the subject had I not breached it first, but since it was now on the table, he began to recount what he knew about her. In the summer after Dave was killed, Charlie was stationed in San Diego undergoing training with the Marine Corps. One day he received a phone call from a woman who identified herself as a friend of Dave's from Thailand and asked if she could meet him. She was living in the Los Angeles area at the time. They set up a time and place to meet. Upon meeting, Valerie stated that Dave had given her a number to contact the family and asked her to contact Charlie if anything happened to him. She then went on to explain her situation.

She told him she was pregnant with Dave's child, and their planned marriage was scheduled for a few days after Dave's death. She explained that she had been discharged from the Air Force and had decided to have the baby and put it up for adoption. Obviously, Charlie was surprised. Dave had never mentioned an impending marriage to him or any other family members, and the thought crossed his mind that this was not a true story. However, as Valerie explained her relationship with Dave and recounted aspects of his personality, it became clear that her story was true. She did not ask him for any assistance and seemed to be satisfied with simply letting the family know of her plight. She told Charlie she would let him know when she had the child and would keep him updated on her location.

Charlie would later find out through Valerie that she had had a baby girl and the child had been adopted immediately after birth. She said she knew nothing about the girl's whereabouts or her welfare.

"Did your parents know about Valerie?" I asked.

"Yes, she wrote them a few letters in which she talked about Dave, but I don't think she ever mentioned the child and I never did."

"Are your parents still alive?" I asked.

"No, they both died several years ago, and the loss of David always haunted them."

"Do you know where Valerie is now?" I asked.

"Ed, I have not heard from her in years. The last time I heard from her, she was living in Salt Lake. I have a phone number and address for her, but I'm not sure either are reliable."

Well, enough of that.

"Charlie, do you know anything about the National League of POW/MIA Families?"

"I've heard of them but don't know much more than that," he answered.

I explained what little I knew about them and also mentioned the annual meeting in D.C. happening in July and asked if he and John might be interested in attending. I told him he could get more information on the League's website, and he said he would check it out.

"Great, Charlie. I'm pretty sure I'm going to go and would really like to meet you guys there if you can make it." Charlie said he would let me know after talking with John.

OK, having cleared the deck a bit on the Valerie issue, I now considered if I should try to contact her and let her know what I was up to. Indeed, I'm not sure I would want to pursue this further if she had huge objections—that is, if she was still alive. I decided to put off any efforts to contact her until after the League meeting. I was hopeful Charlie and/or John would attend and I could discuss the whole issue with them.

It was now March and the meeting was in July, so I decided to make a visit to D.C. before then and visit both DPMO and Ann Mills-Griffiths to get as smart as I could before the meeting. I had a phone number of a contact at DPMO, which I had gotten from Howard Plunkett, so I called to find out what I needed to do to gain access to their empire. Guy Benz answered the phone, and I let him know I was interested in case #1408. He asked me to stand by for a minute while he looked for the case officer.

"Bill Habeeb is the case officer for that case, and I can connect you."

"Bill Habeeb," came an almost immediate response.

I explained who I was and my relationship with Dave, and he quickly responded that he didn't know much about the case off the top of his head, but he would be happy to pull up the file and get back with me.

"Great! How about if I come to D.C. and visit with you and you can brief me then." Bill was amicable to that idea, and I told him I was also going to visit with Ann Mills-Griffiths during the same trip and needed to set up a date when I could do both.

"Glad you're going to visit with Ann. She'll give you some sage advice related to finding the remains of MIAs."

When I called the League, I got the receptionist and asked to talk with Ann Mills-Griffiths. She asked me to repeat my name and told me she would check and see if Ann could take my call.

"Ann Mills-Griffiths," came a hurried reply from someone who wasn't necessarily annoyed, but she seemed preoccupied with something else. I explained who I was and my situation with Dave's case and told her I would like to come visit with her and get her outlook on my efforts. I explained that I had just talked with DPMO, and they had agreed to meet with me, and I would like to meet with her during the same visit. She explained that her schedule was quite busy with the annual meeting coming up in July, but she might be able to see me in early June.

Sold, I thought, and within a few minutes, we had scheduled a meeting on the morning of June 10.

"Are you related to Dave?" she asked.

"No."

"You should come to our annual meeting in July. You're not eligible to be a Family Member, but you can become an Associate Member and attend the meeting. Sign up as soon as you can," she urged.

"Will do."

Although she didn't seem bothered by my request, I felt that she was not all that excited about taking the time to visit. I called back DPMO and asked Bill Habeeb if he could meet the afternoon of the 10th, and he agreed. He would pull Dave's file and be ready to answer my questions at that time.

Once off the phone, I downloaded the forms I needed to become an Associate Member of the League and register for the Annual Meeting and sent off my check the next morning. I also let Charlie know I had registered and that he and John were eligible for Family Membership with the League. The League meeting was to be held at the Crystal City Hilton, and DPMO was located a few blocks away from the hotel. Mary had a first cousin who had a really nice apartment across the river in Foggy Bottom, so I asked them if I could use their guest quarters for both of my trips; they agreed, and I was all set.

So, what did I plan on accomplishing with these efforts? As I had read about the various excursions into the Laotian jungle, I had one idea that most intrigued me—asking DPMO to include me on an effort to find Dave's remains. After all, I was only 68 years old and in reasonably good shape and might be able to add something to the effort with my extensive knowledge of fighter tactics, not to mention my superior skill and cunning.

What the hell, put me in, coach!

***

In early June, I traveled to D.C. to visit the League and DPMO. I spent the night before my meetings with Mary's cousin and her husband in their apartment in Foggy Bottom, right across the street from the Watergate Hotel. I told them about my plans and saw the skepticism in their eyes. So be it.

The next morning, I grabbed a taxi and gave the driver the address of the League office. He drove me to a complex of office buildings on Columbia Pike in Falls Church, a few miles from the Pentagon. We had some difficulty finding the League's office as it was not on the main drag but was in a third row of offices further back in the complex. The office was marked only by a small sign on the window and did not appear to be very large. Upon entering, I introduced myself to the receptionist and told her of my appointment with Ann. She soon motioned me to go down the hallway.

In Ann's office, I found an elderly woman behind a desk stacked high with documents and artifacts who appeared to be preoccupied with the tasks at hand.

"What can I do for you, Col. Sykes?" she asked, glancing away at some papers on her desk.

"Thank you for taking the time to visit with me, Mrs. Mills—"

"Ann," she interrupted.

"My good friend, Patti Sheridan, told me that you were the first person I needed to talk to if I was interested in recovering the remains of my fighter pilot friend, David Dinan." She now became a little more attentive.

"How is Patti?" she asked. I answered that despite suffering from COPD, she was doing OK and was as sharp as ever.

"And how is Frosty?" she inquired. I now sensed that her relationship with both Patti and Frosty was long-lasting, likely through working together with Patti through VIVA in earlier years.

"That old fart is one of the toughest bastards I ever met." I went on to explain the health problems he had experienced, and the whole time she was smiling and nodding her head.

"So, tell me about your friend."

I gave her a short version on the circumstances of Dave's death and followed that with a short version of Valerie's involvement. Ann did not comment, but it seemed that her interest in my story was piqued.

"What about your friend's family? I don't remember any Dinans being League members or involved with the League."

I told her the family had given up on any recovery of remains several years ago, but that I had made contact with them and they were interested in my search. They had also indicated they would join the League and try to attend the meeting next month.

She nodded approvingly. "Good, you'll learn a lot about the process at the meeting. Right now, it's a mess." She launched into an explanation of the problems with the government's efforts and how it was disappointing and in need of new leadership. She vented about how the DPMO office in D.C. was not working well with the operational element, the Joint POW/MIA Accounting Command (JPAC), which was located in Hawaii, and how the identification laboratory was trying to control everything, including field operations, and creating a huge bottleneck for the identification of remains.

Ann's rundown of the situation left me behind in a matter of minutes as I had been unaware of any of this. However, I also realized that I had stumbled onto an intense and sincere woman, whose level of dedication to what she was doing was high. I gained a huge amount of respect for her and put her on a pillar reserved for few. This woman is a fighter pilot.

I did not have any response for her detailed description of the existing challenges and problems. *Why expose my ignorance.*

Realizing I was not going to respond, she asked, "So, how do you expect to assist in this effort, Col. Sykes?"

"I'm going to volunteer to accompany an investigation team to search for Dave's remains."

"How much do you know about the actual incident? Were you there during the attempted recovery?"

"No, but I know a lot about the tactics we used in the War and a lot about the F-105 and its capabilities and equipment," I stated confidently.

"They will never let you be part of a team. DOD is very averse to letting anyone, other than pre-identified personnel, be included in their investigations. They have allowed former military members who were directly involved in the incident to accompany them, but it is unlikely they will think you have anything to contribute."

Ouch. I pondered her direct body slam to my great plan. "I'm going to meet with Bill Habeeb over at DPMO this afternoon, and I'll present my offer to him," I countered.

I got the "you're a dumbass" look from her that I recognized from those times I had made stupid pronouncements to my fellow fighter pilots.

"Good luck and please say 'hi' to Bill. He's one of the good guys over at DPMO."

I thanked Ann for her time and asked her how long she had been involved in this effort. She explained that her brother, Jimmy Mills, had been a back-seater in an F-4 and was shot down in 1966 and had never been recovered. Her efforts, along with others in similar situations, had evolved in the '60s and took on a life of its own when the League formed May 28, 1970. She had been at it ever since, initially at the local, state, and regional level, but then nationally.

As I departed, she asked me to try to get the Dinans involved with the League.

Although I had not accomplished much with respect to my goal of recovering Dave's remains, I had just met a woman who had gained my immediate respect and one I would continue to return to as I needed help with my effort.

I made my way back to Crystal City and, over lunch, considered Ann's viewpoint regarding my desire to join an inspection team. What else could I offer to convince the DPMO to take me on an investigation? *Give it your best shot, Ed.* Surely my superior skill and cunning and dynamic personality will convince them that I would be an essential member of any team they might send out to recover Dave's remains.

At the appointed hour, I called Guy Benz and he asked me to meet him at the Starbucks in the Crystal City Mall. I gave him a description of myself and proceeded to the Starbucks. After a few minutes, I was joined by a middle-aged man of medium build who was with a taller man. He approached me directly as I was one of the few patrons in the shop and the rest were much younger.

"Col. Sykes, I'm Guy Benz and this is Bill Habeeb." After exchanging some pleasantries, the two sat down at the table, and we spent some time discussing my relationship with Dave and the circumstances of his loss. I was beginning to think that all DPMO was going to give me was a coffee shop chat and a handshake when Bill offered, "Why don't we go up to my office and review some of the case information?"

They led me to an elevator and took me up to the 8th floor where we encountered a receptionist with a sign-in sheet. Once I signed in, I was given a badge and told I could not wander around the secure area without an escort, which today was Bill. I was also asked to leave my cell phone in a storage area outside the DPMO offices.

Bill led me through the secured doors and down a hall past several compartmentalized small offices with folks toiling away at their desks. I had visited many government offices like this in my past and was always grateful that I never had to work in such a confined space. My primary office had always been the cockpit of a jet fighter—much more confined but with a hell of a lot more freedom.

Bill led me to his spacious office, which had a pair of windows, so I assumed that Bill was not a standard POG but had an elevated status. Bill reached for a folder from the side of his desk and began to go through what he had on Dave. Most of the things he discussed I already knew but did not reveal that because I wasn't sure Charlie was supposed to share the documents he had sent me. His review did not take long, and following his presentation, I commented, "It doesn't sound like they have ever made a serious attempt to visit the site where Dave died."

"We haven't really done much on this case, Col. Sykes. We don't do a lot of searches in northern Laos as most of the Laos losses were in the south, and other missions have taken higher priority."

"So how do I get higher priority for a mission aimed specifically at Dave Dinan?" I asked.

"There will be a planning meeting later this year in Hawaii, which I'll attend, and I'll try to get Dave's case on the schedule for an investigation. We normally do four JFAs in Laos every year, and normally one is to the north, and I'll try to get him included in that JFA."

I asked him to explain what a JFA was. I had seen the term in the reports, but it was never explained. "That's a Joint Field Activity. Normally 75 or more people divided into three teams that investigate or dig for remains at predetermined sites. These exercises normally last for about six weeks and are carried out under the leadership of Detachment 3, a permanent small team of folks stationed in Vientiane, Laos. The Det. 3 Team is augmented by a number of active duty and civilian participants. It's a very big effort, and it's under the constant scrutiny of the Laotian government," he explained.

It was time for "put me in, coach."

"OK, Bill, if you can get Dave's case on the schedule for next year, I would like to be included on the JFA Team."

Bill glanced to his left, staring at the wall for several seconds, and then he looked back at me.

"I'll try to get Dave's case on the schedule, but it may be hard. There are several other ongoing cases that may have priority, but I'll give it a go. However, it is highly unlikely that DPMO or JPAC will support your request to be part of an investigation team. They never let family members participate as part of a team. Also, unless you have firsthand knowledge of the site and circumstances of the event, they will not include former military members. They have done this on a few occasions in the past, but they don't like doing it. There are all manner of liability and bureaucratic issues to overcome before such an action is taken."

As Bill made his case, I rolled his statements over in my brain. *What now?* A new idea popped up.

"Bill, what if we found the PJ who went in to recover Dave's body? Do you think he would be allowed to go in with an investigative team?"

"There's a chance they would let someone like that go, if he was in good health and had a good recollection of the event. Do you know who the PJ was?"

"No, but if he's still alive, I bet I can find him," I stated, knowing the records I had of Dave's case contained almost no information about the rescue attempt. "Do you have any information about the rescue attempt?"

"I haven't seen anything in the 1408 case file that talks about that," he replied. "All of my researchers are pretty busy on other cases right now, but I could ask one of them to dig into it when they have time."

"How about if I look into it?" I volunteered. "Where's the best place to start?"

Bill explained that the best records of the Special Operations units that carried out most of the SEA recoveries were kept in the AF Library at Maxwell AFB in Alabama. He flipped through his Rolodex, got out a sheet of paper, and wrote something down on it and handed it to me.

"This guy is a researcher at the Maxwell Air Force Library, and he helps us out a lot on cases like 1408. Give him a call and see what he has to say. We send some of our researchers down to Maxwell from time to time, but right now, we don't have anyone scheduled to go down there."

"Is there anything keeping me from going to Laos and doing some looking around?"

Bill studied me a minute. "We can't stop you from doing that, but we can't give you much help either. The Laotians will probably object to your making any visits to this area, but if you can get the right permission, it can be done. Others have done it in the past, but I don't think many have done it lately."

Next question: "Is it possible to get any maps of the area that mark the loss location?"

Again, Bill hesitated. "I don't have a good set of maps right now, but I can send some by email next week."

"Great, is it alright if I share the maps with the pilot who was with Dave when he was shot down? It may kick some cobwebs loose in his mind. If you like, I'll have him call you also."

"Sure, you can share them with him, and I would also like to interview him and add his inputs to the case file."

I then asked Bill about the possibility of any photos of the site that might exist. "There were a lot of RF-4s getting in our way all the time, and maybe one of them did something of value," I added. RF-4s were the photo reconnaissance planes during the War.

Bill smiled and said there were no photos in the file, but he would inquire about the possibility. "Most of the photos from the Vietnam era are of very poor quality and are hard to find because they were not very well indexed—but they do exist. I'll look and see what I can find."

Bill then changed the subject. "Are you coming to the League meeting next month?"

"I plan on coming, and I have asked Dave's two brothers to come also. They talk like they may come."

Bill gave a positive nod and said, "Great. Make sure you tell them to ask for a DPMO briefing on the latest status on Dave's case. Also, tell them to ask that you be included in the briefing. It's normally just given to families, but they may let you attend if the family requests it."

"Great, I'll talk to them about that." We shook hands, and Bill escorted me out of the DPMO office complex.

I jumped on the Metro at Crystal City and thought about my day. I had learned a lot but not sure I had accomplished anything to move the case forward. Ann and Bill had similar views on the possibility of me being included as part of a JFA Team, and knowing my experiences with government bureaucracies, I figured they were correct. Didn't hurt to ask, though. However, this foiled effort might have put me on a better course. Do you suppose the PJ is still out there and willing to go back as part of a recovery effort? His involvement could add more to the effort than my getting in the way. I then remembered the note Bill had given me and reached in my shirt pocket to pull it out. It contained the name "Archie Defante" and his phone number at the Air Force Library at Maxwell.

Flying back to Kansas, I did an assessment of where I was at and what I needed to do next. First on the list was an attempt to get Charlie and John Dinan to attend the League meeting in July and to have them ask for a formal family briefing by DPMO. Next, I needed to plan a trip to Laos on my own to satisfy my curiosity but also, more importantly, to bring attention to Dave's case. Perhaps a nearly 70-year-old former fighter pilot running around the jungles of Laos kicking over cans and doing their job might convince DPMO to move the case to a higher priority. Last, I needed to find the PJ who went down the helicopter cable and attempted to rescue Dave. A popular TV show in 2011 featured comedian Larry the Cable Guy, whose favorite phrase as he began a project was "Get 'r done!" So now I was searching for the elusive "Cable Guy" and knew I had to "get 'r done."

The next day I sent out emails to garner continued interest in Dave's case to some of the players. I sent an email to Bill Habeeb, thanking for meeting with me and telling him how much I had learned from him and Ann Mills-Griffiths.

I also sent an email to Charlie Dinan:

Charlie, sorry I haven't gotten back with you sooner. I just got back from D.C. where I had a meeting with the DPMO guys in charge of recovery of KIAs and a meeting with the Exec Director of the League of POW/MIA Families—Ann Mills-Griffiths is her name—and she is quite interesting. There is a meeting in D.C. on July 20–23, which I am planning to attend, during which I am going to ask for a detailed briefing on Dave's situation from all of the agencies that are involved in this recovery. Would be great if you could attend also. Rather than explain all I have learned by email, it might be best if I give you a call and let you know what is going on. Is there a time Sunday evening or later in the week that would work? I will try to collect all of my thoughts/ideas before we talk. Just got back to Wichita an hour ago and haven't had a chance to organize some of my ideas. I think there is a good chance that we can generate a good deal of additional information in Dave's case and maybe get some results.

Charlie responded within a few hours, and we set up a time for a phone call later that evening. As I shared my information, I could sense Charlie becoming more interested in a cause that, for the most part, his family had given up on long ago. He had discussed the possible trip to D.C. with his brother and thought there was a high likelihood they would attend. He also said he would contact the Air Force Casualty Office and request a detailed briefing. He explained that this office was his primary contact with the government concerning Dave, and he thought they arranged the family briefing. He would let them know that he would like to have "Cousin Ed" included in the briefing. I was elated. Progress was being made!

A few days later, I got an email from Bill Habeeb:

I was out yesterday w/ doc appt. Will send out the maps tomorrow. I'll be at the league meeting too. If this is your first one, I think you'll find it interesting. The mass briefings will help you with understanding more about the overall effort. You'll also get to meet with some JPAC folks. I'll introduce you to a couple of guys that have a handle on what's happening on the ground in SEA.

V/R, Bill

Two days later, Bill sent me the maps attached to an email. His email explained what was plotted on the maps and how. The first maps contained nothing more than the official loss location plotted and annotated. Annotations included the source message and coordinates. Some of the maps had several loss locations plotted on them. He also mentioned that one set of coordinates had been converted incorrectly, so Bill had replotted it. He also explained that JPAC goes out and canvasses villages, interviews, and surveys possible sites. Bill's email indicated he was still searching for photos, and if he found any and could declassify it, he promised to send me a copy or bring it to the League meeting.

As I read Bill's email and scanned the maps, I decided the government was pretty much clueless concerning the exact loss site. It was even more clear that DPMO had not focused much attention on finding it. I forwarded Bill's email to Jonesy to see if he could lend anything to the discussion and also asked Jonesy to call Bill once he had studied the maps. I wasn't surprised Bill hadn't had much luck finding reconnaissance photos. The motto of reconnaissance pilots was, "Alone, Unarmed and Unafraid." In this case, you might add "Useless" to that motto.

The next day I got an encouraging email from Charlie Dinan saying he was available to attend the League meeting, and his brother John was interested in attending. He'd also received all the information he needed from the Air Force.

Shit hot! Since both Ann and Bill had pointed out the importance of having family involved, having Dave's brothers there would add great emphasis to the need to pay more attention to this case. Charlie and I discussed the upcoming meeting on the phone. Charlie also let me know the Air Force was going to set up a briefing on the case for the family, including Cousin Ed. Again, I was impressed by Charlie's enthusiasm in this undertaking.

Within a few days, I had checked to confirm my status as an Associate Member of the League, made hotel reservations at the meeting hotel, and handled my travel arrangements. I decided to put off contacting Archie Defante until after the meeting. However, I did start looking into options for traveling to Laos. I concluded that, with it being about halfway around the globe, it would take at least five flight changes and 30 hours to get to Vientiane, the Laotian capital.

The week of the annual meeting, I got to the hotel on a Wednesday afternoon and settled in. I had agreed to meet the Dinans the following morning when they were scheduled to arrive, a moment I was greatly anticipating.

Although the Dinans were not there for the Newcomer's Reception that evening, I had signed up for it and drifted down at the appointed hour. The only person in the room I recognized was Ann, and she was busy talking with other guests. I would soon discover that most of the people at the meeting were family members or government officials who were conducting briefings. Eventually, the crowd around Ann thinned out and I went over to say hi. She smiled. "Hi, Ed, are the Dinans coming?"

"Not tonight, Ann, but they'll be here in the morning."

"Good. Make sure you introduce them to me," she replied.

"Will do!" I assured her as other guests came up to her. A couple of glasses of wine and several chunks of cheese later, I was ready to get some sleep and prepare for the next day.

I had asked Charlie to call me on my cell phone when they arrived that morning. I had finished registering for the meeting when, around 8:30, I got a call that they were downstairs by the hotel entrance. I said I would be right down and that I would be wearing a New York Mets baseball cap. Before I had left Kansas, I looked at several pictures of Dave, thinking it might be easier to identify his brothers. But at the front entrance, I saw two men standing together but didn't think either one of them looked at all like Dave. They didn't even look like each other. However, the shorter of the two men saw me and my Mets hat and approached me with his hand held out. "Ed, Charlie Dinan, and this is my brother John." The first thought that went through my mind was, "Nutley must have had an active milk man back in the day," but, obviously, I didn't say it.

Charlie was a little taller than I remembered Dave being, and he was more slender with no similar facial features to Dave. John was much taller than either of his brothers with a heavier build. Perhaps John's facial features did resemble Dave's a little.

"Really glad to meet you two. I must say I couldn't have picked you two out as Dave's brothers."

"No, neither of us look much like David. However, David and our father bore a very close resemblance."

They said they had already had breakfast, so I suggested we go upstairs and get them registered. By now the hallway outside the meeting auditorium was filling up with lots of military uniforms from each of the services and several sharply dressed men and women who were probably from the many civilian agencies represented.

However, the largest group of people were, for the most part, casually dressed and fairly elderly. These, I guessed, were the family members of the missing. Many of them were greeting one another. They often embraced and greeted each other like long-lost friends. I assumed (correctly, I would discover) that many of these people attended the League meeting every year to get updates on what was being done to recover their loved ones' remains.

Charlie and John got their Family Member registration packets and were asked if they had attended previous meetings. When they replied that this was their first meeting, they were asked to give a DNA sample at a desk located further down the hall. Charlie stated that he had given someone a DNA sample several years ago but that he would give another one, and John nodded that he would do the same. On the way to the DNA desk, we passed Bill Habeeb, and I introduced him to the Dinans.

"Have you set up a family briefing?" Bill inquired.

Charlie said we had a briefing scheduled for the next day at 2:30. Bill said that he might be able to attend the briefing, but it would probably be given by someone from JPAC.

After Charlie and John both gave their hair as DNA samples and filled out the associated paperwork, it was time for the meeting to begin. We found a table in the back of the room and watched as Ann kicked off the meeting by introducing the distinguished guests. She then gave a summary of what the League had accomplished over the past year and expressed her displeasure with the slow progress being made on the overall effort to account for missing service people. She then introduced Richard "Dick" Childress, the League's senior policy advisor, who spent a full 30 minutes giving a scathing rebuke of the bureaucratic incompetence within the ongoing effort.

What a great comprehensive statement about the inefficiencies the League and the individual families were encountering. If what he was saying was factual, my previous suspicions were confirmed. I joined the vigorous applause and noticed many of the bureaucrats there were also applauding vigorously. My sense, based on this reaction, was that there was a leadership problem at DPMO. I filed that tidbit, reinforcing my earlier discussion with Ann, in my "brain file" and decided to spend more time looking at leadership in this effort.

I didn't have to wait long as one of the next presenters was the head of DPMO, a retired two-star Army general. His speech, packed with the necessary stuff, was clearly aimed at the family members. It all sounded good on the surface. We sat through several more presentations about the ongoing efforts in SEA as well as some efforts related to Korean War and WWII recoveries. As I had learned during my visit with Ann, the League was originally formed to focus on the unaccounted-for in the Vietnam War, but over the years, its success was recognized by families of other conflicts and they piggybacked on.

The afternoon session included briefings by some of the folks who conducted investigations in the field, and one of the members of a panel discussion was the Defense Intelligence Agency (DIA) Stony Beach investigation specialist in Laos, Dustin Roses. I added meeting Dustin to my to-do list. If I was going to travel to Laos, he would be a good guy to work with. By midafternoon, the brothers and I were suffering from briefing fatigue and decided to go get a beer.

We found a quiet spot in the hotel bar and ordered some drinks. It was the first time I had really had a chance to talk at leisure with the Dinans. We covered much of the history stuff: family, our personal backgrounds, and reminiscences of Dave. I outlined my plan to travel to Laos and to find the Cable Guy. They expressed interest in these objectives and asked me if I needed any financial support. "I won't turn it down, but I think I can manage this on my own," I told them. I didn't say I had decided to take $10,000 out of my IRA to finance my adventure.

The next morning, we again attended the meetings, and on one of the breaks, I found Bill Habeeb and asked him to introduce me to Dustin Roses. Dustin was in a quiet area down the hallway, and Bill introduced me and the Dinans. Dustin, tall and in his mid-30s, was well dressed in his dark suit and soft-spoken but straightforward. He said he worked in a one-man office out of the U.S. Embassy in Vientiane. He often worked with Detachment 3, the JPAC Team also located in the Embassy, and frequently accompanied them on JFAs as a worker bee. He said the Laotians were reluctant to let him do any "lone wolf" excursions into the countryside, but he was trying to get permission to have more freedom of movement. I told him of my plan to visit Laos and look for Dave's site and would like him to accompany me on the visit. He said that he doubted the Laotians would allow him to do that and that I was probably on my own. However, he told me to contact him concerning my movements, and he would give me any help that he could. Somehow, I had established in my mind what a field investigator should look and act like, and Dustin filled the image to a T.

Lunch was an official function with individual meetings of the services. Only family members were invited. However, the Dinans, calling me "Cousin Ed," took me to the luncheon, and we were lucky to sit with some active-duty representatives of the Air Force Casualty Office who talked about the ongoing programs to provide more information to family members, including a new database program that would soon be available to families. However, Cousin Ed was too distant a relative to be included for access. Charlie assured me he could pass any information on to me.

After lunch, we attended the family briefing Charlie had arranged. It was in a semiprivate room in the hotel with several briefings going on at the same time. The guy who briefed us identified himself as Bob Maves. He was a short, slightly overweight man wearing a black baseball cap. He told us he was an analyst at JPAC in Hawaii. It quickly became apparent that Bob was not into small talk and was focused on his work. I was impressed by his short, staccato answers to our questions. There was no attempt to baffle us with bullshit, and without a political bone in his body, Bob simply told things the way he saw them. His conclusion on #1408 was that not much had been done and it was doubtful that anything would be done soon.

In the course of his briefing, he alluded to some difficulties that existed between the DPMO and JPAC efforts. He also had some harsh comments for the scientists and how slow they were to identify remains. I asked him about my being allowed to accompany a JFA Team to the loss site of #1408, and after asking me a few questions, he bluntly replied, "You have nothing of value to contribute, Mr. Sykes." As this was the third time I had been told this, I figured I had struck out.

"Is there any reason I shouldn't attempt to visit the site myself?" I asked.

"You can do whatever you want, Mr. Sykes."

"Can I call you for information concerning questions I might have about a trip?"

"If I can answer your questions, I will."

Despite his blunt responses, I liked Bob from the beginning and got his contact information. "Make sure you consider the Hawaiian time difference," he reminded, then he was off to his next briefing. He left us a packet of information on Dave's case, but as we went through it, we realized there was nothing new in it.

As we exited the briefing, we were met by Bill Habeeb, who asked us how it went. The Dinans told him they were happy to get the briefing, but they did not get much new information. Bill nodded.

"We'll begin working this case a little harder," he replied.

Just then, a well-dressed young Asian man passed us in the hall.

"Hung!" Bill exclaimed, and the young man turned and joined us. Bill introduced Hung Nguyen to us.

"Hung works for JPAC and is stationed in Vietnam. He researches the records of NVA troops in the War and might have records regarding Dave's case. The North Vietnamese kept very careful records of their activities, and if they found bodies, they would normally bury them and record the location. None of the other parties in the ground war kept good records of their activities or buried bodies. Hung could review the records of NVA movements in Laos during the time of Dave's death and see if there is any mention of the event."

Hung replied, "I'll be back in Vietnam in a few weeks, and I'll begin looking for information at once." He opened a small portfolio he was carrying and wrote down the case number and promised Bill he would get back with him soon.

Long shot, but what the hell, I thought.

A banquet was scheduled for that evening, but Charlie and John had decided not to attend. They had planned to head back home after the briefing. I realized I had not introduced them to Ann, so we went back to the meeting where we found them in the middle of a break. Ann was at the head table chatting with other presenters in preparation for the next part of the program. I took the brothers up to the front, and noticing a break in her discussions, we walked over. I was greeted with, "Hi, Ed. Glad you could make it." I quickly introduced her to the brothers, and she thanked them for coming and urged them to get involved. She was soon back to her conversations with her presenters. I felt bad that Charlie and John had not had more time to chat with Ann and get a flavor for what a dedicated soldier she was. However, I wanted to let her know that Dave Dinan now had family members involved, which would have a much greater impact than some old washed-up friend of a fighter pilot.

When Charlie and John were ready to depart, I walked them to the hotel lobby and told them I would keep them posted. They wished me well, and I concluded that they were not sure Dave's remains would ever be recovered, but they were willing to support my efforts. I'm not sure they realized how important their involvement was, but I was glad to have them as active players.

That evening's banquet was a neat event, highlighted by a candlelight ceremony during which the names of all the missing whose families were attending were read. As Dave's name was read, I decided to make sure I asked the Dinans to attend next year's banquet. This would have been the first time Dave's name had been included in this ceremony. Goose bumps! I was now fired up concerning this effort, having found some new folks who might be able to help me along the way.

I was also painfully aware that the "coach" was not going to let me play, so I was on my own as far as the Laos investigations.

What the hell. This looks like a mission for a single-seat fighter pilot—those other guys haven't gotten much done in the last 40 years!

Chapter 15

Finding the Cable Guy — Get 'r Done

When I returned to Kansas, an email from Jonesy awaited me.

Ed, I have looked at the maps, and while my memory is far less than perfect, I am surprised that the official loss location is that far from the road. It seemed to me that Dave ejected quite soon after he was hit. My guess would put the location on a line from the official loss location to what I recall as the target area, which I think was in the area of UG375 695. Because he ejected so quickly, I would have put the location between five and ten miles from the target area on that line. I know this is not much help. Sorry, I'm afraid too much time has passed.

Please pass my respects to Dave's relatives and give me a call when you return with whatever news you've found. I continue to share your interest in doing what we can to bring Dave's remains home. Jonesy

As I reviewed Jonesy's information and tried to place his provided coordinates on the map, this only confused the situation further. I called Jonesy, briefed him on the meeting, and again asked him to call Bill Habeeb with his inputs. I was pretty sure Bill would add Jonesy's inputs to the #1408 file. If nothing else, this would show that there was ongoing interest in the case. Making bureaucrats bring your file to the top of the stack is always a good thing. Since the actual location of Dave's death was only becoming more confused, I now decided to shift my focus to finding the Cable Guy.

I dug out the information Bill had given me and gave Archie Defante a call.

Getting no answer, I left a message and he called me back later that afternoon. Archie was a real conversationalist. I suspected that spending all of one's time in the Archives of the Air Force Library would make one a little stir-crazy. When I told him what I was looking for, he told me he might have some information on the incident. When I mentioned I was a former F-105 pilot, however, the conversation went far afield. He was soon telling me about airplanes, beer, food, and women, and I realized that Archie was quite a character. After a while, I was able to get the information I needed to plan a trip and gain access to the Archives. I would not be able to actually go in the Archives, but I could request the records I wanted, and he would go look for them. He also let me know I could view only records that were unclassified or had been declassified. I could make copies of the documents for a nominal fee.

"Great, Archie. Thanks for taking the time to talk to me, and I should be out in a few weeks. I'll give you a 'heads up' message as soon as I have the visit set up."

I sensed that Archie was excited about having a visitor. I got busy setting up the details and sent an email to Archie when my plans were completed:

Archie, as requested, this is a "heads up" that I will be at Maxwell AFB this coming Monday and Tuesday (22 and 23 Aug). I will arrive on Sunday PM and should be at your office fairly early on Monday. I am researching the ejection and loss of 1Lt. David Dinan on March 17, 1969. Dave was flying an F-105D and was assigned to the 34th TFS at Korat RTAFB at the time. He was shot down in NW Laos. I would also like to access any records that may be available on the FAC and SAAR operations associated with this event.

It was fun swapping lies with you over the phone a few weeks ago, and I look forward to hearing some more of your jive next week. Hope you can join me for lunch or a beer at the bar during my visit. Cheers, Ed Sykes (wgfp)

Meanwhile, I got an email from Bill that included this email he had gotten from Hung Nguyen:

I just got back from leave today. Mr. Forsyth and I have been discussing Case 1408, see emails below. I think Dinan is closer to the target area of Route 7, but Bill thinks he is near the record location. The target area is approximately 24 kilometers northwest of the record location. Anyway, Mr. Teel told me he has IAW leads to do in Laos, which he is coordinating with you.

Case 1408 occurred on 17 March 1969 in Xieng Khouang Province. According to the Vietnamese history book of the 148th Regt (PAVN 316th

Div), at the end of February 1969, the Vietnamese Northwest Military Region HQ tasked the 316th Div and 5th Volunteer Youth Group to coordinate with the 701st Pathet Laos and local Xieng Khouang militia units to start the Xieng Khouang-Muong Soui campaign. Muong Soui is at 48Q TG 795 598, which is right on Route 7 but far west of the target location of Case 1408.

This is all I have right now. I will let you know when I find out more info.

Based on this input from Hung, the location of the actual site of Dave's demise was even more uncertain. I wasn't surprised. Nor was I surprised by the information regarding the location of NVA troops in the area. I was sure there had been a lot of NVA presence along the route structure where Dave had been first hit as there was a lot of high-caliber guns along the roads. However, I couldn't ever remember enemy activities in the general area where Dave ejected. The terrain, for the most part, was mountainous and covered with heavy jungle. There were few, if any, strategic points of value in the area, so I felt it was unlikely NVA troops would have ever ventured there, meaning Dave's body was probably never buried.

Years ago, a friend introduced me to Frontera merlot, a cheap but pleasant red wine from Chile. I was told it contained a number of antioxidants that helped brain function. Seemed like a good enough excuse for me, so I consumed a good amount of it at times like these. With my brain hitting on all cylinders, I decided to "get 'r done" with respect to finding the Cable Guy and planning a trip to Laos in 2012. I already had my trip to Maxwell set up with hopes of finding the Cable Guy. However, the only thing I knew about travel to Laos was that it was a long way away. I also knew that I did not speak the language or know anything about the culture. Otherwise, I was well prepared. *But right now, I have bigger fish to fry.*

My trip to Maxwell was less than two weeks away, so I reviewed the documents to make sure I asked the right questions and asked for the right documents while at the AF Library. I had visited Maxwell AFB on numerous occasions so was comfortable with getting around Montgomery, Alabama, and the base.

When the day came, my taxi delivered me to the Base Billeting Office where I checked into a "space available" room. I then took a short walk around the base to make sure I knew where to go the next day. Maxwell's primary role is to provide "professional education" to AF officers and senior enlisted airmen for the entire USAF. There were few classes in session at this time, so the base was quiet.

The next morning, I made the short hike to the library where I was greeted by a receptionist. I identified myself and told her I had an appointment with Archie Defante. She led me to a large room with study tables and told me she would let Archie know I was there. I was the only person in the room until Archie showed up. He told me I had come at a good time because during periods when classes were scheduled, this room would be packed and he would be much busier.

He then explained the rules. I would ask for the documents I wanted, and he would go to the Archives to see if they were available in unclassified versions. He would then bring them down to me and I could peruse them at my leisure. I would not be allowed to take any documents from the room, and if I wanted copies of any of the documents, he would make them for me at a nominal price. Once I was done, he would collect the documents and return them to the Archives and retrieve any additional documents I requested. Easy enough.

Based on our previous discussions, he had already gathered several documents concerning the SAAR effort. Before he left to retrieve them, I suggested that I would buy him lunch at the Officers Club. This launched into a discussion of what we would find on today's menu and touched on several far-ranging topics Archie wanted to chat about. After what seemed like a long time, Archie retrieved the folder of documents.

"Check these out and let me know when you're ready for more," he said and then told me where to find him. Most of the documents in the folder had message traffic from the period. The messages told the story of the rescue attempt as information was sent back and forth from incident participants to their own bases and then retransmitted to each of those base's headquarters. It was slow work, but I eventually began to put a picture together. Shortly after the ejection, two A-1Es were scrambled off alert from Nakhon Phanom (NKP as it was officially called, or "Naked Fanny" as the fighter pilots called it) and arrived at the scene within an hour. The FAC at the scene let them know where the downed pilot was and turned control of the SAAR over to them. The A-1E was a large turboprop airplane with the ability to carry a large load of munitions along with a cannon. They used the call sign "Sandy." They played many roles in SEA, but as far as fighter guys were concerned, their primary job was to get our ass out of the jungle if we got shot down. We also knew they had a large set of balls and were good at their job.

Shortly after the ejection, two HH-43 "Jolly Green Giants" were launched from a remote Lima site. A number of these sites scattered around Laos were used for all manner of activities in the "secret" war in Laos. One use of these sites was to forward-stage Jolly Greens to allow for quick response if an ejection occurred. In this case, it worked as advertised, and the two Jollys were on the scene quickly. From the messages, I now had the call signs of the two rescue helicopters: JG-9 and JG-16. I knew the Cable Guy was on one of those two choppers. As I read the reports, I discovered that JG-9 was the primary rescue bird and JG-16 was the backup. However, JG-9 experienced some problems with its hoist system, and the primary role was turned over to JG-16.

JG-16 then hovered over the site and the PJ (the Cable Guy) descended the cable, where he found the body of the downed pilot and reported him deceased. Earlier, one of the Sandys reported the possibility of gunfire close to the scene. However, there was no indication from the messages that the Sandys took any action against approaching forces. The Cable Guy experienced difficulty removing the body as it was wedged between a tree and a rock, and the decision was made to get the hell out of Dodge. The Cable Guy ascended back up the hoist as the Jolly Green was departing the area, which ended the SAAR mission.

The whole episode had taken place over a short period of time, and no reference was made to fighters being called in to support the rescue. (I remembered that I had been flying in southern Laos that afternoon, and there was never any indication of a SAAR effort being undertaken that needed fighter support.) Also, there were no reports of any shots being fired to protect the rescue site or rescuers. Of course, what I was looking for was information about the Cable Guy. I now knew that he was the PJ on JG-16 and was stationed at NKP, the home base of the unit where the Jollys were assigned. However, message traffic sent during combat activities never included names of individuals, so more digging was necessary.

Archie and I went to lunch, and I got the lowdown on much of his life history—more than I really needed to know. During a break in his narrative, I told him that what I really wanted to find was the name of the PJ on JG-16. He suggested we look at the Unit History Reports for the special operations wing (SOW) the Cable Guy was assigned to. Sometimes these histories would include names. We decided we would look at the History Report the next morning and I would continue to study the message traffic for the rest of the day. Not much new was uncovered that afternoon, and as I returned the folder to Archie, he told me he would not be able to join me for dinner—something to do with a girlfriend.

The next morning, Archie met me with the Unit History Report for the 56th Special Operations Wing for the period of Dave's demise. Much of it was boring stuff about how many missions they had flown and how many bombs dropped, but eventually I found a recount of the attempted rescue of my friend. The first variance I found in this report was that the Sandys had encountered light ground fire and had "worked the area over." That was a little different from the messages I had read the day before. Then came the shocker. The report stated that the PJ had found the deceased crewman under his chute, and "the PJ and the dead crewman were then picked up by Jolly Green 16 and the SAR forces then returned to Base."

I read the statement several times not believing what I was seeing. I then checked the last page of the report to see who had signed it off. It was signed by the colonel who commanded the 56th SOW. This had to be a mistake. If this were true, what had become of the body? Is Dave buried at NKP? Had they misidentified Dave's remains and returned them to the States under the wrong name? I knew this could not be correct but felt I should give Bill Habeeb a call at once.

"That can't be correct," was Bill's reply when I gave him the news. "Fax me a copy of the report, and I'll follow up on it with the Air Force."

I had Archie assist me with a fax to Bill. Archie did not seem overly surprised by my discovery. He said that these kinds of mistakes were common. As I thought about it, I reflected on my experiences in the War and realized that the continuing chaos of the sustained battle would result in numerous mistakes. It was always the next day that was most important; you had already survived to the present.

But then something more important dawned on me about this finding. Why had this report not been discovered and investigated by DPMO previously? It shouldn't have been that difficult for the professional researchers to find this mistake and make a note of it. Bottom line, it was another example of the inefficiency of this part of government and how little had been done to research Dave's case. As I said goodbye and thanked Archie for his help, he handed me a mug with the AF Library inscription on it. Archie told me it was a real "chick magnet." He then handed me a red paperback book. "This might help," he explained. "This guy knows as much as anyone concerning PJ efforts in Vietnam. I'll bet you can get some answers here."

The book was titled *PJs in Vietnam: The story of airrescue in Vietnam as seen through the eyes of pararescuemen*. It had been written by retired AF SMSgt. Robert LaPointe. "Thanks, Archie. That looks like a new book. How much do I owe you?"

"It's on me, Ed. We have several copies here at the Library."

As I was waiting for my airplane at the Montgomery Airport, I called Charlie Dinan. His response to the possibility of Dave's body being recovered and returned to NKP was immediate. "I don't believe that, do you?"

"No, I don't, Charlie, but I'm a bit upset that this information had not been discovered earlier by DPMO. It was right there in plain sight."

I gave him a few more details on the SAAR attempt and told him I would be in touch. I scanned the *PJs in Vietnam* book on my return trip to Kansas and found it to be an interesting review of many of the PJ rescues during the War. They were listed in chronological order, so I hastily skipped forward to the spring of 1969 and found no mention of the attempted rescue of Simmer 2. I thought that strange, so I went to the

Index and looked for "Simmer 2" and "Dave Dinan," but there was no reference to either. I decided once I was settled back in Kansas, I would attempt to locate SMSgt. LaPointe and see if he had further information. If he had information on the incident, he might be able to tell me more about the Cable Guy.

First thing I did on my return the next morning was email Archie:

Archie, you were right about the mug you gave me. Going through security at MGM this morning, TSA opened my bag to see what was in it and several beautiful female onlookers would not leave me alone before my departure. You are so clever.

Thanks for all of your help the past two days and hope our paths cross again. Ed Sykes, (wgfp)

Next, I sent an email to Charlie, John, and Jonesy with a copy of the History Report attached:

This is part of a declassified document that I found (actually dug out by a very able archivist named Archie Defante) during my research at Maxwell AFB on Monday and Tuesday. It is part of the history report of the 56th Special Operations Wing (SOW) that was stationed at NKP. I have my doubts about its accuracy, but we must follow up on this report. I faxed a copy of this to Bill Habeeb at DPMO yesterday, and he has already begun an effort to establish the report's accuracy. (There are several reports by message traffic that refute this claim.) We agree that we must attempt to find the crew members of Jolly Green 16 and 9. I spent several hours on Tuesday attempting to get information on these crews without success. One of Bill's folks, Guy Benz, will be at Maxwell next month researching records for DPMO and he may be able to help. Although it is not his responsibility, I will also bribe Archie (he likes to eat a big lunch) to give us an assist in his spare time.

Bill is also beginning a search through the mortuary records for that period.

Will keep you posted on any progress and would appreciate any information you might come across.

Cheers, Ed Sykes (wgfp)

The next day I got an email from Charlie:

Ed, it's an interesting document, but it's at odds with virtually all the information I received. It is inconceivable that David's body had been recovered in light of the information that we received.

I haven't been able to connect with the Randolph computer. There's either a security block in my firewall or I am missing a critical program to enable access. Their IT person couldn't figure it out, although he said that he could access it from his home computer. He uses Comcast, as do I. Anyway, they sent me a disc with all of the information including the information my parents released to the Library of Congress. I'll burn a copy and send it to you in the next day or two.

I have an old friend who recently visited Thailand and tried to get access to the area that David went down. I'm not sure how hard he tried, but he emailed me that he was told by the American Consulate that the Laotians would only allow family members into the area.

Charlie

I had come to the same conclusion as Charlie concerning the removal of the body from the jungle. It simply could not be factual. Charlie's reference to Randolph was his attempt to obtain the data files that families had been promised would become available. I wasn't sure how to deal with the part about his friend trying to gain entry to areas in rural Laos. Along with my discussions with Ann, Bill, and Dustin, it sounded like permission to enter these areas was as solid as Jell-O.

***

I decided to relax my search for the Cable Guy so I could better prepare for my hoped-for trip to Laos. During this time, I also took a break to attend Homecoming at the University of Wisconsin (UW). My fraternity and the UW AFROTC Detachment (Det. 925) had both invited me back for Homecoming activities, and I decided to go and have some fun.

Visiting the fraternity house during Homecoming was always a blast. The place hadn't changed much since I was a student over 40 years earlier. As usual, the tap was flowing in the bar in the basement, and it was fun talking to the current members and finding out what the job market was like for young electrical engineers. Very good, thank you! It was also fun learning the newest change to my fraternity—women were now "brothers." In fact, a few of the "brothers" had married each other.

The AFROTC Detachment now had a new home. Gone from the Mechanical Engineering Building where I had begun my Air Force journey, it was now housed in a new building only a block or two from the football stadium. They put on a light lunch and a status briefing for alums, so I wandered around the gathering to see if I could find any of my old friends. Bingo! I recognized a name on one of the nametags: "Len Knitter." I was pretty sure we had graduated in the same year, and once the introductions were over, we talked about our Air Force careers. I couldn't believe my ears when Len told me he had become a Jolly Green Giant pilot after flight school and had been in Vietnam at the same time I was.

I immediately zeroed in on his adventure in SEA and hoped to identify ways that he could enhance my search for the Cable Guy. However, Len had arrived in SEA just after Dave was killed and could not recall anything about the incident. I explained what I was doing and asked him about the JG community.

"Do the Jolly guys have reunions?"

"Yep, damn good ones. Just had one a few months ago."

"Do PJs show up?" I queried further.

"Some do, but normally it's mostly pilots."

Since he still had some good contacts within the Jolly community, he said he'd be happy to assist me with my search.

The next week I sent an email to Len to follow up:

Len, was great meeting you last week at Homecoming. The 925th has much better digs now! Too bad about the Badgers' fortunes Sat.

I just got back to Kansas and would like to start working with you concerning the death of my good friend, David Dinan, on Mar 17, 1969, in northern Laos. His call sign was Simmer 2, and the call signs of the JGs that day were JG 09 and 16. 16 was the crew that sent down the PJ and reported that Dave was dead. However, I also have a report from the 56th SOW that Dave's body was recovered by 16 and "returned to base". My desire is to confirm or deny this report with someone that was on the scene and might remember the incident. The DOD MIA folks say he was left at the scene, and there are a number of reports that verify that. However, I would feel better if some of the JG folks could back up the report.

Len's quick response mentioned that he hadn't arrived in Da Nang until July 1969. Since he'd heard nothing about the incident, he forwarded my email to John Flournoy, the Jolly Green Association president, who might have a contact on his JG roster who could help.

A day later, I was cc'd on a message sent by John Flournoy to several of his Jolly contacts, including Bob LaPointe, the author of the book that Archie had given me at Maxwell. The email was succinct:

Any help, guys? I drew a blank. John Sr.

I soon got a message from Bob LaPointe. It was an informational email he sent out to all involved:

I reviewed all the mission reports for that time period and then looked at the unit histories for the Thailand-based Jollys. There was no reference to Simmer 02 or David Dinan. I looked a week back and a week forward but still no luck. Last I ran it through the SAR database that contains a listing of all known SARs but still no hits.

If the Jollys were involved in this mission, there should have been some sort of record of it.

The last place to examine, probably, is the best source of reliable information. The RCC SAR logs are kept at JPAC (old JTTFA). In those books are handwritten entries of every flight of ARS/ARRS aircraft in the Vietnam War. The entries included takeoff time, why they were airborne, time arrived on scene, and in many cases dozens of entries listing every significant radio communication with SAR aircraft. Had a Jolly gone out to find David, it would be listed in this logbook. Unfortunately, the only copy of those logs is at JPAC. On a case by case basis, they may assist in copying the relevant pages and providing them to the requester. If the requester is a family member, they will likely do the research for free. Otherwise, you can ask and if they say no, you can still get the information using the Freedom of Information Act (FOIA). If you have to go this way, it may take a year and they may charge a fee for the research.

I would suggest a friendly email asking for help and providing the same information listed in the email we received stating that you would like to avoid a FOIA request if possible. JPAC also likes avoiding all the paperwork associated with FOIA so they may provide the information just to avoid the extra paperwork.

Well, crap, the Cable Guy is sure elusive. If Bob LaPointe had never heard of him or even the incident, my mission was even tougher. I wasn't much closer to finding the Cable Guy, and I wasn't sure I wanted to spend several days in Hawaii in archives that appeared to be hard to access. Also, going through an FOIA request was not something I wanted to take on right now. My planning was moving forward for my trip to Laos, and I decided to move the Cable Guy search forward to a later date. "Get 'r done" might still happen but not anytime soon.

Chapter 16

Making the "Show"

I had become serious about a trip to Laos in the spring of 2011 as it became evident that the government wasn't investing in Dave's case. When I expressed my plans of this adventure, all four of my children embraced the idea with a "Do it, Dad" attitude. Mary, however, was not sure it was a good idea, but realizing it would be difficult to dissuade me, she held her fire and made only mild objections.

My oldest son, Bartz, a lieutenant colonel in the Air Force and within a year of retirement, was especially enthused about the project. At the time, he was serving as an AF liaison at the U.S. Embassy in Moscow, and I told him it would be fun if he would accompany me on the journey. He was in favor of such an adventure, but breaking loose from his duties in Russia and his wife and six children—"punks," as he referred to them— might pose a problem. Bartz had spent most of his AF career pushing F-16s around the sky, so we shared a sense of what it means to be part of the fighter pilot brotherhood.

The week before I left for the 2011 League meeting, I had the good fortune of talking to a member of the church I attend in Derby, Kansas, about my plans to travel to Laos. He told me that one of the members of the congregation had a nephew who was a missionary in Laos. So I attended a Sunday School class with Evelyn Kersting, and following the class, I asked about her nephew.

"Yes, Kenton is a missionary in Laos, and he teaches English in a university there. He is currently in the U.S. traveling around the country raising money to continue his ministry. I think he is going to visit Kansas sometime next month. I have his cell phone number if you would like it."

I accepted the number and called him that afternoon. Reaching him on my first attempt, I explained my plans and asked if we could get together if he made it through Kansas. He agreed. He gave me his email

address and asked me to give him more information prior to our meeting.
I got right on it:

> Kenton, it was great talking to you today! My interest in Laos concerns my
> friend and roommate during the Vietnam War, Dave Dinan, who was shot
> down and killed NE of the Plain of Jars in 1969. His body was never
> recovered, and it may not be possible to recover. However, I would like to
> give it a try and give him some type of proper burial in his homeland.
>
> If you think you could make it through the Derby area later this month or
> in Sept., I would be interested in meeting with you and giving you some of
> the details of my quest and see if you could give me some advice and an
> assist. As I said earlier, I would be happy to spring for a meal if we could
> make it happen.
>
> Let me know—hope your trip in the U.S. is going well and hope to see you
> before you wander back to Laos. Cheers, Ed Sykes (wgfp)

About a week later, Kenton replied, laying out his upcoming travel
itinerary. He planned on passing through Wichita/Derby near the end of
the month. Hoping to help in my quest, he said he'd check in with me
closer to the date so we could meet in person. *Great!* If I'm lucky, I may
be able to find a guide to help Bartz and me on our adventure.

When I returned from the 2011 League meeting, along with what I
was doing to find the Cable Guy, I decided to see if I could get Bartz to
hook up with Dustin Roses and get some embassy-to-embassy intel. I
figured he might get a feeling for what the journey would entail from a
diplomatic standpoint since they both worked in U.S. embassies and
knew the "secret handshake" (i.e., how to get something done without
pissing too many people off). Following Bartz's initial call to Dustin, they
began to correspond by email. Only a few days after calling Bartz, I got
a reply back that included a lengthy email from Dustin:

> Dad, I followed up with Dustin Roses …
>
> If you want me to go along, I'll waive my normal consulting and wilderness
> guide fee (you couldn't afford it, anyway) and I'll charge only for expenses
> incurred. Save the money from Ez's wedding, this would be money better
> spent.
>
> Getting tired of my punks being there?
>
> See ya, Bartz

(Three of Bartz's "punks" were spending the summer with us in Kansas and avoiding the Moscow heat. Ez is my youngest son who was getting married the following summer and was hitting us up for a good amount of cash for the party.)

In contrast, the email from Dustin to Bartz was much more helpful.

Lt. Col. Sykes,

I definitely remember meeting and discussing this with your father at the League Meeting.

Let me give you a little bit of background on this case (you should already have access to all this, but just in case): Ref. no. 1408, David Dinan, F-105, 17 March 1969.

The circumstances of the loss incident make it hard to pinpoint a spot on the map where you can go see the "site." The pilot bailed out, and so the crash location of the F-105 is not relevant from an MIA standpoint. The parachute/ejection seat location is of more interest and is what we've been looking for with no luck. I have three grid coordinates of interest (forgive me for no consistent format, this is how I get them in from JPAC): the "record loss incident" official GeoCoord is 192338 N 1033736 E—this could be the crash site, or a village nearby, or some geographical feature associated with the case. The plane crash site is believed to be at GC UG 558 448, but to the best of my knowledge, we haven't found the plane, and witnesses say it was on a cliff, then fell off, then got scavenged and there's nothing left. The SAR report in 1969 placed the body at 48Q UG 547 446, and pieces of parachute and military material were found at GC UG 496 472, but later determined not likely the remains site. The pilot was seen dead on the ground by SAR forces, who couldn't recover the remains due to hostile fire in the area. Since it was on the surface, it is likely it decayed in place, washed away in the rain, carried off by animals [etc.]. If someone buried it, as was typical of these types of cases, we would have likely found a witness who knows about it. We haven't heard any stories of a burial associated with this case. Another problem with this area is there are so many crashes on top of each other that locals get things confused. We recover materials and get reports all the time of stuff people saw in the jungle out there, but it is very hard correlating to a case due to the number of cases it could potentially be.

The good news is that you can go to that part of the country very easily, and all the associated villages and locations are easy to get to. Once you get to Vientiane, you can take a short commercial flight to Phonsavan, Xieng Khouang Province. You can drive, it's a beautiful drive over somewhat treacherous terrain, but about 8 hours in a car. This is where you will lodge overnight. From there it is a 30-minute drive over nice roads to Muang Khoun (so you know when searching maps, "Muang" means district capital, district, or city. Ban or Baan means village. Every village is Baan something.) Muang Khoun is the district the loss location is in. From there you will drive over pretty good roads to get as close to the villages you want to get to as possible. Villages associated with this loss are Ban Xan Noy, Ban Khap, and Ban Paka Noy (you may have to try all kinds of alternate spellings). Roads in this part of the country are good, and if you are going further out, an SUV will be beneficial. I drove my 4-door sedan to Muang Khoun, and the standard vehicle in the area is a minivan. Renting an SUV is possible but will cost you. We have a good reference for a car/driver company up in Phonsavan.

Legally speaking, if you just show up and head out there yourselves, nobody is going to think you are any more than a couple [of] western tourists. This province gets a lot of tourism due to the Plain of Jars world heritage site, and it has some very nice natural areas with hiking, waterfalls, etc. People also come to see the ethnic hill tribes prominent in this area. If you get a driver, he won't really care what your business is, he'll just drive. If you get out at some small village and head off into the mountains, you might get some weird stares, and the village chief will find out and wonder what you are doing, but nobody will likely approach you. If you are approached, it will be because someone thinks you are lost or somehow turned around and wants to make sure you don't really think there is some great tourist attraction down the trail. Explain you don't care, and they won't bother you. The language barrier actually helps in this instance, where if you spoke the language, they could tell you to just go away, but if you don't understand, then they will tolerate you being there. You shouldn't pay anyone, give gifts, or otherwise make a scene. You could try to ask people about the case/site/crash location, etc. People will tell you all about it but won't likely want to get into details without a central government official present. You may get lucky and someone will have a personal item, know of something they can take you to, etc.

The geography in this area is mountainous, but uniquely most of the area is thin pine forest (think rocky mountain pine forest, northern Idaho, Eastern Oregon...). This is a unique climate zone in Laos and only covers a 30x30 mile area in Xieng Khouang province. The weather tends to be cooler, and the rainy season hits around April–September, but you can still visit any time, just may have a few rain delays per week. In January–February, it is really cold for the locals, which means quite pleasant for us, like October weather in the U.S. Xieng Khouang province is my favorite part of Laos. The people there are much more friendly and industrious, the city is not over touristed and developed, but there are enough decent hotels and eating venues.

There is a lot of history and beautiful pine forests (a change of scenery from the heat and dense jungle everywhere else).

Costs: Plan on flying into Bangkok and buy a separate ticket to Vientiane, or travel overland to save money. Bangkok to Vientiane costs $250 roundtrip, it's usually more if purchasing a ticket from the U.S. all the way to Vientiane. Vientiane lodging about $40–$100 depending on how nice of a place you get. Transport and food in Vientiane, not more than $20–$40 per day depending on your preferences. Flight to Phonsavan, I think it is $175 roundtrip from Vientiane. It's something like that. Transport in Xieng Khouang will be your biggest expense, I'm thinking around $60–$80/day. Lodging in Xieng Khouang is variable, from $20 for an OK place to about $50 for this French-style lodge on a hilltop (that's where we stay, it's atmospheric but still just as hygienic as the $20 place). Food in Phonsavan will be around $10–$20/day depending. You should plan for being totally self-sufficient as western-type supplies/gear, etc., are not readily available anywhere. You'd be hard pressed to find AA batteries let alone a new GPS, digital camera, hiking boots, etc. There are no exceptional immunizations you need. There are no serious diseases that would put a westerner at risk, just do the standard stuff the doctor recommends. Malaria risk is low, and whether you want to take antimalarial pills is up to you, but I don't when I go up-country, and the only people I know who do are the JPAC guys because they are there for 35 days and are forced to. Dengue fever is a problem, but the only defense is mosquito repellent. Foodborne illness is by far your biggest threat, but that's just playing it safe and having some pills ready when it hits. Older people tend to have more problems with this.

Ok, hopefully that's enough information to make an informed decision. I'm not allowed to officially help you, but I can definitely point you in the right direction and give POCs/recommendations. If you are actually coming, I'd love to see you guys, have you in for an office visit (since you are in the DAS), and we'll see about me accompanying you up-country. Not sure, but it might be something we see as worthwhile. Although me being there might make it harder for you to fly under the radar. We'll see.

Thanks, Dustin Roses, USDAO Vientiane

Wow, what a great primer. He could market this as *Traveling in Northern Laos for Dummies.* I was not surprised by the confusion over the loss location (what else is new).

Bartz's response to Dustin was included in the chain. Besides echoing my thoughts on the thoroughness of the information provided, Bartz mentioned that he wouldn't be able to go on a trip like this until January or February 2013. With that, I realized Bartz would probably not be going to Laos with me on this first trip. I felt that I had a little momentum going and needed to get the trip in as soon as I could. Never let the bureaucrats relax!

I sent Bartz an email in which I hinted that I would probably make this trip without him since I was looking at going in January/February of 2012.

Bartz and Dustin had given me a lot of insight into the challenges I was facing. Bartz had described it as a "tiny, tiny needle in a haystack," which was reinforced by Dustin's inputs. In aviation, when you fly directly over a navigation aid, you enter what is known as the "cone of confusion," an area where your navigation instruments become unreliable. This search was clearly in the cone of confusion. However, the travel information was helpful, and I excitedly began looking at airline schedules and costs. I hadn't heard from Kenton in a while, so I sent him an email concerning his status. He replied when he would be in the Wichita area, and we set a time to meet.

Mary and I met Kenton over lunch, and he told us about his efforts in Laos. He said the teaching of Christianity was not condoned by the Lao People's Democratic Republic, and consequently, he did what proselytizing he could in a rather covert manner. His primary duty was teaching English for the University of Laos, and he discussed religion with his students outside the classroom. We discussed my desire to find an English-speaking guide from the area where Dave was killed, and he said he could probably find me one. He was going to return to Laos in a few weeks and would keep in touch.

A few weeks later, I sent Kenton an email asking if he had made it back to Laos and I got a reply around the middle of October:

Good morning (my time), Edward,

Yes, I did make it back in good shape, but have hit some snags with my work visa and the equivalent of our green card being denied—at least through the public education sector.

I'm presently working privately as a teacher at another school. So, I'm still in Laos but at a change of venues. Though I've got minimal students presently, my teammates are still at the university and collectively know over 1,000 students. I bet amongst us we can find some assistance!!!

Do keep me abreast of your plans and any "advance information" that you can send so I can assist you in this. I'm happy to serve in whatever capacity I can.

Warmly, Kenton Kersting

Reading between the lines, I assumed the problems he was having with his visa had to do with his proselytizing but thought it best not to ask via email.

I did not hear from Kenton for about three weeks and sent another email to check on his progress. I told him my tentative plan was to travel to Laos in March and asked if that timing would work out for one of the students to help guide me. I informed him that the crash site was in Xieng Khouang Province and in Khoun District, and although I was trying to isolate the search to a particular village, I hadn't found out anything solid yet. I asked in the email if he had any suggestions regarding that search.

***

As 2011 came to a close, it was a time to do an assessment of how far I had come. I had found and met the Dinan brothers, gaining their complete support for my efforts. I had met and formed a relationship with Ann at the League and felt she would be somewhat supportive of my efforts. I had also met and worked with Bill Habeeb who introduced me to some prime players, most notably Bob Maves and Dustin Roses from DOD. And let's not forget Kenton Kersting.

Bottom line, however, was that none of my efforts had resulted in the movement of a single finger to get anything done on the ground in Laos. By 2012, I had decided that my continual prodding of the DPMO and JPAC was not going to have much effect. The bureaucracy was too cumbersome and driven by too many factors to get much action on Dave's case. I would be told prior to planning sessions that Dave's case would be reviewed for action, but following each session, I was told that Dave's case was not on the list of events for the Joint Field Activity (JFA). Depending on funding, there are normally four JFAs each year, mainly to southern Laos. Most years there is one JFA in the North, and Dave's case has never been a primary site on any of those exercises.

Before the end of 2011, Bill Habeeb had told me that he was going to attend a meeting in SEA at the beginning of the year and he would push for action on Dave's case. A week or so into 2012, I sent Bill an email asking how things were going and if any attempt to take an on-the-ground look at Dave Dinan's case was being contemplated.

Three days later, I received a reply from Bill:

Hi, Ed, I'm in Hanoi right now. Just finished mtg with Vietnamese Govt officials on select cases. I did raise your case during mtg with JPAC recently. Will call you when I get back next week. Hope u had gr8 holiday. Bill

It was not hard to read between the lines on this one. No good news undoubtedly meant bad news. A few weeks later, I spoke with Bill on the phone and he confirmed my suspicions: there was nothing on the calendar for 2012 in support of Dave's case. He promised to continue bringing it up, and I thanked him for that. Time for me to get involved on a more physical level.

I had also given up on being included on a JFA as a team member. I understood why they didn't want family members to be included, but I felt that, as a crew member, I might be able to lend some insights concerning what could have happened in certain situations. But none of my arguments for inclusion in the process would be accepted. Therefore, I decided to do my own JFA. I had some information concerning where Dave might have died, and I decided I would get as close to the site as I could without the assistance of the government. If the government found out a 69-year-old fighter pilot was running around the jungle doing their job, maybe it would stimulate some interest in Dave's case. Even if it didn't, it would allow me to better assess the difficulties of the effort and see for myself the aspects of the terrain we were dealing with. Where to start?

I decided to change plans and make my trip following the next League meeting in June to garner as much attention from DOD as I could as well as see what kind of help I could get from some of the players. I knew who some of the primary players were but hadn't met them face-to-face. I also let the Dinan brothers know I was planning the trip. No problem with them. Finally, I decided to start a "Dave Account" on my financial records to my primary checking account. I took $19,000 out of my IRA, stuck it in my account, and kept a log to see how much it was costing to undertake this effort.

I called Bill again to tell him I was going to make a trip to Laos after the League meeting in June. I also mentioned I was going to make a trip to D.C.—the first week in February—to visit him at DPMO. I told him I wanted to get as much information as I could in preparation for my trip, which was true. But I also wanted them to realize that I was not going to stop harassing them. Once again, it was time to get the bureaucrats to put Dave's case at the top of their stack.

The next day I was copied on an email from Bill to a guy I didn't know, Al Teel, who worked at JPAC and was an expert on Laos:

Al, I spoke with Ed Sykes (cc'd). He was an F-105 pilot in SEA and a squadmate of Dinan (1408). He is a great guy and sensible. I've been keeping him updated on the case. He is considering a trip to Laos, and I thought it might be useful if you would advise him on the area considering your experience. Thnx. Bill

Well, I was glad to know I was a "great guy." A few days later, I got an email from Al Teel asking for more details on my needs for the trip and offering his assistance. Although I had never met Al, it looked like I had another source of help at JPAC, which I appreciated. I wrote an email to Al with my plans for the summer and a rundown on Dave's case.

The next week I made a short visit to D.C. I had a great lunch with Ann Mills-Griffiths and told her what I was up to and listened to her describe her dissatisfaction with the leadership at DPMO. This woman was a dynamo who never seemed to run out of energy and worked to tweak everything until it suited her (and that wasn't easy). She informed me of her connections with members of the Laotian government and told me she would help me as much as she could, but she also said the Lao folks were often difficult to deal with and would find ways to make you pay for any services rendered. My admiration for Ann was only enhanced by this meeting.

The visit with Bill was about what I'd expected. He talked about his visits to SEA and his request to move Dave's case forward but indicated there would be no change to the current schedule for 2012. The case may be on the schedule for 2013, most likely in the spring as that was when the JFA normally took place in northern Laos. I asked him to continue pressuring to put the case on the schedule. I discussed my likely trip to Laos, and he offered to give what assistance he could, but it would be better to rely on folks like Al Teel and Dustin Roses.

I also went by the Laotian Embassy to see if I could get a visa on my passport during July. It was a small building in the Embassy Row of D.C., and I was one of only a few people waiting for a service window to open. After taking a number and waiting for over an hour, a window finally opened and a man who I assumed was Laotian gestured me to come over. As I approached the window, I smiled and greeted the man, "Good afternoon."

He looked at me and, without smiling, said, "What you want?"

I told him I wanted to get a visa to enter Laos in July. He noticed I was holding my passport in my hand and motioned me to give it to him. He glanced at it briefly and said, "You get Laos."

"Can I get it now while I'm in D.C.?" I asked.

"You get Laos!" he stated, shaking his head in annoyance at my question. In broken English, he explained some procedure that would require I leave my passport there and it would be mailed to me. No way, Jose! I would get my visa in Laos. This was the first meeting with a Laotian government official, and it brought back memories of other government officials I had dealt with in such garden spots as East Berlin, Oman, Russia, and China. Why must they be so unpleasant and condescending?

So, what concrete accomplishments came out of this trip? Not much except they knew I was still around and would most likely keep coming back. Major League Baseball was about to start spring training, where all the young talent and even some old veterans were hoping to impress their teams and make it to "The Show." You make it to the Show when you get on a Major League roster and play in the Big Leagues. For me, getting to the "Show" was making it to the official DPMO annual schedule for that year's JFAs. After 42 years, it was finally time for Dave to make it to the "Show"!

Over the next couple months, I continued to plan for the trip and look for any new, helpful information. In 2005, I had traveled to Sri Lanka as part of a consulting contract and had used a travel agent to make many of my arrangements. I soon found that a ho-hum flight to Columbo, Sri Lanka (the capital city) was much easier to plan than a trip to far-out Laos. The agency I had previously used had some good ideas about getting to Bangkok, but it got tougher beyond that. I also hadn't heard from Kenton in a while and wondered if I would have to make the trip solo. That would add to the excitement!

In addition, Charlie had finally gotten access to the data files out of the Air Force and was getting a little new information. In one email, he discussed a data plate that had recently been recovered around Dave's ejection site:

The only information that they had was the data plate from the canopy hatch assembly. The message stated the data plate "correlated the item…to an F-105D type aircraft, but not to the specific aircraft involved in the case 1408 incident." It went on to state that there were five operational F-105 losses "and one other F-105 incident (currently on the JPAC excavation list) that were within 25 KM of the 1408 record loss location."

I've received a few notifications like this over the past several years, mostly reports from Laotians who found pieces of parachute, survival gear, or witnessed the NVA salvaging wreckage. As far as I can recall, none of the information was definitively specific to Dave's incident.

Not a lot of new information there, except it again highlighted the confusion around Dave's site.

In late March, I awoke one morning to find an email from Kenton Kersting:

I don't know where you are in your plans for a Laos visit, but I remain available to assist in whatever manner I can from this end.

I faced a radical change in my teaching situation last fall in my release from the university for conducting religious activities with students. I was able to remain in-country working for a private medical education firm, but access to the myriad students at the National University was curtailed. So, though I don't know students/people from the area where your friend's plane was shot down who can assist you, I DO have a young man from another province who would be an EXCELLENT travel companion/

translator/assistant available should you still be planning to make this trip. He has excellent English skills; I often have him accompany me to translate for me.

Drop me a line when you've the chance and let me know what's stirring in your plans. Warmest regards (from hottest Laos; we're entering the hot season now. Temps in the 90s in the days, spiking to triple digits some afternoons!).

I immediately replied to Kenton:

Kenton, great to hear from you also! I had thought I would have already made my trip to Laos, but it is being dragged out by my work with the agency that makes visits to Laos to recover bodies. I was in D.C. last month working with DOD. I now expect to be in Laos during the Oct/Nov time frame. Let your friend know. I would be very much interested in obtaining his services.

I stopped at the Laotian Embassy in D.C. and found there was a waiting period for a visa. Also, was told I can get a visa when I enter the country— easiest if I land in Vientiane, I guess? Do you have any advice on this matter?

This time I got an almost-immediate reply from Kenton:

It is quite easy as American citizens to enter Laos. You can get a visa-on-arrival at the airport or at the bridge here at Vientiane (and, I presume, at other border-crossing points; I've never entered anywhere other than at Vientiane). It costs $35 USD and you need a passport photo. The visa is good for one month, which, I presume, would more than cover your planned stay here. If not, you simply cross over the bridge to Thailand and return and fill out another form. Quite simple really for a generally red-tape-happy communist nation!

One thing is to make sure your passport has more than six months remaining on it. Most all nations are sticklers on this. You'll be denied boarding in the States should you have less than six months remaining on your passport—or worse, should this not be checked stateside, you'll get all the way over here and be denied access to Laos and be returned to the States straightaway.

My friend I mentioned who could possibly travel with you is a college graduate and seeking to enter the job market. I can't discourage his doing so based on a few days' work sometime in the fall, but we'll see where he is/where you are later in the ballgame. Maybe he'll be available after all. If not, we'll find someone else.

Warmest regards, Kenton

I barely was able to digest this information when another email from Kenton appeared:

Something to consider in your travel plans, Ed, would be the rainy season—generally from late May to late October/early November. Many remote areas would be inaccessible during the rains, so a November or later trip would be more ideal. Let me rephrase—obviously places are accessible all year, people have to live, but ease of accessibility would be better in the dry/drier seasons. Kenton

P.S. For what this is worth, December/January/February might be even more ideal in that the weather will be cooler and there'll be no rains to contend with. Everything will be very dry, making remote sites more accessible.

If you're able to work out something with the DOD, then, of course, you'll work on their timetable, but if you come personally, I'd recommend the time frame mentioned here. March is doable, but you could face days of 100 degrees pretty easily.

After reading this email, I decided it was decision time. I decided I would attend the League meeting in June and then make my trip to Laos in July. The rainy season might present some problems. I wasn't sure I could accomplish much on the ground in Laos anyway, but I was fairly confident I could stimulate some action on the part of DPMO to put case 1408 on the 2013 calendar. And since I wanted to take advantage of Kenton's efforts to get me some local help, it seemed like the sooner I moved, the better.

So, I began to let everyone know I was headed to Laos in July. I also let Charlie and John know that I would attend the League meeting in mid-June. The Dinans both said they had conflicts and elected to not attend the League meeting. I would have enjoyed spending more time with them, but the biggest drawback was that there would be no family briefing for Cousin Ed to attend. I decided not to push the issue since I wasn't sure there was much more information they could share that I didn't already know.

The next step was to reveal my plans to Mary. When I did, I was not met with a lot of encouragement. She asked a lot of questions I couldn't answer about almost every aspect of my trip. One thing she made clear was that she was not going to make a trip to Laos to recover MY remains.

So, to calm these waters, I found an insurance policy that would haul my ass back to the U.S. if I expired somewhere en route. It was not cheap, but it was damn well worth it!

As I looked for flight options to Vientiane, Laos, I found several ways to get there. The cheapest was through China, but it meant an overnight in western China, which sounded like a goat rope as I read the reviews of folks who had done it. Other options went through Korea or Vietnam, but I finally decided to go through Bangkok. As I talked with my old Asia travel agent, he said he could assist with the flights and some of the hotels, but once I got outside Vientiane, I was on my own.

Before the League meeting, I tried to get as much information concerning a trip from every available source. I had already been in touch with Al Teel, so I sent him an email letting him know of my plans. I soon got an email from him that confirmed my conviction that the time was right:

> I am in Nong Khai at this time taking an immersion course in Lao. It might not be so easy to get to Udorn; however, would be happy to meet you here in Nong Khai, which will be your overland jump-off point to Laos via the Lao-Thai Friendship Bridge. Let me know if that would be okay with you.

Great! Meeting Al in Nong Khai would be a nice way to begin my adventure. It would also allow me to revisit Udorn. Years before, I had diverted into Udorn Royal Thai Air Force Base with battle damage. At the time, it was one of the primary F-4 bases in Thailand along with Ubon. I soon found that it was now a municipal airport and had a number of daily flights out of Bangkok. I let Al know I was going to be there the last week in July, and he replied that he would still be there. Good timing.

I also consulted with Dick Messce, whom I had met through Charlie and who had traveled extensively in Laos. He sent me an interesting anecdote about his attempt to get to the area where Dave died:

> I rented a dirt bike while in Thailand to cross into Laos to reconnoiter the Plain of Jars hills in search of information, etc. The Laotian Embassy shot me down right away. Only supervised tours are permitted with a strict format. Foreigners are not allowed to travel inside the country unescorted. I have a trip to China in mind. Maybe rules are different entering Laos from the other side?

Dick's description of obtaining access to the Laotian back country was the most severe I had heard thus far. Only one solution—go find out for myself!

One humorous incident occurred just before I left for the League meeting. I was celebrating the annual "Riverfest" celebration on the Arkansas River with an old wealthy friend, Ron Ryan. I told him of my plans and why I was making the trip. He offered to help, and I told him I would let him know if I ran into real trouble.

The next day I sent him an email:

Ron, was great to see you all on the River Sat. I very much appreciate your offer to help with my efforts to help recover Dave's remains. I really expect the U.S. Government to do the heavy lifting on this effort. I leave to meet with several of the DOD folks tomorrow and hope I can get some action on this case. I will probably travel to Laos later this summer and will keep your offer in mind if I run into difficult circumstances—not unknown for U.S. citizens to get kidnapped. How much do you think I'm worth—don't answer that Ed Sykes (wgfp)

I soon got a response from Ron:

If they do kidnap you, it will only be a matter of days when they will agree to pay to give you back. Seriously, good luck and let me know if I can help in any way. Ron Ryan

I also decided I should try the political angle. My congressional district in Kansas had a fairly new representative in Congress, Mike Pompeo, who was a graduate of West Point (number one in his class) and seemed pro military. Mary and I had been the Republican chairman and chairwoman for the precinct we lived in for many years, but I had not yet met Mike. I decided I would meet him on my trip to D.C. I asked the current Republican county chairman, Bob Dool, who I knew well, to let Rep. Pompeo know I would like to visit with him while I was in D.C. the next week. Bob said he would do that.

A few days later, I got an email from Rep. Pompeo:

Ed, I got a note from Bob Dool indicating that you will be in D.C. on Thursday and Friday of this week. I am actually back in Kansas all week.

I am happy to talk with you by phone. Or, if face-to-face matters, I am back in town the week after next (as I can best recall) and would be happy to find a time to get together. Mike

It was my intention to ask Mike to give DPMO and other associate folks a gentle nudge of encouragement to pay more attention to Dave's case. I would have to put this off until after my trip to D.C. as I wanted to meet face-to-face.

***

It was now mid-June and time for the League meeting. Having been to a previous meeting, I knew the drill and was interested in the briefings but was more interested in letting some of the players know I was still around. The first morning I took in the briefings by Ann and Childress and, once again, heard them voice a good deal of displeasure with the bureaucracy and the inefficiencies of the organizational structure. Although interesting, it didn't affect me much.

After lunch, I sat through a few more briefings, but when my information overload light came on, I headed to the bar. I ordered a beer and noticed Bob Maves walking in. He looked around and, not seeing the folks he was looking for, turned to depart.

"Hey, Bob!" I shouted as I headed in his direction. "Ed Sykes here. We met at last year's meeting. I'm working on the Dave Dinan case."

"I remember you, Mr. Sykes. I was just looking for some folks who are coming up to my room to slam down a few beers. I'm in room 408. Why don't you come up and join us? I'm headed that way now."

"Let me drop off my stuff, and I'll be up shortly."

I grabbed my beer and headed to my room to drop off my meeting bag and handouts and made my way to 408. The door had a waste can holding it open, and I could hear voices inside. I knocked and someone yelled, "It's open." Inside were several men I didn't recognize as well as Bob. He introduced me around the room, and I noted that most were DOD guys, but one was an author.

"Beer's in the cooler," Bob said as he pointed in its direction.

I listened to the talk but much of it had to do with stuff I knew nothing about. Eventually, Bob looked at me and said, "I've heard you're going to make a trip to Laos. What are the dates?"

"I'm looking at the last week in July. I just want to get a flavor of the environment. Don't think I'll stay more than a week."

He asked me if I was working with Al Teel. I replied that I was planning to meet Al in Thailand prior to crossing into Laos. "Great, Mr. Sykes."

"When do you think we might get 1408 on the JFA schedule?" I asked.

I could not believe his response: "You're on next year's schedule, Mr. Sykes."

He pulled out a card from his shirt pocket and said, "1408 is on the JFA schedule for the March JFA in 2013. It won't be a dig, just an investigation team."

I couldn't believe my ears. "Is that schedule firm?" I asked.

"As firm as anything can be in this business. I expect it to happen."

I decided at once not to ask him how, or for what reason, the decision had been made. I was simply elated there was finally some action. I stayed around and had a few more beers with Bob's gathering, but this news had moved my focus elsewhere. Perhaps being the "squeaky wheel" and continually moving the issue to the top of the stack had had an impact. I was also convinced that the active participation of the Dinans made a huge difference.

I left that room knowing we had made the "Show." I decided a drink of fine Scotch was in order and found a quiet place near the bar. I celebrated by myself, quiet on the outside but yelling "Shit hot!" on the inside.

I was surprised that the Dinans had not been notified. I decided not to call Charlie because I felt DPMO should notify them first, if they were going to. I suspected the decision had probably been made just this week. The excitement made it tough to get to sleep.

I now considered putting off my trip, which was only a few weeks away. Or should I press on? As I weighed all the factors, I decided to push on with the trip and then make a second trip in March 2013 to be around when the JFA did its investigation. *Don't let the momentum slow down. This is kick-ass good news, and there is no sense in letting the bureaucrats relax!*

The next morning, I found Bill Habeeb in the hall and asked him about the Spring JFA, which he confirmed. He apologized for not letting me or the family know, but the decision had only been made that week and he had been busy getting ready for the meeting. I told him I would let the family know and he nodded. Our discussion was brief as he had to be somewhere.

The rest of the meeting was anticlimactic. Without Dave's family there, I didn't get a briefing from DPMO. I briefly said hello to a busy Ann and told her the news. "Good for you. Keep pushing," she replied with a warm grin.

Dustin Roses was part of a panel discussion in which he talked about many things, but I focused on the difficulty of working with the Laotian government. I looked for Dustin after the panel finished, but he was nowhere to be found. No big deal. He already knew I was coming to Laos and had already provided me with valuable assistance and a willingness to do more.

Back in Kansas, I sent Charlie an email after a few days home. I had given DPMO enough time to notify the family and had not heard from Charlie or John, so I wanted to send the news.

Charlie, John, just returned from the League meeting in D.C. Dave's case is finally on the official investigation schedule. JPAC should visit the site area in March '13. I am trying to put together a short trip to Laos during the week of 22–28 July. There is a JPAC investigator in Thailand (near the border) during that time, and my plan is to meet up with him and get some insights and then fly up to the area where Dave ejected and do a little look-see. Will keep you posted.

Best news is that the Feds are beginning to show some real interest in the case. Cheers, Ed (wgfp)

I got a call from Charlie within a few hours. He was excited to see some progress being made on an actual investigation. I explained how the whole thing unfolded and mentioned that their participation had made an impact.

As we rambled on about the case, we eventually came around to Valerie. I hadn't thought much about her in the last several months and told Charlie that I might attempt to link up with her after my trip. Charlie told me he would try to contact her using the last phone number he had for her.

"I'll let you know what happens," he promised as we concluded our conversation.

A few days later, I received an email from Charlie:

I contacted Valerie (Galullo) Zoolakis. We spoke for about an hour. I was surprised that she knew so little about the actual crash and recovery attempt. She's a retired grammar school teacher and is living in the Salt Lake City area. She said that she has followed the case and had been pretty sure that Dave's remains had not been recovered. Val did most of the talking and gave me a rundown of her jobs, marriage, divorce, and frustrations. She was very interested in Dave's case, and I described the coincidences that led me to speak with you. I'll be calling her back today or tomorrow. I needed to end our conversation after slightly more than an hour because I had a pressing babysitting commitment. Val said that she'd be very interested in speaking with you.

Charlie gave me her phone number, but I decided not to call her until after I returned from Laos. I would have a lot more to talk about then and maybe some of it good news.

Note to self—try to meet with Valerie following the Laos trip. Over the years, from time to time, I remembered our last encounter and reflected on the terrible impact the event must have had on her life. What a burden to carry.

Right now, I need to get this trip right. I sent an email to Dustin telling him I was coming to Laos for sure at the end of July and got a lengthy reply within a day:

Good hearing from you. I will be in the office in Vientiane when you are here. I would love to get together for lunch or something. You are trying to get to Khoun District, Xieng Khouang Province, and you will certainly have to spend time in Vientiane on your way. You can't get there from anywhere else. I spend a lot of time in Xieng Khouang Province, and I can tell you everything you want to know to have a successful trip up there. Just let me know what your questions are.

I can't tell whether your intention is to go out to the 1408 loss incident location and try to drum up information. Please let me know exactly what your intention is. Don't worry, I can't prevent you from going there and doing whatever you want. I would, however, like to give you a few things to keep your eyes and ears open for that may help you be more successful.

Stay out of trouble—just be a tourist. You will have no problems with the authorities, cooperation, etc., unless you get yourself out of the tourist box. This is why it would be very hard for me to go with you to the site, then it would be official and next to impossible. So wherever you go, just say you are there to see the war history, you're just a harmless old guy, want to get some Beer Lao, see a pretty temple, and oh by the way, can you help me find more info on this crash incident I've heard so much about. If you get in the situation where a village chief or the police approach you because they are suspicious about you going a little too deep off the beaten path, just back off, and lighten the mood with lots of beer and silly conversation about girls, sports, the officials' relatives in the U.S.

For as few family members we get out here, I'm happy to give you some extra attention. Please keep me in the loop and I'd love to set you up for success. Dustin

Harmless old guy, my ass! The last time I was in Laos I wasn't so fucking harmless. Guess I can pretend. Of course, the last time I was in Laos I wasn't old either. Perhaps this is some rather sage advice.

I emailed Dustin and thanked him for his input and told him I would get back with him as soon as I had a better grip on my schedule. I emailed Kenton to see how he was doing on a guide. I could do this on my own but would be much more effective if I had some "adult supervision," someone that knew what the hell was going on when I got myself in the cone of confusion.

At the end of June, I got a promising email from Kenton:

Ed, I'm going to be home from mid-July to mid-September. I leave Laos on the 15th of July, so our paths won't cross on this side of the Pacific.

The original fellow I was hoping to connect you with is now employed full-time, so I doubt he can go with you now. Though, with this being Laos and life being lived in flux here, maybe he can arrange to be free for the time you'll be here. Another option is a fellow who has EXCELLENT English skills and recently moved to my village. He's seeking employment. He'd be ideal were he still free come the time for your visit, but with it being three weeks

269

or so in the offing, I don't know that he'll still be free or that he can afford not taking a job now in order to be employed by you later in July.

I'll inquire of both today and see what the options are. If neither would be willing to commit, let me look at other options for you. Kenton

I sent Kenton an email and asked him to nail down the new guy he had found. He said he would, but I was getting nervous as the time approached for my trip and I had heard nothing further.

Finally, a couple weeks before I was due to depart, I got a promising email from Kenton:

As expected, my first guy (who's now employed) will not be able to assist you, but the other young man is VERY keen on the prospect. He'd be a great liaison for you and with the right advance from you would book hotels at whatever price range you suggest, could take care of flight or bus details and tend to your needs at restaurants and in interviewing/visiting with nationals.

As soon as you get your travel plans ironed out, contact me and I'll put him on whatever needs doing that can be done before you're here. I've got his number but not his e-mail address. He's to send it to me when he's at an internet cafe next time and then I'll put you two in direct contact with each other. His name is Phet (pronounced /pet/; the "h" in transliteration implies the aspiration of the letter), which means "spicy" in Lao.

So I nailed down my flights to Bangkok and to Nong Khai and then returning to Bangkok from Vientiane. No easy task. Booking international travel can be pretty tedious, especially when I needed to use multiple airlines. I attempted to book flights from Vientiane to Xieng Khouang and back but found it impossible to use the Lao Airlines site. Screw it. That's a detail I will have to take care of with Phet's help.

Now with a firm schedule, I sent it to Kenton:

Kenton, just completed my ticketing. I will leave Wichita on the 23rd and arrive in BKK on the night of the 24th. Expect to be in V Laos sometime on the 26th after a visit with a U.S. Gov guy in Nong Khai earlier that day. Arrangements are still pending. Let me know the best way to hook up with Phet Ed Sykes (wgfp)

Within a few hours, I got an interesting response from Kenton:

Coming in from Nong Khai, you'll be crossing on the bridge. Phet could meet you there, but timing is what's going to be awkward. If you'll be in a

U.S.-based office (I didn't know there was any U.S. presence in Nong Khai), perhaps they would let you call Phet from their office as you're ready to leave for the border. He could head to the bridge at the same time you do. Or you could cross the bridge and take a bus to Talat Sao (the Morning Market) and give someone 5,000 kip (80 cents) to call Phet on their mobile phone from there. (EVERY bus at the bridge heads to the Morning Market as it's the main terminal for both the city and the nation.) Actually, as I consider options, the latter would probably be the better option. You get yourself to Talat Sao and call Phet from there via anyone with a phone (which will be everyone—even the farmers). The challenge will be to find an English speaker to whom you can explain your situation and offer to pay to use his phone. Once you and Phet connect, have him take you to a phone shop so you can buy a mobile phone ($30 for a phone that's just a phone, no bells and whistles, and $5 to set up an account), so you always have recourse in being able to call him or someone else for assistance in tight spots. (I can forward some numbers to you.)

Before I say anything to Phet, we need to determine a price to pay for his services. I'd recommend $30 a day plus the covering of his expenses—meals, bus/air fare, hotels/guest houses. The Lao would be fine sharing a room with others; if you don't mind, then you halve your hotel expenses. If you prefer privacy, then, of course, hotel expenses are doubled. $10 will get you into a lower end guest house—safe, but minimal amenities; $35–$50 a night puts you in hotels comparable to a budget hotel (Red Roof Inns, Budget Inn, etc.) stateside. The latter are without ANY chain connections; they're just nicer hotels that are privately owned/run here. All this in Vientiane; I've NO idea what the countryside will hold for you. It might be $5 buys the best room in the town and that will be with an outdoor toilet, etc.

You'll not want for quality food in Vientiane. That's one of the perks of being here, but outside the city, it'll be bowls of noodles and plates of fried rice OR plates of fried rice and bowls of noodles. Lao cuisine is not fit to be called cuisine.

A bus up north would probably only be $30 a ticket; flights in the vicinity of $200 and then you'd have to hire a tuk tuk or some form of transit to get to the sites you're wanting to visit. I'm NOT adventurous, which is what the bus takes—mountainous roads with poor suspension on the buses. I'm subject to motion sickness. Friends went to the Plain of Jars in Xieng Khouang and sat next to a man who started throwing up 1 hour into their trip and didn't stop for the next 12 hours. They would've had two people on the bus doing that had I been with them. Attendants on the buses hand out plastic bags for such events as you board the bus; personally, I'd fly, BUT you're looking at considerable expense then with Phet accompanying you.

So, kick these ideas around and write me back with your thoughts. If needed, we can talk before I leave; I have Skype on my computer and can connect from over here for about 2 cents a minute. Some bargain. Warmest regards, Kenton

Phet could turn out to be a German shepherd and he would still be worth more than $30 a day! I got right back with Kenton:

Kenton, thanks for all of the great information. The rate suggested for Phet is fine, and if he does good, there should be a nice tip at the end of the adventure. The friend I am meeting in Nong Khai is a U.S. Gov employee who is there doing immersion in the Lao language. He is also a member of one of the teams that search for POW/MIA remains. I plan on spending one night there and arriving in Vientiane on the morning of the 26th. I will probably stay there for one night at a hotel near the U.S. Embassy. That might be a good place to meet Phet. I will meet with a member of the U.S. Embassy staff that day to get a briefing and some advice and may want to include Phet in that briefing—not sure yet. Would like to use Phet from the 26th to the 30th—will make decisions on the bus vs plane later but think I would like to experience the bus at least one way. How long is the trip to the Plain of Jars by bus?

Please let Phet know I would like to use his services and look forward to meeting him. Thanks so much for all of your help! Ed Sykes, (wgfp)

I love it when a plan comes together. Now two weeks before my departure, I needed to nail down my visits with Al Teel and Dustin. I first shot off an email to Al:

Al, got my act together finally and was wondering if you would be available for some lunch/dinner on July 25th in Nong Khai? Am coming into Udorn that morning and hope to make it up to Nong Khai by bus later that day. Any ideas on the best place to meet? I plan on spending the night in Nong Khai— any good hotel recommendations?

Following our meeting I am planning to go to N. Laos and meet with others there and possibly try to get up to the area of the ejection site. I have no intention of looking for any remains but would like to get a feel for the challenge. Bob Maves says that Dave's case (1408) is on the investigation schedule for next spring, which is pretty exciting. Any thoughts?

Would like to get together and take advantage of your knowledge of the best way to proceed. Cheers, Ed Sykes (wgfp)

I also fired off an email to Dustin:

Dustin, … Will try to meet Al Teel of JPAC in Nong Khai on the 25th and then be in V'tin on the 26th. Would like to meet with you that day if you will be available. I have asked a travel agent to book me a hotel in V'tin near the U.S. Embassy and we could meet there—or anywhere else you choose. I have hired a guide, Lao who speaks English, who will accompany to the area of the ejection—don't know if you would like to meet with him as well? He would be more familiar with some of the stuff we might discuss. Your thoughts?

Look forward to getting together! Cheers, Ed (wgfp)

The almost immediate response from Dustin told me to call or email him when I arrived in Laos. He'd meet me and my guide to tell him about getting to the site. Perfect.

The next day, I got an email from Al:

I'll still be here on the 25th. If you arrive in Udorn that morning, I would recommend taking the airport shuttle to Nong Khai—about 200 baht ($6) and about a 45-minute trip up the main drag to Nong Khai; may have to share with a few other airline pax, but that's okay. If you come solo, will run you from 800 to 1000 baht ($24 to $30), no need since the shuttle is a comfortable, air-conditioned Toyota or Nissan van.

When you get on the van, tell the driver you want to go to the Royal Nakhara Hotel. The van will probably take the paved detour just outside Nong Khai to the Lao-Thai Friendship Bridge. If he goes to the bridge, he will have to turn around and about 100 yards back make a left-hand turn down a wide paved drag. He will cross the railroad tracks and then the hotel is about 150–200 meters past the tracks on your right-hand side (there is about a 6ft X 6ft brown sign with yellow lettering on it; sits about 10–12 feet off the ground, next to the large drag you're on). The Hotel is cream colored with a red Thai-style roof. (That's where I am staying, but since it's Wednesday, I will still be in class until about 1600.) Recommend you get settled into the hotel and we can hook up there after I get out of class. I am in room 417. In case you want to stay here, I already reserved a room in your name and should not cost more than $45 for the night (probably less), which includes breakfast next morning. If you don't want, let me know so I can cancel.

Any particulars on dinner? There are a few relatively good restaurants down along the waterfront with an assortment of foods, to include western. Chat with you later and see you on the 25th.

I sent an immediate reply:

Al, thanks for the great assist and thanks for reserving the room. I will use it. Makes no difference on where we do dinner. I'm sure you have some great places in mind. I look forward to getting your advice and downing a Singha or two.

I took care of a few more details, booking hotel rooms in Bangkok for two nights and at the Mercure Hotel for my first night in Vientiane.

A week before I was to leave, I got an email from Kenton:

Mr. Phet is at my house right now and we've reviewed the information together before I leave tomorrow for the States. He's excited about being your guide and assistant.

I'm cc'ing him on this e-mail so you'll have his e-mail address and he'll have yours. He'll be happy to make advance plans for you (i.e., a hotel reservation near the embassy) or other things in advance that you think of. His full name is Khamphet Keosiripanya. (Blessedly he goes by Phet, pronounced /pet/ like a house dog! He's laughing over my shoulder now.)

Wishing you well on this venture. I really wish I could be here for your trip, but I'm leaving you in good hands. Warmly, Kenton

I replied to both Kenton and Phet, letting Phet know I looked forward to meeting him and telling him my plans to stay at the Mercure Hotel July 26. I also told him about the meeting with the embassy employee (Dustin) I wished him to be a part of and how long we'd have to travel throughout northeast Laos. I concluded by telling him I'd send an email with more specifics later.

The next morning, I received my first email from Phet:

Dear Edward,

Thank you so much for emailing me. I just have read your message this evening was so great idea to meet you at the hotel or some other convenient place. For me always free during these days so I can go and help you at any time. I don't know if you have put my number correctly or not, here an example, 856 20 …

I hope to see you on 26th,

Have a nice day

take a good care

phet

That looked like pretty good English to me. It was now five days before my departure, and I tried calling Phet using his phone information. However, I was never able to complete the call. I assumed that phone service in Laos is not the best. This suspicion was confirmed when I looked for an ATT access number in Laos that I could use to call back to the U.S. Laos did not have one. I had traveled all over the world, and this was the first country I had found without this service.

I didn't expect to get all the way to Dave's ejection site on this visit, but I thought it might be a good idea to record some GPS data if I did get in the area. I discussed the possibility of buying or borrowing a GPS system with Al Teel and Phet. Al said the GPS used by JPAC was quite expensive, and he was not sure if I purchased a system in Vientiane that it would work in Xieng Khouang Province. Phet found a merchant that sold such devices, and we decided we would visit the store after my arrival.

Phet and I also discussed the trip to Xieng Khouang, and he said he had never been there. Not a problem. As we sorted through the schedule, it became obvious that we would have to fly to Xieng Khouang because a bus trip would take too much time. As it was, we would only have one day, July 28, to try to make a journey to Dave's suspected site.

I also brought up the question of the best beer to drink in Laos. "Beer Lao is everywhere in Laos and it is a very good beer," was his response. I remembered that Bob Maves had told me earlier at the League meeting that I had to try Beer Lao. So, if I accomplished nothing else on this trip, I'll at least get to experience a new beer.

Also before departing, I wanted to update the Dinans, so I sent them an email with my schedule. First, I again stressed my elation with finally having made the "Show":

> It means they will send an investigative team to the site to evaluate the case for an actual "dig." Your involvement in the case is probably one of the primary reasons it is on the schedule.

> I am going to be in Thailand/Laos between July 23–31 doing a look-see of my own. I am going to visit with Alvin Teel of JPAC on the 25th in Nong Khai, Thailand, and with Dustin Roses of the U.S. Embassy in Vientiane on the 26th.

I have hired a Laotian, English-speaking guide to accompany me up to the area of the ejection site. Am counting on Alvin and Dustin to give me advice to keep me out of trouble and my guide Phet to make sure I only drink the best beer (actually, he is probably the key to my success). I would like to get as close to the site as I can without upsetting the locals or Laotian government to obtain a personal assessment of what the challenges are in making a recovery. Cheers, Ed (wgfp)

The day before I left, I got the following message from Charlie:

Ed, here's a footnote on my trip to Singapore and Indonesia in 2009. The beer selection hadn't changed very much since my prior visit to SE Asia 40 years prior. Tiger, Tsingtao, San Miguel, and the Australian brands seemed to be the beers of choice in both Singapore and Bali. There were also the usual European brands, but I stuck to the regional brews and even bought a Tiger T-shirt.

I replied:

Charlie, I will try one of each and rate them for you. The beer we all drank in Thailand was Singha, the most common Thai beer—I will throw that into the mix. Cheers, Ed (wgfp)

The time had come to "stoke the burner" and launch this adventure. In the Thud, you stroked the burner as you released the brakes and moved the throttle to the afterburner position, waiting for the burner to give you a hard nudge and move your body down the runway. As my flight departed Wichita, destined for Laos, I sensed a hard nudge that brought a level of excitement comparable to launching a Thud on a combat mission. Why is the unknown so exhilarating?

Chapter 17

Xan Noy

When traveling to Asia, it has always been my preference to break up the journey if I can. Therefore, I spent a night in Los Angeles en route. A lazy day spent at the pool, exercise facility, and bar always seemed to energize me, making the long journey across the Pacific more bearable. I had a couple thousand dollars in cash, four credit cards, and a carry-on suitcase filled with casual clothes. If I was supposed to look like a tourist, I fit the bill.

During my two long connecting flights, I worried about the details of the trip. Although I had a well-planned entrance and exit strategy, the four days in Laos were rather murky. The different advice I had been given about getting around outback Laos, dealing with the government, and the weather had not been consistent. I would have to adopt a "make it up as you go" strategy and push hard to get as close to Dave's site as I could. This should be fun!

Before boarding my flight in Los Angeles, I found a communications kiosk that dealt with iPhones. I asked the representative if I could get a sim card that would work in Laos, but after examining my phone, he told me it had not been unlocked and so he couldn't help me. It looks like I would be buying a phone in Laos as Kenton suggested.

After spending the night at a hotel near the Bangkok Airport, I caught an early morning flight to Udorn. I didn't recognize a thing at Udorn. My previous trip here was when I had diverted into Udorn in 1969 with suspected battle damage and somewhere on this ramp had shut down my stricken F-105 and thought seriously about kissing the ground.

I followed Al Teel's instructions and caught a shuttle to Nong Khai, a large Thai city on the south side of the Mekong River with Vientiane across it. The shuttle dropped me off at the Royal Nakhara Hotel as I requested, and I checked in using the reservation Al had made for me. I had never met Al before but had been told he sounded like Burl Ives and was of similar build. I got to my room and attempted to call him but got no answer, so I left a message for him to check in with me when he got back.

I laid down for a nap and got a call in the late afternoon from Al. He asked me if I would like to join him and a friend for dinner. I agreed and went to his room, where he gave me some information about a Mr. Yui who lived in Xan Noy who might know the location of Dave's demise. Apparently, they had gotten his name from another Laotian on a different investigation but felt Mr. Yui's information might be related to Dave's case. Great, I now had a destination. I'm headed to Xan Noy.

Al's official JPAC business was in Naung Kham taking a language course. Leaving the hotel, we crawled into a tuk tuk, which is basically a motorcycle in front and a covered carriage in back with seats on both sides with the capacity to carry four average-sized people. We ended up at an open-air restaurant in downtown Naung Kham, and the atmosphere of Thailand made me recall times in downtown Korat. The flavor of the food also brought back nice memories.

Early the next morning, Al met me for breakfast. He explained how to get to Vientiane. Al also gave me a laptop and asked if I would drop it off at Detachment 3 at the U.S. Embassy for him.

Following his instructions, I caught a tuk tuk to the Friendship Bridge that crosses the Mekong River. There I boarded a bus for the short trip across the river. A couple of buildings sat at the other end of the bridge, with some serious-looking folks behind their windows. I presented my passport at one of the windows and was given a form to fill out to obtain a visa for entry into Laos. I sat down at a table to fill out the form and discovered it was going to be a bitch to complete. The form had small boxes to enter information, which would be difficult to fill in given my eyesight. About that time, a young fellow walked up and, in pretty good English, asked if I needed a taxi to Vientiane. I told him I did, but it was going to take a while to fill out the form because of my vision problems.

"Let me see the form, sir," he replied. He took the form from me and asked me for information requested on the form. In a matter of a few minutes, the form was complete. I gave it to the official at the window with

$35 USD, and he stamped my passport and gave me the part of the entry card to turn in when I exited the country.

"Where are we going, sir?" my new handy friend asked.

"Mercure Hotel," I responded. About 30 minutes later, I was in the capital of Laos facing the Mercure, a nondescript building on a wide street. The small lobby area contained what appeared to be a temporary bar and a few lounge chairs for visiting. At the reception desk, I was able to communicate with the clerk well enough to get checked in. I guessed this hotel dated back to the period when the French still occupied Indochina. The hallways and room had a slightly musty smell, but the room was nice enough. I entered the room using a large metal key, which was something I had not seen in a while.

Knowing that my iPhone would not work in Asia, I went to the front desk and had the receptionist call Phet's number for me. After a few rings, he picked up and said he would come by the hotel in a few hours. While I was waiting for Phet, I explored the Mercure and discovered that behind the older lobby area was an addition with a nice pool and pool bar as well as an up-to-date fitness center. Peeking into the fitness center, I discovered an elliptical machine and knew this was the hotel for me. To celebrate, I went to the pool bar and ordered a Beer Lao, sitting next to the pool and settling into my new surroundings. *Pretty good beer.*

After a few celebratory beers, I retreated for a short nap before returning to the hotel lobby at the appointed hour to meet Phet. At precisely the correct time, a short young man entered the hotel and, after looking around the room and spotting me, approached. "Edward?" he asked.

"Yes, have a seat. And we can figure out what we're going to do." He asked me several questions about my trip and how I liked the hotel and the weather. I then explained some of the things we would have to do the next day: meet Dustin Roses for lunch, get airline tickets and make hotel reservations in Phonsavan, and buy a GPS for positioning ourselves in the countryside. He said he would bring his laptop the next day so we could make many of the arrangements on it. I asked about buying a phone, but he told me to depend on his cell phone.

He seemed comfortable with all the things I had outlined, so I asked him if he would like an early dinner.

"Sure, Edward." I took him back to the pool bar and ordered another Beer Lao and asked if he would like one. "No, Edward, I do not drink," he stated before ordering tea.

Oh crap, I've found the perfect guide, but he won't be much of a drinking buddy, I thought. "We'll have to change that," I offered.

He laughed and gave what would turn out to be his standard reply: "Yes, Edward."

After some sandwiches, a few more drinks, and a discussion of some minor details for the next day's activities, Phet left and headed back to his room somewhere in the city. One of the topics we had discussed was what type of currency we needed. He told me that many places would take U.S. dollars and most would take Thai baht. I had converted several hundred dollars into baht when I entered Thailand, so I had that covered. Some places in the countryside would require Lao kip. He also said most larger cities and villages would have ATMs. After finishing my beer, I went to the front desk and asked about the nearest ATM.

It turns out it was on this street a short walk from the hotel, so I wandered out in the twilight, found the machine, and attempted to use it. I put in my request for the equivalent of $200 USD in Lao kip and discovered that amounted to 1.6 million kip. The machine churned out a large wad of bills in several denominations. I was barely able to get the folded bills in my pocket much less my billfold, so I decided to carry kip in one pocket, baht in the other pocket, and dollars in my billfold. Mr. Moneybags.

The next morning, I met Phet in the hotel lobby and asked him to call the driver I had met at the Friendship Bridge and ask him if he could stay with us all day. Within a few minutes, he arrived to pick us up. We first stopped at the airport to purchase tickets for the trip to Xieng Khouang the next day. Only one aircraft flew in and out of Xieng Khouang each day. This concerned me, but as I would later find out, it was very dependable.

After Phet successfully booked our flights, we headed to the U.S. Embassy to drop off the computer Al Teel had given me and to have lunch with Dustin. Upon reaching the Embassy, I asked for instructions to Detachment 3. I was met by a young sergeant who seemed to be expecting it and took the computer to deliver it. I then asked her if she could ring Dustin Roses for me, and I got a blank stare. "Who, sir?" I repeated myself, and after consulting with others, she returned and told me that Dustin was in another part of the embassy across the street; I should go to the side door and use the phone at the door to ask for him.

This jarred some cobwebs loose, and I now remembered that Dustin was not part of JPAC or Detachment 3 but was assigned to Stony Beach, a part of the Defense Intelligence Agency. No one should be surprised that there is significant duplication of effort within our government. Stony Beach had a similar mission to JPAC but, from what I could gather, were mostly intelligence gatherers who spent a good deal of time in the field looking for MIAs. From the Stony Beach reports I had read, they operated in small teams.

We went to the door as directed and soon Dustin made his appearance. He suggested we eat lunch at a little open-air café a short walk from the embassy. We ordered drinks, and once again, I found myself the only one ordering a Beer Lao. I introduced Phet to Dustin, and after they spent a few minutes talking to each other in Lao, Dustin looked at me and asked, "Do you have a plan?"

"Sure," I replied. "I thought we would hike up to Dave's site and dig up his remains. Want to go with us?"

Dustin smiled knowingly and in a more serious tone replied, "Sorry, but I can't make this trip. I can't go anywhere without the permission of the Lao government, and on an investigation, I have to take one of their agents with me."

I told Dustin we were going to visit the village of Xan Noy and attempt to locate Mr. Yui to see what information he had.

"Good luck with that. I'll email you some maps you might find useful in getting to Xan Noy and getting your bearings to the suspected crash site once you get there." Phet gave him his email address, and Dustin said he would send them later that afternoon.

"Got any good hotels to recommend?" I asked. Dustin told us about the Phouphadeng Hotel, which overlooked the Plaine des Jarres (PDJ) from a hill and was made up of wonderful guest cabins and a nice lodge.

"It's off the beaten path and costs a little more, but it's worth it." We wrote down the name of the hotel and decided to try and make reservations on the internet that afternoon. Phet put Dustin's cell phone number in his phone, and after a nice lunch, we bid Dustin adieu and headed off to purchase a GPS.

Our driver hauled us around to a couple shops that sold handheld GPS machines, and I eventually purchased one using my credit card. I wasn't sure I would need it, but if we got close to a suspected site, I wanted to be able to get the most accurate coordinates possible. At the hotel, Phet was able to make reservations at the Phouphadeng Hotel in Phonsavan. Cost was about $75 a night, which was over twice what most hotels in the area charged, but I decided to take Dustin's advice. Phet also got an email from Dustin with some Google Earth maps with the site and village markings superimposed over them, which was nice of him to supply. Phet said he could get them printed overnight.

Figure 7 - Map of Xan Noy and crash site supplied by Dustin Roses during Ed's first visit to Laos, 2012.

As we got ready to pay our taxi driver, I realized that counting out all that kip was a real pain in the ass for someone not familiar with the currency, so I made an executive decision. I took the entire wad of kip and gave it to Phet, telling him that from now on he was responsible for the kip we spent and his official title was "Kip Boy." He took the kip and smiled approvingly. "Yes, Edward."

"Just let me know when you're running low, and we'll find an ATM and recharge."

"Yes, Edward."

I ate at the restaurant at the hotel that evening. It was a buffet that was not all that good, but I didn't really care. I had gotten a lot accomplished on my first full day in Laos. I was especially happy with Phet. His personality was perfect, and I could tell we would have some fun together. His laptop and computer skills were an additional benefit. If he would only learn to drink beer. Sleep came easily after a few Beer Lao. Its flavor was growing on me.

Phet and I got to the domestic airport an hour before our flight the next morning, got our boarding passes, and waited for our flight to arrive. As it pulled up, I didn't recognize the twin-engine turboprop plane, but it turned out to be a French ATR-72, which had a comfortable passenger area with gracious attendants and a light snack provided. The flight to Xieng Khouang took a little over 30 minutes. As we sat in our assigned seats, Phet by the window, I asked him if he had ever flown before. "No, Edward, this is the first time." If he was frightened, he didn't show it. He did seem to be fascinated by the view and how quickly we reached our destination. "The bus ride would take between 5 and 8 hours and is very scary," he stated. "The mountain roads are very steep, and the roads turn back and forth in sharp curves. This is much better."

Upon landing at Xieng Khouang, we waited for our bags to be delivered from the plane to the tiny terminal. It was little more than a shack with concrete floors and a few hard benches. With almost no overhead baggage space on the airplanes, bags had to be checked.

As we walked into the lounge area, we spotted an unexpected sight—a man holding a large white card marked "SYKES." I hadn't requested a taxi, but the hotel owner had sent one to pick us up. He introduced himself as Mr. Somphong and escorted us to his van. He drove us to the eastern edge of town and then up a rather steep hill to our hotel. The owner greeted us, checked us in, and escorted us to our cabins. They were quaint little buildings that could have been set in the Colorado Rockies.

Through Phet, I had asked Somphong to stick around, and after we left our bags at the rooms, we met him at the lodge and moved to a wooden deck with a fabulous view of the PDJ and Phonsavan. I asked Somphong about taking us to Xan Noy. We negotiated a price for an all-day trip and settled on $90 USD. I told him we would plan the trip the next day and make the actual trip the following day. He did not mention any possible travel restrictions, so I assumed getting to Xan Noy would be a straightforward process. Somphong departed, and once again, I felt a lot had been accomplished in a day's time and nothing could go wrong.

That evening we ate at the hotel restaurant where I had a really good curried chicken. Phet told me a little about himself. He had been brought up in a tiny rural village in northwest Laos west of Luang Prabang. When he was 12 years old, he had been placed in a Buddhist monastery by his parents and spent 12 years as a monk, getting his education through their programs. He had departed his Buddhist order only six months earlier. I asked him many questions about Buddhism, and he described the levels of growth in the religion, terminating in the state of Enlightenment, which sounded like what Westerners call "self-actualization," where you are at inner peace with the world. I told Phet I wasn't "enlightened" yet, but one more Beer Lao and I would be getting close. Getting my joke, he laughed.

I awoke the next morning to the sound of several roosters crowing. I wondered which of the roosters would be in my curried chicken tonight. Somphong arrived after breakfast and told us we would first have to get permission from the Tourism Ministry to visit Xan Noy. Apparently overnight, he had considered his exposure to danger with the government officials if he violated any restrictions and had decided to play it safe. So, we first went to the Tourism office, but it hadn't opened yet. A large collection of inert bombs and jettisoned fuel tanks were displayed around the office, so we spent time looking at them as I answered Phet's questions about what they were. After about an hour, the office had still not opened, so we decided to tour the PDJ while we waited.

I had always wanted to see the Plaine des Jarres up close. We discovered there were four different areas to tour, so Somphong took us to the largest area. Phet and I spent a couple of hours looking at the large mysterious jars that dotted the landscape. Some of them appeared to be 8 feet tall or more. It has never been determined what the jars were used for or what ethnic group made them. Touring the area, we found several trenches and revetments remaining from the Laotian Civil War. Memories of bombing and strafing passes made in support of this effort came back to me. How many Thud drivers had died near this place?

After finishing our tour, we found Somphong waiting for us. He told us the Tourism office had opened, and they had told him that we could not go to Xan Noy as it was in the Khoun District, which was closed to tourists. This was not what I wanted to hear, so I asked him to take us back so we could talk to them ourselves. Phet and I entered the small office where two solemn men in dark suits sat behind a table. The did not smile or greet us, so Phet said something to them in Laotian, and they motioned that we should sit in two chairs across from them at the table.

"They want to know why we want to go to Khoun District," Phet informed me.

"Tell them I want to visit the site where my friend died," I replied.

As Phet explained why we wanted to go there, the men shook their heads and made negative gestures with their hands.

"They say we cannot go there, Edward. You cannot visit such sites."

"Tell them we want to meet Mr. Yui and talk about the history of the War in Xieng Khouang Province."

Once again, the officials demonstrated their displeasure with our plan, their tone becoming more stern. I was steaming inside, and as I looked at the two men, who appeared to be in their 40s, I thought, *God, I wish I had killed your fathers!*

Persistent, I told Phet to tell them we only wanted a short visit to Xan Noy. I was beginning to fear that I was getting Phet in trouble with these officials. However, he persisted, and the tension eased as it appeared they were entering into negotiations.

"They say we can only go for 8 hours, and we must take a policeman for our protection, and we will have to pay him for his work."

"How much?"

"150,000 kip," Phet replied.

That's a big number, but as I did the calculations in my head, I realized we were talking only $20. "Ask them if we can meet the policeman at 9 A.M., and we will depart and be back at 5 P.M."

Phet explained our plan, and after more discussions, Phet finally said the plan would work; we could pick up the policeman here the next morning.

Phet then stressed, "We must be back by 5 P.M., Edward."

"Roger that, Phet."

I did not thank or even smile at the two men, but Phet did, and I was glad he did for his own safety. We ventured outside where Somphong awaited us with an eager look on his face. Phet explained the plan to the driver, who nodded and discussed something with Phet. Phet told me, "He says it may take as long as 2 hours in each direction. He says the roads may be very bad since we have had a good deal of rain lately." Well, at least we were going to make the trip, which is more than our government has done in over 40 years. Phet and the driver talked some more, then he let me know the driver would pick up some provisions in the morning so we would have food and drink for the trip.

"What would you like, Edward?" Before I could answer, he suggested we take some chicken and sticky rice as well as several bottles of water. "Make sure we take some Beer Lao also," I added. He communicated this to Somphong, and we agreed to meet at eight the next morning to pick up the provisions and then meet the policeman as agreed upon.

Phet and I ate at the hotel restaurant. I again had chicken curry. I noticed that the flock of chickens that ran around the hotel was a little smaller than the day before. I thanked Phet for his perseverance with the Tourism officials, and he said it was no problem. I was becoming very impressed with Phet and his willingness to try as hard as he could to satisfy my requests. For the first time since I began the trip, I had trouble sleeping. I guess there was some risk involved. Why had the two men insisted on protection? What if the roads were impassable or we had an accident and couldn't make it back? What if we got back late? Would those bastards throw me in some Laos jail and toss away the key?

The next day yielded picture-perfect weather. Somphong was right on time, and we went to a shop to get the food. Somphong had brought a cooler with ice to keep the beer and water cold. Phet paid the bill in kip, and I asked him if his "kip gun" was running out of bullets. He looked at me, questioning what I was talking about.

"Do you need more kip, Phet?" I explained. "Will you have enough to pay for the policeman?"

"Yes, Edward, but we may need to pick up some more after that."

We drove to the Tourism office, where a thin young Lao man, dressed in loose-fitting clothing that looked like pajamas, waited. He really didn't have the appearance of a protector. I tried to determine if he was armed, but it didn't look like there was anywhere to conceal a weapon. He was definitely not what I had expected, and I surmised that maybe he was part of an "employment" plan for one of the Tourism official's family members. At any rate, he was pleasant, and he and Phet got along well from the start. Phet introduced him to me as Mr. Bua Sone.

We were off almost precisely at 9 A.M., and the first part of the trip was along a well-maintained road that headed out of Phonsavan. I had studied Dustin's map and estimated the distance to Xan Noy to be around 22 miles. We had not traveled far, however, before we made a left turn and suddenly were on some pretty rough roads—more like jungle trails. I was sitting up front, so I had a good view of the terrain, and as we made another turn, I observed that we were about to cross a fairly wide streambed. Somphong hesitated for a moment and then pushed on. The bed seemed to be solid, and we passed through it without much difficulty.

As we progressed slowly toward Xan Noy, we encountered some challenging switchbacks as Somphong carefully drove. I guessed he had experienced these conditions before, but he must be nervous risking his van on these roads. About an hour after leaving Phonsavan, the road leveled out and became easier to navigate. Before long, we came to a small village with only a few huts along both sides of the road. We drove on, so I assumed it was not Xan Noy. We finally arrived at a bigger village with several dozen huts scattered around the hillside to our left. When we came to a junction in the road, Somphong pulled over to the side and parked the van, letting us know through Phet that we had arrived.

It was probably around 10:30 in the morning, and as we walked along a road into the village, I spotted a large grassy area about half the size of a football field where numerous blankets were spread on the ground along with a few tents. Several people, mostly women, occupied each spot, displaying various trinkets, household goods, and foods.

"It is the village market, Edward," Phet explained, and we wandered through the little "shops" with the women exclaiming, "*Sawadee*," and smiling at us. After walking the length of the market, I asked Phet to find out where Mr. Yui's house was. He asked some of the women, who pointed up the hill and gave Phet some instructions.

We arrived at a typical village house, which was built on stilts, putting the living quarters about 5 to 6 feet above the ground. The area below the living area was fenced in, mostly with wooden sticks, and as I looked closely, I realized this was where each family kept their livestock. It was mostly pigs, but I saw a few small cows. There were chickens everywhere and many dogs. Most looked ill-kept, and some showed signs of serious dog fights. I wondered right away what it must smell like in the upper chamber when the livestock had the occasional upset stomach.

At Mr. Yui's house, we were greeted by several people at the top of the stairs. Phet asked about Mr. Yui and was told that he was in Luang Prabang working on a construction project and wouldn't be back for six weeks. Chalk that up to bad luck. I asked Phet to ask if Mr. Yui had a phone and, if so, see if he could get his phone number. After a little conversation, I saw Phet get out his cell phone and put in some information, so I assumed he got it. He returned to me, telling me he had the number and that the family insisted that I go visit the village chief. Proper protocol had been violated by not going there first, it seemed. So the four of us hiked the 100 yards or so back toward the van where we had been told the village chief lived.

When we arrived, it appeared no one was home, but Phet climbed the stairs and knocked to announce his presence. Soon, a thin middle-aged man appeared wearing a plaid shirt and jeans, nothing to indicate royalty. Phet spent some time talking with him, and he then invited all of us to climb the stairs to a balcony where there was a table and several chairs (probably where the Chief met with the village elders).

Phet introduced me to Mr. Khammerng. The Chief greeted me with a warm handshake and a smile, and I liked him at once. We sat around the table, and Phet explained why we were there and that we were disappointed that Mr. Yui was not there. They chatted for a while, and finally, Phet told me that the Chief did not know where the site was, but he did know which hill it was on and that he would be willing to show us.

"Great!" I exclaimed. "How do we get there?" Phet posed the question to the Chief and told me that we could take the van up the road a ways, and he would point it out from there. "Let's do it. Is he ready to go now?" Phet nodded, indicating that issue had already been settled.

We got in the van with the Chief in the front passenger seat and drove up the road for what I guessed was about two miles. The Chief had Somphong stop the van, and we all got out by a harvested rice paddy. As we walked into the paddy, the Chief pointed toward the closest hill. I noted that the hill was completely covered with trees, but there was one tree on the north side of the hill that towered well above the rest. Phet had been talking to the Chief and told me that the site of the American's death was on the other side of the hill.

"How long would it take to get there?" I asked. After some discussion, Phet said there was no way to get closer by vehicle and it would take at least 5 hours in each direction to get there by hiking through the jungle. Given this information, a trip of at least two days would be necessary to get there and look around and would entail spending at least one night in the jungle.

Figure 8 - Phet Keosiripanya and Ed Sykes in front of the karst where Dave died, July 2012. Author's photo.

Phet volunteered immediately, "We could sleep in the forest, Edward. I have done it many times and could show you how."

"We may need to come back at another time and do that, Phet."

As the rest of the party and I headed toward the van, I suddenly turned, as if being drawn by some spiritual power, and walked back across the rice paddy toward the hill. Although I could never fully explain it, I felt Dave's presence in the surrounding atmosphere. I had never experienced anything like it in my life. My skin tingled and erupted in goose bumps.

"We're coming to get you, Dave," I whispered, and just as suddenly, as if Dave had heard my promise, the feeling went away. Crazy!

Finally, I returned to the van where the others stood by, wondering what I was doing. As they saw me, they climbed into the van and I jumped in behind.

"Are you OK, Edward?" Phet asked.

"Yes, I'm fine, Phet. Why don't you ask the Chief if he would like to eat lunch with us?"

Phet discussed the proposition with the Chief and said we could bring our food to the Chief's house and eat it there. Once at the Chief's house, Phet and Somphong grabbed the cooler and the chicken and sticky rice and entered the central room of the Chief's house. He had the guys give the food to a woman, who I assumed was his wife, and she took it back to a kitchen area. Meanwhile, we got out the beverages. Somphong and Phet pulled out bottles of water, but the Chief, the policeman, and I all grabbed a Beer Lao. It tasted so good, and you could tell the Chief was pleased to be drinking the beer. I was surprised the policeman took a beer, but I wasn't convinced we would need protection in Xan Noy.

Soon, the woman brought out metal plates of food and we dug in. With a good deal of laughter, Phet and the others talked with the Chief while I was busy concocting a plan with my beer-enhanced brain. About halfway through the meal, the first round of beer was gone, and we distributed the last three beers to the drinkers.

"Phet, ask the Chief if I can stay in Xan Noy if we come back later to search for Dave's site."

After a good amount of discussion with smiles and laughter, Phet finally answered, "Of course, you can stay in Xan Noy, Edward, and the Chief has invited you to live in his house when we come back."

Wow, I should have been a diplomat. As the discussion resumed, I began to work out the terms of the agreement, but it sounded like the rent would be free as long as I brought more Beer Lao. Unfortunately for now, we were out.

I had Phet apologize to the Chief for being so shortsighted, and he waved his arms in a sign that it mattered not. He then got up and wandered back to what appeared to be his bedroom. He soon reappeared with a clear glass bottle containing a clear liquid. He yelled something to his wife, and she brought us three small glasses. He filled them and kept one, giving the other two to the policeman and me.

Oh shit. Am I still a real fighter pilot? "What is this stuff, Phet?"

"Edward, it is rice whiskey the Chief has made, and it is the best in the village."

Figure 9 - From left: Mr. Somphong, our driver; Chief Kammerng; the Laotian security guard, and Ed Sykes at lunch in Xan Noy, July 2012. Author's photo.

OK, fighter pilot it is. I took a short sip and, finding it tasted somewhat like the moonshine whiskey the folks around Kentucky used to sell, chugged the rest of it. It went down hot and made me sit up straight with a start. The biggest mistake, however, is that my empty glass caused the Chief to reach for my glass and fill it again. No more chugging. I sipped it slowly and asked Phet to compliment the Chief for his excellent—no, outstanding—whiskey! The Chief nodded his approval, and for the second time in two days, I was rescued by Phet. "Edward, if we are going to be back in time, we had better leave soon."

"Right you are, Phet. Please apologize to the Chief but explain our problem and let him know we must leave." Phet explained our dilemma to the Chief. I could tell he would rather have us stay, but we slowly got off the floor and headed for the door. My unsteady legs got me down the stairs and to the van, but I no longer sat up front. I didn't want to see a thing of the trip back—and I didn't.

I awoke as we were entering Phonsavan, where we delivered a somewhat unstable policeman to his starting point. Phet paid him in kip, and we headed back to the hotel. By now, I was starting to feel better, and I asked Somphong to come back and get us in an hour (enough time for a shower and short nap), and we would all go downtown to eat. I also announced that tonight Phet would drink his first beer. The look of distress on Phet's face could not be concealed. "I will try, Edward," he allowed.

Once we arrived at the hotel, I took Phet to the balcony overlooking the PDJ and asked him to call Dustin on his cell phone for me. He handed me the phone as it was ringing, and after a few rings, Dustin picked up.

"Dustin, Ed Sykes here. I think I'm going to need your help. I'm in a Laotian jail up in the Khoun District, and I may need some help from the embassy to get me out." I was proud of myself for getting it out without laughing.

"No, you're not," he quickly stated. "If you really were in jail, I would have heard about it. Did you make it to Xan Noy?"

I gave him a brief rundown of the trip and told him I could get with him in Vientiane the next day and give him more information. He mentioned he wouldn't be around the next day, so I agreed to send him a full report of my adventure. "Thanks for all your help, Dustin. It would be nice if the Lao would give you a little more ability to move around on your own," I concluded.

"I don't see that happening anytime soon."

"Why did you tell him you were in jail, Edward?" asked Phet when I had hung up.

"I was only pulling his chain," I replied, which led to more questions about what that meant. I told him I would discuss it after he had a few beers in him that night.

When Somphong came to the hotel to get us, he took us to a restaurant near the center of town where we ate and I tried to get Phet to be a beer drinking buddy. He did choke down one but allowed that was enough. I thanked him for all his help and let him know I would probably be back the next year to make another effort in my search.

"Make sure you let me know, Edward, because I want very much to help you." It was great knowing that Phet was willing to be involved again, lending me his capable computer skills and cell phone for communications.

"Next time you come, I would like you to bring me a better computer," he said, and I told him I would do that. Back at the hotel, I gave him several hundred dollars for his efforts and told him he could have the GPS since we had discovered it wasn't very effective for what we were doing in the Khoun District.

The next day we flew back to Vientiane, and he left to visit his family. I spent another night at the Mercure followed by a night in Bangkok at a hotel Dustin had recommended. Back in the States after nine days on the road, I felt more comfortable about my ability to make future visits to Laos. But I felt much less confident about recovering Dave's remains. It would be difficult.

Following my Xan Noy adventure, I put together my trip report, complete with photos, and sent it to everyone I could think of who might be able to influence additional emphasis to case 1408. I titled my report "Out of Laos," and I began to get comments and inquiries almost at once. All the prime players responded. Al Teel and Bob Maves both applauded my report and asked a few questions about my findings that were not included in the report. JPAC covered! Also, Bill Habeeb was upbeat about my trip and the way I handled working with JPAC. DPMO covered! I also got a nice letter from Ann Mills-Griffiths, who applauded my working with the official government agencies. Ditto Dustin Roses. And, of course, the Dinans were happy and told me they would make sure they were at the next League meeting. By then, they would be able to get a briefing on the results of the spring JFA.

Chapter 18

A Broken Lady

Quite proud of myself, I spent some time basking in my own glory. The trip had gone well, and Dave's case had made the Show for the 2013 spring JFA. I could sense the momentum building. When I finally got tired of basking, I decided it was time to contact Valerie.

So much for a perfect world. I attempted to contact her on the phone number I had used earlier in the spring and found the phone had been disconnected. I asked Charlie to attempt a call, and he had the same result. He also did a Google search and came up with two addresses in North Salt Lake City for Valerie Zoolakis, but that was all he could find.

Through Charlie, I had told Valerie that I would try to visit her after the Laos trip, and she had agreed to that meeting, so I decided to make a trip to Salt Lake during the last week in September and see if I could find her at one of the two addresses I had gotten from Charlie, which were only half a mile apart. I had found a hotel not far from the two addresses, and after landing at the Salt Lake Airport, I took a taxi there. Now late in the afternoon, I decided to wait until the next morning to give it a go.

The next morning, I got a taxi to take me to the closest address. The small ranch-style home looked quiet. I asked the driver to wait as I went up and knocked on the door. Nothing. I waited a bit and knocked again then heard rustling within the house. After the door was unlocked, the door opened a crack and a woman's face appeared. I knew at once it was not Valerie by the woman's age. She said nothing.

"Good morning. I'm Ed Sykes, and I'm looking for Valerie Zoolakis."

After a long hesitation, the woman asked, "Why do you need her?"

I explained that I had known Valerie years ago while serving together in Thailand. I had tried to call her, but apparently her phone number had changed and I'd been unable to contact her.

"Just a minute," she said and disappeared, closing the door.

She eventually returned and said, "I just talked with Valerie. She lives a short distance from here." She handed me a slip of paper with an address on it. I thanked her, and she closed the door. I checked the address, which was the same as the second address Charlie had provided. The taxi driver took me to a duplex with a central garage area, and I again asked him to wait. I would be back in a minute to pay him or depart.

As I knocked on the door, I could hear a woman talking inside and was not sure she had heard my first knock, so I knocked a little harder. The door opened, and through the screen door, I tried to recognize the woman I had last seen in my hooch, over 43 years earlier. It was a stretch, but as I studied her face, I was confident to whom I was talking. "Valerie, we met several years ago at Korat. I was Dave's roommate."

Before I could go any further, she said, "Just a minute," and went inside the house to finish her phone call. When she returned, she invited me in, and I told her I needed to release my taxi driver and would be right back. I paid the driver and got his phone number and hustled back to Valerie's door. She let me in the house and invited me to sit down on a couch in her living room. I was almost immediately joined by a large long-haired cat who decided she needed to cuddle up to me. Valerie introduced the cat as "Puss" and asked if I wanted her removed from the room.

"She's fine," I stated, making a close friend for the remainder of our visit.

The shades were drawn and it was a little dark, but I'd noticed she walked with a slight limp as we entered the house. I also noticed that her youthful figure had followed the same course as most women who were approaching their 70s. I first apologized for the no-notice arrival but said that Charlie and I had both attempted to contact her with the original phone number we had, but it appeared to be disconnected. She said she had recently moved out of a house with her friends, Robert and Carol, and now used a cell phone. She gave me the number.

We began by talking about our time at Korat, and she admitted that she didn't remember me, which made sense as we had never really had any contact other than that meeting in the hootch after Dave's death. Also, I had only been on base for about six weeks at the time.

When I described the meeting, she said she didn't remember it at all. I was not surprised as she'd been distraught at the time. However, when I described my meeting with Father Gene Gasparavic, she acknowledged that our meeting probably did occur. She then went into her memories of Father Gene and the marriage counseling sessions they had had before Dave's death.

She also described the beautiful wedding dress she had had tailored for herself. As she described her relationship with Dave, I noted that she demonstrated no emotion or distress, with one exception. When she described returning to the United States after losing her fiancé and being discharged from the Air Force for being pregnant, she revealed to me that at the San Francisco Airport, she took her wedding dress out of her suitcase and threw it in the airport trash can.

I could tell that she had not enjoyed her assignment to Korat with the exception of her relationship with Dave and a few of the other lieutenants. Her disdain for almost all senior officers was obvious. She felt like they had afforded her little respect, simply because of her youth and gender. She named several of the lieutenants she chummed around with but did not mention Bob Zukowski, Dave's good friend and roommate who was killed six weeks before Dave. Reflecting on this, I remembered that I had rarely seen Bob at the bar and perhaps that's why Valerie didn't mention him. I had concluded long ago that one of the reasons Dave hung his ass out so far on that fatal day was related to Bob's death.

Since Valerie left my hooch, two days after Dave's death, I had not seen her at Korat and had no knowledge of what had happened to her. She said she went to Personnel and explained that she was pregnant, and after confirmation by the hospital, she was scheduled to be separated from the Air Force upon her return to the U.S. a few days later. It all happened quickly. During the few days she had remaining at Korat, she spent most of her time in quarters but did venture out for meals with a friend from the C-121 Unit.

I told her that I had never revealed the meetings we had at the hootch or the fact that she was pregnant and about to marry Dave to anyone at Korat, and I assumed that no one, not even Col. Prosser, ever learned about it. Again, during a combat exercise, everyone involved is looking forward and little time is spent dwelling on those who "went somewhere else." For the most part, combat is a gossip-free zone, especially for the most involved combatants. As I explained this to her, she simply nodded her head in understanding.

After her return to the U.S., she elected to stay in California and have her baby there. In 1969, being an unwed mother was a mark of personal irresponsibility, regardless of the circumstances, and her mother indicated she did not want her to return to Connecticut to have the child. She entered an adoption process, and after she delivered the baby girl, the child was given to an adoptive family.

"Have you ever seen the child since then?" I asked.

"Never."

"Do you know where she is?"

"No, I don't, but my sister does. Several years ago, after she was older, I sent a letter to her through my sister. I tried to explain why I had put her up for adoption and why I now felt remorse for having done that. I got a blistering letter back from her, again through my sister, that let me know she hated me and never wanted to meet me."

Ouch! Again, Valerie passed this on to me without emotion. It appeared that she had long ago accepted that fact and moved on. I could only imagine how devastating this letter would have been at the time.

Dave had given Valerie Charlie Dinan's phone number while they were together at Korat in anticipation of just such an occurrence. She had managed to contact him and found out he was a Marine stationed in San Diego. Shortly after the baby was born, they met in southern California where she updated him on her situation. Other than an exchange of information about Dave and their related families, little else resulted from this meeting. Charlie was soon on his way to service in SEA, and Valerie got her act together to qualify for being a grammar school teacher in California. Sometime during that period, she sent Dave's parents a letter expressing her sympathy for their loss and giving them a glowing picture of Dave. She mentioned the impending marriage but not the child.

She then described her ongoing years in California to me. She had spent many years as a public school teacher. She was involved in relationships with a few men, but none measured up to Dave. Although she experienced a few fulfilling relationships, most of these flamed out. She did have one marriage, which lasted a few years and then hit the rocks. She'd had no more children.

She eventually migrated to Utah and was befriended by a married couple, Carol and Robert. Carol was the woman who met me at the door at the first address. Valerie spoke highly of Carol and Robert and said she was going to move to southern Utah with them in the near future. She was not happy with the cold winters in Salt Lake and did not think her Mormon neighbors treated her well.

Another issue was her health, both physical and mental. She had undergone a hip replacement surgery a few years earlier and was still experiencing a good deal of pain. A second hip replacement was currently being considered and might be scheduled in the next few weeks. She'd also been battling breast cancer. In addition, she had, on many occasions, sought help for depression and mental trauma and was currently working with a therapist. She felt her mental issues were initiated by the loss of Dave and had been a continuing problem.

As she was telling me this, I thought, This woman was within a few days of becoming Mrs. David Dinan, but because of this misfortune, she lost her future husband, her career, her child, and the ability to be a respected member of her family. Little wonder she had experienced these problems.

I also realized that in today's Air Force and society, she would be celebrated as a courageous woman and patriot. She would also still have her career, her child, and the respect of her family. Sadly, in this case, timing was everything—and the timing did not favor Valerie.

Finally, she told me about her attempts to save stray cats around the neighborhood and how she had gotten some resistance from some of her neighbors. From the conversation, I sensed her strong affinity for these animals, especially those that were imperiled.

By now, we had spent nearly 4 hours chatting, and I suggested I should return to my hotel but we could continue our conversation later that evening at a little restaurant I had eaten at the night before. I told her I would call for my cab driver, but she volunteered to drive me to the hotel instead. When she dropped me off, I told her I would call her later that afternoon and set up a time to eat. She agreed.

However, when I called later, she said she was feeling pretty tired and declined my offer. She talked around the subject a bit, but I sensed she didn't want to discuss this part of her past any further. I accepted her decision without complaint.

The next morning, I reflected on our visit on my return flight to Kansas. There are many ways to be a victim of war, but what had happened to Valerie was devastating. Dave, of course, was the biggest victim of this war, but his suffering was short-lived. I am certain that Dave's parents and the rest of his family suffered immensely, having been close to the suffering experienced by my sister and mother. However, all the survivors except for Valerie had some safety nets to help them regain strength. Valerie had to dig out of this hole by herself, and I sensed she was still digging.

It was at this time I decided to compile my notes about this adventure and write a book. The book would detail the undertaking to recover Dave's remains and the eventual achievement of that goal. It would include a burial at Arlington with Dave's family and many of his old fighter pilot comrades in attendance. But, most importantly, it would be attended by Valerie, who would be reunited with her daughter, and they would finally reconcile their differences, and we would all live happily ever after.

What a great saga. However, the reality of being able to give this story that fairy-tale ending seemed difficult to impossible. I had seen firsthand the "needle in a haystack" aspect of the search itself. And now I had discovered that Valerie was not inclined to have anything to do with what I was dreaming about. *Oh well, I'll push on and see what happens.*

The following day I called Valerie, and we spoke for a short time about our visit and she updated me about the doctor's appointment she had told me about.

I summed up our visit and phone call in an email to the Dinan brothers the next day, including her current address and new phone number:

Charlie and John, had the opportunity to visit Valerie in North Salt Lake, UT, on Wed. She plans on moving to southern Utah (La Verkin) in the next few months with some friends. She does have some health issues. The day I visited with her she was talking about a second hip replacement in the very near future. She also told me she had some other problems including nerve damage to her leg and severe depression.

Other than that, she was very chatty, and we discussed all kinds of things over the course of four hours. I phoned her after her doctor's appointment on Thursday, and she said she was putting off the hip surgery for now and had enrolled in a swimming class beginning Friday morning. Think it will do her a great deal of good for both the exercise and social contact—seems like she has not ventured out of her house much in the past several months. She does not yet have a computer so does not use the 'net. Says, however, when she moves, she will get established on the 'net.

If you get a chance, you might drop her a note or give her a call. I know it would mean a lot to her. Cheers, Ed (wgfp)

Two weeks later, I received a letter from Valerie. As I read it, I realized that the likelihood of her involvement in my fairy tale was unlikely.

Dear Ed and Mary,

Ed, when you showed up on my doorstep, I was very surprised!!! I commend you for your persistence in your quest to bring closure to David's family.

I am currently seeing a therapist as I have been plunged into a deep depression. I am not looking for sympathy. I need for you to understand what I'm going through.

You and Charlie made good lives for you and your families. I lived a life of guilt and shame and coped by ignoring everything. I made a decision 43 years ago to adopt out our child. I had and still do have no family support. So I made a decision, good or bad, based on the circumstances at the time and tried to forget a lot of my past. Charlie, you, unresolved medical problems, and isolation tipped me over the edge into a huge depression. I was vigilant enough to recognize all the red flags and reached out for help.

301

I have to learn to forgive myself for life's choices and love myself, without the guilt and shame. I also have to learn to lower my anxieties so I can heal mentally and to move forward with my physical ailments. The anxiety is unhealthy and has worn me down.

I am doing my part, in small steps. I go swimming twice a week and like it. I'm doing a little walking 2 other days. I am trying to eat more and will be working on goals with my therapist. She specializes in special depression, stress management, and women's issues, and I am all of the above! I don't have the skill sets to see the "glass half full." I'm a "glass half empty," so practicing positive affirmations is new and hard for me. You want to tell David's story, I am a "reluctant" part of that story. I don't want contact via phone or knocks on our door from our daughter. Carol and Robert support and understand my not wanting to revisit that part of my life. I have no motherly urges or curiosity to do so.

I also don't want any phone calls or anyone to visit me. I have to heal mentally so I can face my right hip replacement as a strong person.

I will drop you two a note and let you know how I'm doing and when I move. If you recover David's remains, please don't ask me to join any reunions or gravesite because I don't want to grieve again. I've had a lifetime of that and didn't know how to get or seek help.

My therapist is very supportive of Carol, Robert, and myself pursuing a downsize of households to a warmer climate. Carol actually reconnected with one of her sons in S. Utah. Robert is a gentle soul and has sisters nearby who never reach out to him. He is the brother I never had. He even took care of my cat when the "girls" went to S. Utah on vacation 3 years ago. He is clearly a dog person, but he took care of Puss when I was recuperating at their house. You experienced how Puss could win over a guy by being a cute pest.

I feel good writing this letter. When we all get settled, "two or three months," I do know how to email you, two fingers and I won't disappear in "plain sight," you will know where I am, but please let me grieve and heal in my own way. I wish I could have remembered you and lots of other things, but they are gone. What you are doing for families is truly a labor of love. Please share this with Charlie.

Sincerely,

Valerie Zoolakis

The letter didn't sound all that good and I wasn't sure what to do next. I did share the letter with the Dinans, and a week or so later, I got the following reply from Charlie:

John and I both read Valerie's letter. It was pretty disturbing, especially since I spoke with her for about half an hour two days before and she seemed upbeat. The first time I contacted her, she seemed to be extremely upbeat and genuinely happy to hear from me. We spoke for more than an hour, and she kept me on the phone even as she negotiated an arrangement with a lawn service. I was really glad that I had contacted her. Now I'm not so sure. I phoned her when I was on my way home. There was no answer, and Val called me back a few minutes afterward, told me that she was trying to contact a physician and promised to call me back later (she hadn't recognized my phone number). She didn't call back and didn't pick up or call back after two more attempts. I'm not going to call her again, at least not for a while. According to her letter, it's apparent that my prior calls upset her and I'm afraid that my additional unanswered calls may have upset her more.

Best Regards, Charlie

My interaction with Valerie in March 1969, following Dave's death, was one of the prime movers in my decision to search for Dave's remains. Having visited her now, 43 years later, my desire to find Dave's remains was not diminished at all. Even though she did not seem enthused about a recovery, I felt more strongly than ever that Dave must not be left behind.

About this time, Charlie sent me some articles and letters concerning Dave that had been compiled by the Homecoming II Project. Valerie had written a couple letters to Dave's parents after his death and one of them read:

David is truly a son to be proud of. I only knew him for nine short months, but every moment we had together was a cherished one. He was a man and a gentleman in every sense of the word. No higher compliment could have been paid to my womanhood than when he asked me to be his wife.

A few years after my visit with Valerie, she sent me a letter describing some of her memories:

I have a few memories, bits and pieces but no chronology of the past anymore. David was "my first." No one ever measured up to his talent and intellect. No one ever appreciated me for my smartness the way he did. I was always one of the guys but didn't fly a plane! I wore a skirt and that did not bother my 2nd Lt. friends at all!

I had taken my sewing machine with me overseas. I made a blue silk scarf and a yellow one for David. I can't remember which one he was wearing the

303

day he died. He used to wear a different color depending on when he flew a mission. I don't remember which color on what mission.

Every month, he would help me balance my checkbook. I could never find my own mistakes. For almost a year after his death, I could smell his body odor and felt his presence when I would sit down to balance my checkbook. The hair on my neck would stand up. I also heard his voice for months later while sleeping.

I vaguely recall him being shot down and recovered his first time. I don't recall any emotions being attached to him on anyone shot down. I had a job to do, crews coming in to be debriefed, and there was no time to think about our friends who were now gone and [would] not be at the O'Club anymore. In the back of my mind, I knew David could be next as we lost friends daily. It was a crap shoot and a lot like Catch-22.

I never knew any details about his being shot down other than Search and Rescue could not recover his body. I was already pregnant then, and our plans for me to separate from the Air Force and be married and able to accompany him were all canceled. I had a beautiful Thai silk street-length dress in off-white made for our marriage. I picked it up after his death. When I arrived at Travis Air Force Base, I took the dress out of my suitcase and dumped it in a trash can in the airport in San Francisco.

Chapter 19

Xieng Khouang

My visit with Valerie had not gone as well as I had hoped. Her personal situation was brutal. Perhaps her move to southern Utah with her friends would bring her some comfort with a reduction in her anxiety and depression. Perhaps someday she would want to look upon her time with Dave as a precious memory, one she would like to revisit.

Following my visit to Xan Noy and with Valerie, I began to focus on the upcoming JFA to northern Laos in March 2013. We had made the "Show!" I still wanted to be part of the JFA Team but realized that was unlikely. However, after receiving my "Out of Laos" report, Ann Mills-Griffiths sent me an email that gave me a glimmer of hope:

> *Col. Sykes, thank you for sending this report. Naturally, I read it with great interest, and it does sound as though the names and information you came back with more than justified the data you were provided by friends who decided to help, for which I'm most grateful.*
>
> *I applaud your sense of responsibility and your dedication to recovering your friend, a commendable frame of mind from where I sit, of that you can be sure. I hope that JPAC will consider asking you to accompany them when they go to the site, as you have proven your ability to make that trip, though I readily admit to being skeptical about most such endeavors, especially by family members.*
>
> *Best wishes, Ann*

Despite Ann's feelings, I was pretty sure that JPAC would not allow me to be part of their official team. However, I planned to be in the area when they did their investigation and give them a gentle nudge if I felt they needed it. I realized that my biggest challenge would be getting access to the area during the period of the investigation. I wasn't sure how to do that, but I could always fall back on bribes if all else failed.

***

Another opportunity presented itself in the fall of 2012 that was triggered by an email from my sister-in-law, Ret. Col. Claudia Bartz. Claudia had been an Army nurse and, since her retirement, had kept up with general officer moves within the military. She sent me this announcement:

> The chief of the National Guard Bureau announced today the following assignment: Maj. Gen. Kelly K. McKeague, special assistant to the chief, National Guard Bureau, to commander, Joint Prisoner of War/Missing in Action Accounting Command, U.S. Pacific Command, Joint Base Pearl Harbor-Hickam, Hawaii.

A Guard guy running JPAC. I immediately pulled up his biography and found out that he was a real "Guard Baby" who had always been part of the ANG. I did not know him personally but knew some of my friends would know him well, so I began trolling. Two of the best officers I had working for me when I commanded the fighter group in Wichita had become major generals, and I was sure they would know him.

I sent an email to one of these guys, Tod Bunting:

> *Tod, how's it going? I can get you into the swine business if you are interested. Big bucks to be made.*

> *I am interested in knowing more about MG McKeague and figured you would know him. I am working to recover the remains of a former roommate and friend, 1Lt. David Dinan. We were Thud drivers together in 1969 at Korat RTAFB Thailand when he was killed. MG McKeague's new job is as Commander of JPAC, the organization primarily responsible for recovery of POW/MIA remains. I was in Laos a few weeks ago and was able to make some judgments about the recovery of my friend's remains. I have attached the report I sent to JPAC operations folks and others. As you can see, I want to be part of the JPAC recovery effort. Any ideas? You were always excellent at ideas—some of them I even liked Ed Sykes (wgfp)*

I did not hear from Tod over the next several days and assumed he must be preoccupied with general officer bullshit. Then a few weeks later, I got an email from Tod:

> *ETG, have not heard from you. I have talked with Kelly and he started that new position this week. He and I are standing by to hear from you.*

> *His email is … – TB*

Tod had apparently replied to my original email and I had either misplaced it or it had gone to my Spam folder. ETG was a term Tod had used to address me during the nearly 7 years he had been my chief of personnel back in the day. It was short for "Ed The Great." I never asked him why he called me that.

I immediately replied:

Tod, great to hear from you. I think your reply to my first email must have ended up in the spam. I am gathering some intel from my friends in Laos right now and trying to decide my next move. I do know that I would like Gen. McKeague to put high emphasis on Dave's case (1408). In the 43 years since Dave ejected, there has never been an attempt to visit the site where he died. I will send an email to Gen. McKeague in the next week and ask for his support and also ask about some of my ideas on how I might help them resolve this case (if that is even possible).

Was really elated to get your email today—thought you were ignoring all of my crazy ideas like you did back in the 184th glory days. I get a smile on my face every time I think of your successes. Who would have thunk it? I'm enjoying my retirement a lot—cattle, hogs, chickens, a great dog and lots of grandkids. Mary is doing very well and is still convinced she made a big mistake by hooking up with me—and she is always right. Hope your family is doing well and you are figuring out what you are going to do when you grow up.

I look forward to working with Gen. McKeague and appreciate your help. I'll be in touch. Cheers, Ed (wgfp)

I wasn't sure how I would compose my email to Gen. McKeague. How would I ask for inclusion on the JPAC Team? However, Tod had already forwarded my earlier emails to Gen. McKeague, so he already knew what I wanted. Before I had sent an email to the general, I got an email from him addressed to me and Tod, first apologizing for his delayed response, caused by the time it'd taken to get his email account set up. He continued:

Leaving today with current commander and JPAC J5 on a 7-day visit to Thailand, Laos, Cambodia and Vietnam, then take command on the 31st of Oct. Will look into this case both on the trip and afterwards. VR Kelly

Getting to know Gen. McKeague hadn't been that hard. It helps to have friends in high places. I decided to send him an introductory email to keep the conversation going:

Gen. McKeague, my good friend, Tod Bunting, encouraged me to send you an email and introduce myself and also give you some background on my good friend/roommate 1Lt. David Dinan III. Tod and I go back a long way. He was the most able of all of the officers on my senior staff while I was the Commander of the 184th TFG (1986–1992). I was not surprised by his success in all of his endeavors following his outstanding service to the 184th.

Let me congratulate you on your appointment as Commander of JPAC. I know a good deal about your mission and have had the opportunity to meet and work with several on your staff. Two of your civilians, Bob Maves and Alvin Teel, have been especially helpful. Your mission is very important, especially in the eyes of those families and friends of those who died fighting our Nation's battles and have yet to be returned home.

I then described the circumstances of Dave's death:

A few years ago, I decided to make an effort to bring additional attention to Dave's case. During the course of these efforts, I have formed relationships with many individuals and groups associated with the recovery of POW/MIA remains. I have also made contact with Dave's family and close acquaintances. This past July I made a trip to Laos and was able to visit the village of Xan Noy, which is very close to the ejection site. I am currently considering a second journey to Laos around the end of this year. I want to be careful to not jeopardize any of JPAC's efforts in Laos. (I discovered there are some sensitivities on the part of the Laotian Government.)

During my visit in July, I got some great assistance from Dustin Roses, the POW/MIA Officer at the U.S. Embassy in Vientiane. When you visit with him, please ask him about Dave's case. I look forward to hearing from you after your visit. Safe journeys! Cheers, Ed Sykes (wgfp)

As I was composing this letter, I decided not to ask him again to be included on the JFA Team in March, now only a few months away. He had already seen my desire to be part of the team, and I was hoping for an invite.

A day before Gen. McKeague was to take command of JPAC, he sent me a long, detailed email concerning Dave's case:

Greetings, Ed. Had the privilege of working with Tod when he was the ANG/DP. Many thanks for the update, but more importantly, for your staunch support of Dave's recovery.

Transiting Narita now and back home in 12 hours. A very successful trip to all 4 countries. My two main takeaways are the incredible dedication and talents of the JPAC in-country teams as well as the sustaining support of our host nation partners.

In Laos, I was able to meet with Amb Stewart and her team (a huge advocate of our MIA mission), and senior Lao Ministry of Foreign Affairs officials from their MIA Committee. Dustin was not there as he was in the south supporting the first Joint Field Activity (JFA) of the FY [fiscal year].

I inquired of Dave's case (1408), and Bill Goudry, a long-time Det member and resident case expert, provided this update.

Bottom line, the JPAC Det is tracking this case, received Lao consent, and will be investigating it during the third JFA of the year (7 Mar–10 Apr 13). They have a reliable fix on the grid coords of the site from time of incident.

A JFA in 1994 reported the team going to the site of record and finding some possible related material evidence but could not confirm the last location of his remains. They noted the possibility that the body could have been moved from the parachute location.

A 2010 report noted a villager claiming to have seen a possible American body and helmet near his village (5 miles south of the area) during the general period the incident took place, but association with 1408 could not be confirmed.

Part of the Spring 13 JFA will focus on following up on this report. Like most of the remaining Laos cases, a significant challenge for the team will be to locate/prep an Mi17-capable LZ or insert close enough to the incident site in order to conduct both interview and site survey requirements.

Of the 3 SE Asia countries, the Lao are the most challenging. We finally were able to obtain permission to drive to accessible sites and to set up remote base camps where practical. These were concessions long sought but finally secured after Sec Clinton's recent visit. They are also famous for circuitous negotiations and unreasonable demands, but we are making progress with them.

Given the upcoming JFA, I'd recommend holding off on a trip back. I'd be glad to provide you fidelity from the JFA as soon as I receive it and then you can better assess the utility and timing of a trip.

Best regards,

Kelly

Obviously, I was elated that Gen. McKeague would spend this much effort in composing this response. Based on my research, I was convinced that the folks who briefed him on #1408 had painted JPAC's efforts in the most positive light possible. I had studied these reports and was not sure they were really related to Dave's case. All the shoot-downs in the area seemed to make the cases overlap. However, the last paragraph of his email was a little tough to take. Not only was I not invited to participate as part of a team, I was urged to not even attend and wait for the JFA results.

The military and political side of me told me to follow his advice and wait until after the JFA to make a trip. The fighter pilot side of me told me to get my ass in gear and be in Laos in March. I began to plan the trip at once.

While this was going on, I was invited to attend an ROTC meeting at MIT. Mark Rockoff had sent my "Out of Laos" report to Frances Marone at MIT, and she coordinated the whole thing through her office. Frances was the woman who had helped me locate the Dinans through the MIT alumni records. She set up a date of November 14, and we got the word out to the Dinans who decided to attend. Mary wanted to go with me as our youngest son, Ezra, and his new bride, Jill, lived in Boston. It was a great visit. I briefed the ROTC group on the status of Dave's case.

We also had a chance to meet the dean of undergraduate education, Dr. Daniel Hastings, and give him an update on our quest. He seemed interested and asked me what he could do to help. After visiting with him a bit, I discovered he had spent a good deal of time around the Pentagon working for the military and probably had some good contacts.

"Right now, things are going pretty well, but I'll let you know if I find an opportunity to get you involved."

The best part of the visit was spending some quality time with the Dinans and outlining my plans for March. I would try to be in Laos during the beginning of the JFA and would attempt to go back to Xan Noy and be in the area when the investigation took place. My biggest obstacle would be getting permission from the Laotians, but I told them I would go to D.C. in December and try to figure out some angles to make it possible.

Phet and I had kept in contact since our Xan Noy excursion, and he had found a part-time job but seemed anxious to work with me again. I had sent him some money to help his younger sister with college, and he had let me know his computer had quit working.

I tried to call him but was unsuccessful, so I sent him an email:

Phet, I am doing well, thank you. And I hope you are doing well also. I just tried to call you on the number below but there was no answer. I will try again tomorrow. I am interested in finding out when Mr. Yui is going to be in Xan Noy and also would like to find out if the Xieng Khouang Province authorities would let me stay in Xan Noy for 4–5 days. I think I would like to go camping in the forest and see what we can find. What do you think?

Sorry to hear about your computer. When I come again, I will try to bring you a replacement. Ed (wgfp)

A few days later, I got a reply:

Dear Edward, how are you doing these days?

I have contacted with Mr. Yui about the crashed site in Xan Noy village, he is living at his home for the moment. He said that the place where he found the crashed area is the way we went to last trip.

I ask him to go with your next trip, he agreed with that but we have to work with the provincial authority. To stay 5 days in the village, some say yes, some say no, but I want you to work with U.S. Embassy also. Take a good care, Phet

Reading between the lines, I guessed Phet did not want to be put in a situation like on the last trip, and I did not want to put him there. I needed to see if I could find a way to obtain permission to visit the Khoun District without using Phet as the messenger. I hoped my short trip to D.C. could result in some solutions and also bring Dave's folder to the top of the stack.

I sent Bill Habeeb a short notice that I was on my way:

Bill, I will be in D.C. next week—afternoon of the 29th and all day Friday. Would like to stop by and discuss Dave Dinan's case with you. I am interested in the 1994 visit to the site and also interested in discussing some personal planning issues revolving around my next visit to Laos. Was at MIT in Boston last week and met with some folks there that are interested in Dave's case. Dave attended MIT. Cheers, Ed (wgfp)

Bill called me the next day and said he would be in those days and I could come by anytime. As I prepared for my trip, I went through my old emails and found some of my Cable Guy messages. Might as well be an equal-opportunity irritant. I forwarded them to Bill as well as to Bob Maves and Al Teel:

Bill, looking forward to seeing you in D.C.—Thursday PM probably works best for me right now. This email summarizes an effort I made a year ago to dig up information on Jolly Green 16, the JG that attempted to rescue Dave. I started my effort by talking to an old ROTC buddy who flew JGs, Len Knitter. I was reviewing some of my old emails on this subject last night and found I had forgotten the reference to the JG records at JPAC—see the Oct 26, 2011 email from Robert LaPointe below. Have you ever used this source of information?

Bob, Alvin, do you know if the exploits of JG 16 on March 17, 1969, have ever been reviewed wrt David Dinan (1408) using your JG logbooks? Also, can these records be used to obtain names of crewmembers who might be interviewed? Would I be allowed to obtain this information, or would it be more helpful to have a family member request it? I would be willing to travel to JPAC and do the legwork if necessary (Hawaii in the winter is not a bad gig!).

Thanks for all of your help—see you later this week, Bill. Cheers, Ed (wgfp)

My subsequent visit to D.C. and DPMO was not all that exciting. Bill and I discussed the 1994 report and concluded it was unclear if it even involved Dave's case. He told me he had sent an inquiry to one of the investigators at JPAC and asked him to check the SAR logs for March 17, 1969, to see if he could find any information.

When I outlined my plans about being in Laos in March during the investigation of Dave's case, he did not encourage or discourage me from going. He said it would be unlikely they would include me on their helicopter sorties, but I would be free to do what I could to get to the site on my own, once again confirming what I had suspected. The biggest obstacle would probably be the Laotian government.

"Even if we decided you could go on the helicopter, it would still have to be approved by the Laotians. There is a Laotian military officer on every flight, and he must approve all participants," Bill explained.

So, what to do about that? I had not told Ann Mills-Griffiths I was coming to D.C., but I decided to pay her a no-notice visit and seek her advice. I stopped by the League office, and as luck would have it, she was in. Her assistant told me she was busy, but she walked back to Ann's office to let her know I was there. I heard her tell the assistant to send me back, and I headed that way. Her desk was stacked high with papers, and she was hard at work at her computer.

"What brings you to town, Ed?"

I explained my plans to go to Laos and my dilemma with the Laotians. She nodded affirmatively.

After a moment, Ann said, "Why don't you have the Dinan family write a letter to the Laotian government requesting you be included in the Spring JFA?" A brief pause. "Why don't you write a proposed letter to be signed by the Dinans and send it to me for review, and I will edit it and we can then have the Dinans send it. I will get you the address and name of the individual to send it to later. I'm not sure it will work, but it might be our best chance."

We talked a little about what to include in the letter then I bid her goodbye. I thought about the officials from my Xan Noy visit and knew I didn't want to put myself or Phet through that again. Ann's idea might be the solution. Back in Kansas, I wrote a draft letter on behalf of the Dinan brothers and sent the draft to Ann for comments or revisions.

Since it was the Christmas season, I waited until after the chaos was over and sent an email to Ann on December 30:

Ann, attached is a draft letter which, I think, addresses the issue we discussed a few weeks ago when I made my unannounced visit to your office (old fighter pilot trick). Please review the letter and juice it up anyway you see fit.

You have much more experience with the Lao and may have some insights on how to ring their bell. Also, need the formal station and/or names for the address block and a fax number for the Dinans to use.

I very much appreciate your help with this. I am also considering having the family write a letter to Gen. McKeague asking him to give me some help while in Laos. Good idea/bad idea? Cheers, Ed (wgfp)

I don't think Ann ever sleeps or takes a day off. The next day, I got a reply:

Ed, draft received, and I'll look it over and edit later today, hopefully. Things have been a bit demanding of late, and those demands have included letters to foreign governments due to failure by Congress to pass the FY13 budget, much less prevent the sequestration from happening. We'll see, but there are already disruptions to the JPAC schedule and, unless funding is released, more to come.

As to your approach, are you trying to omit telling the Lao government that you went previously? Don't you think they know that? You can be certain they do. Also, you were a friend of Dinan's in the military, fellow pilots; Laos is a military-based government; they understand military mentality and share it.

As to the correct person to whom the letter should be sent, I'll think about that. On Gen. McKeague, it is he and his Det. 3 CO, LTC O'Brien, aided by Bill Gadoury, who need to be convinced that you actually can assist them in locating and recovering LT Dinan's remains. It isn't simply a matter of you WANTING to go, but for them to understand you could help and NOT be a hindrance.

I'll get back in touch and, by the way, I don't agree with dropping in unannounced to try to ensure you can talk with whomever you please. I would have scheduled a time despite my having so little to share. I consider you a serious, responsible person; that is the reason I want to help. Ann

Ouch! Ann, of course, was correct about my unannounced visit. I will not do that again! The good news was she was still willing to help despite my bad manners. I am so grateful that she grew up with a fighter pilot brother and has seen this type of behavior before.

As 2013 rolled in, I began the year with a good deal of optimism. My excursion to Xan Noy the previous summer had resulted in a much more relaxed attitude by DPMO and Ann Mills-Griffiths concerning my efforts. At least they sensed I was not a complete cowboy and would not knowingly step into a big pile of cow shit. Comfortable with Dustin Roses, I would call him occasionally to see if he was making any progress with his efforts to have the Lao lengthen the leash they had on him. I also kept up a frequent conversation with Phet. He was anxious for me to return to Laos for another adventure, happy to be my Kip Boy, guide, and interpreter. If he felt he had gotten in trouble with the Lao government because of our dealings with the provincial officials, he hadn't told me.

Unfortunately, other obstacles popped up. As a budget compromise, President Obama and the Republican-led House came up with a proposal to sequester the budget with massive cuts to the discretionary social issues along with cuts to the Defense budget. At first, it was unknown if the defense cuts would affect the JFA schedule, but eventually it was determined that the funds existed to keep the schedule intact.

Following my initial trip to Xan Noy, I knew I would be hard-pressed to achieve my goal on my own and would need the help of DPMO. Therefore, I decided it was more important than ever to find a way to be involved with the JFA efforts and learn how the JFA worked. Perhaps Ann's assistance with a letter to the Lao would help me be in the area at the time of JPAC's investigation.

Early in the month, I got a response from Ann. She had revised my draft letter quite a bit, and as I read it, I recognized her skill as a diplomat. The suggested letter from the Dinans to a high-ranking Lao official was attached:

Dear Sir: Our brother, First Lieutenant David Dinan, USAF, was killed in Xieng Khouang Province on March 17, 1969, following an unsuccessful ejection from his F105D aircraft, but his body was not recovered.

We are aware that the U.S. and Lao governments plan to conduct a joint investigation of this incident (REFNO 1408) in March of this year. We are very grateful for your government's willingness to assist in this effort that means so much to our family and our community.

Our brother's close friend and fellow pilot, Colonel Edward Sykes, USAF (Ret), hopes to return to the incident location during the investigation. Col. Sykes will again be using a tourist visa, as he did during his first visit in July 2012. We hope both Lao and U.S. officials will permit him to be involved with the investigation to the fullest extent possible. He is not only knowledgeable but a trusted friend of the family who has volunteered to assist in this effort.

Once again, please accept our sincere appreciation for your government's assistance to the U.S., in particular, the JPAC Teams who pursue answers on our missing loved ones, in this case in locating the remains of our brother.

The honor of a proper burial for our brother Dave in his homeland would bring relief and closure at last to this family tragedy, and we would be forever grateful to the Lao and U.S. governments and to the Lao people for their willingness to help. Respectfully, [to be signed by Charlie and John]

She also made some recommendations concerning the circulation of the letter as well as my "Out of Laos" Report from the previous summer:

You may also want to send the earlier report to McKeague and [Lao Official]; I would, as being open and helpful goes a long way, especially with the Lao. I think if Det. 3, primarily LTC O'Brien, the CO, agrees you could be helpful, the Lao will likely not object, but hold the U.S. Team responsible for your welfare and behavior.

I've been to Phonsavan long ago and more recently, and on further N/NE to the former Pathet Lao headquarters and even to Dien Bien Phu. It is a beautiful, complex and underdeveloped country, that's for sure, but keep me posted and I'll try to help. Ann

I got the proposed letter to Charlie and John right away and asked them to put it in a format they felt comfortable with and mail or fax it to the Lao official Ann had suggested should get it. Within a few days, the letter was on its way to Laos as well as to several DOD officials Ann had suggested. After waiting a few weeks, I asked the Dinans if they had heard anything. "Not yet." So, using additional contact information I received from Ann, we sent the letter out again. I also asked Dustin Roses to see if he could do anything with it through the U.S. Embassy.

After a few hiccups with the format to be used in the attached letter, I got a response from Dustin:

Ed, I was able to pass the PDF on to folks in the Embassy. Right now, the Political Section doesn't think they can do anything with it, and I'm still waiting to hear from JPAC.

I can work on at least getting you a different FAX number and/or snail mail address.

Still no word on the investigation status, I have a feeling the decision will come on very short notice, like everything else expected to transpire about that time frame. Dustin

Now mid-February, I finalized my travel plans for early March. Phet was more than willing to get involved and had set aside some vacation time from his job to be available at that time. He was working near Phonsavan and said he would meet me there.

Then, as I completed my travel plans, I began to catch rumors that the #1408 part of the JFA would be canceled because the Laotian government was considering canceling travel to the Khoun District for undefined reasons. Despite this uncertainty, I decided to do the fighter pilot thing, put my head down, and charge forward.

I had touched base with some of my JPAC contacts to find out where the JFA Team would be staying. When I found out they'd be staying at the Vansana Hotel on the edge of Phonsavan, I immediately tried to book two rooms at the hotel over the internet, but it was fully booked for the period I requested. The Team had booked every room. I worked with Phet to locate another hotel close to the Vansana and eventually found two rooms at the Mali Hotel, a short drive away.

Two days before leaving Wichita, I was informed that the 1408 investigation to the Khoun District was definitely canceled. After thinking less than a minute, I decided to press on anyway.

I sent an email to the Dinans letting them know what I thought I was going to do:

Charlie and John, hope you are getting less snow than we are. I leave KS on Mar 2 for a day of spring training baseball in Phoenix and then on to LA for an evening departure to Vientiane on Mar 4. All kinds of issues have been coming up. At first, it looked like the sequester would shut down the investigation (LA 3-13). Now it appears to be on, but for some reason, the Laos government is shutting down the visit to Dave's site. JPAC thinks it may be due to objections by District Officials. Do we have any idea if the letter to [Lao Official] ever made it all the way? Have you heard anything from him or JPAC?

At any rate, I will meet the team in Xieng Khouang on Mar 7 and begin working on them and the Laotians to let us go to the site. I will also have my guide, Phet, with me, and we may make a trip on our own if I can't get an official trip to go. Obviously, I would rather have the JPAC team go in with me. However, if they can't go in, I have been assured that they will give me assistance with the case from the hotel they are staying at in Phonsavan.

(May have to put in some serious bar time with these guys and spend a fortune on Beer Lao to get what I need). Cheers, Ed (wgfp)

Spring training was just getting good, and it was refreshing to stop in Phoenix and watch the 2013 Chicago Cubs trying to figure out which of the young stars were going to make the "Show." So although I thought we had made the "Show" in pursuit of Dave's remains, I was now like many of the young players on the field, hoping for the best but aware I might have to spend more time in the Minors.

Then I was reminded how long a trip it is to Laos. Once again, I spent the night in Bangkok, staying at the Novotel at the airport, which was great—especially the swimming pool complete with a poolside bar and fitness center. Nice way to start the trip. The next day, I flew to Vientiane and sat next to a woman from Egypt whose last name was Morsi. She was related to the new president of Egypt, Mohamad Morsi. She talked about the hope for the Egyptian people now that they had a newly elected democratic government. A few months later Morsi was overthrown by a military coup. I would wonder what happened to this woman as I felt lucky to be born American.

Landing in Vientiane, I got my visa and went to the Air Lao desk to get a ticket to Xieng Khouang the next day. I managed to get a taxi to the Mercure Hotel, where I checked in, went to the pool bar, and ordered a Beer Lao. I was proud of myself as I had gotten all of this done without Phet's assistance. However, when I went to an ATM to get some Lao kip, I missed my Kip Boy as the 1,600,000 kip came out of the machine. I couldn't wait to hand this large wad of cash over to Phet. We were to meet the next day in Xieng Khouang as he was coming by bus from a neighboring province where he was working.

Early the next morning, I headed to the airport for my flight to Xieng Khouang. When I walked into the waiting room at Xieng Khouang, I saw a young Laotian standing in the reception area holding a placard with "Sykes" written on it. Surprised to say the least, I identified myself, and he told me he was from the Mali and was there to take me to the hotel. His English was excellent, and he did not have the standard-issue appearance of a Laotian (he was much taller with more Western features). We loaded up my stuff in a nice van and proceeded to the hotel.

"Do you work for the hotel?" I asked.

"No, my stepmother owns the hotel, and she sent me to pick you up."

"So, what do you do?"

"I'm a consultant and contractor and soldier of fortune, and I'm involved in all manner of work escorting tourists and adventures into the outback areas of Laos," he stated in almost perfect English. He sat in the passenger seat in front as another person was driving.

When he asked what brought me to Laos, I told him I was going to try to join the U.S. Team that was at the Vansana Hotel and see if I could assist in the recovery of my friend's remains. I asked how far it was to the hotel, and he said not far but too far to walk. He volunteered to take me there, and I told him I would be ready to go the next day. We agreed that he would take me there midmorning. I introduced myself, and he said his name was Vilan Phetrasy as he handed me a business card.

Phetrasy—wasn't there a leader of the Communist resistance during the war named Phetrasy? I was sure I'd heard that name included in some of our intelligence briefings back in 1969. I asked Vilan about it, and he said that man was an uncle, his father's brother. He told me that both his father and uncle were now deceased but that he still had relatives within the central government in Vientiane.

Arriving at the Mali, I checked in and, despite a plain facade, found the hotel to be nice inside. Within the front door, I could see an eating area and, on a wall over the entrance, an old POW/MIA flag. What a coincidence. I would later find out that small U.S. teams had stayed in the hotel previously and had given the owner the flag.

The woman who checked me in was a pleasant Laotian who also spoke pretty good English but not as good as her stepson. She introduced herself as Sivone Phetrasy. She took me to my room. I had reserved a second room for Phet and let her know he would be arriving by bus later that afternoon. I told her when he arrived, he could wake me when it was dinner time. I settled in for a good nap to see if I could get my time clock right. Since my retirement, I almost always took a midafternoon nap.

I awoke to a knock at the door and opened it to a beaming Phet. "Edward, it is good to see you!"

I replied, "Phet, my friend, it is great to see you also!"

We exchanged some small talk about our travels, and then I informed him that the JPAC missions to the Khoun District had been canceled by the Lao government. Then I told him that if they did not go, we would probably try to go on our own. I gave him a reconditioned computer I had brought from the States.

"Will the Lao officials let us do that, Edward?"

"I don't know, Phet. We will have to work that out."

I could tell he was not happy with my reply, so I told him about the letter the Dinan family had written to the Lao government. However, no one in the government had responded. I told him about Vilan and his frequent trips to the Khoun District as well as his contacts within the government. Perhaps he would be able to help us out.

We wandered across the street to a nice outdoor restaurant and had some really good Lao noodles. I drank several Beer Lao and I even got Phet to sip one. The setting was perfect: lovely Lao music and a table by a pond with several large goldfish swimming around in it. I'm sure it was caused by the Beer Lao, but I was starting to feel good about this trip despite the obvious obstacles.

Breakfast was included with our stay, so Phet and I ate in the hotel restaurant the next morning. I think we were the only guests there. Sivone made us fried eggs and noodles for breakfast, and I was ready for a successful day. I was anxious to find out as much as I could about Dave's case and held out hope that the Lao officials had changed their mind about allowing visitors to the Khoun District.

"Phet, I'm not going to take you to the Vansana Hotel with me this morning."

"What is the matter, Edward?"

"Everything is fine, Phet. I will be working with the American Team, and I want to make sure they're willing to share their information with me."

Phet seemed a bit hurt, but I didn't take the time to explain the reason. A few DOD folks had told me to not bring Phet into such meetings because the briefers might be suspicious that he was on the take or an undercover agent. They also told me it was better that Phet not be identified and targeted by the Lao officials that were always present around the Team.

Vilan was ready to go with the van and driver, and we took off for the Vansana Hotel. The Hotel was in a strange location—on a hill that had an extensive residential village on the approach. The road was rough and not well kept, but the hotel itself was obviously upscale. However, Lao soldiers were guarding the approaches to the hotel, and a number of tents were set up in front. I had Vilan drop me off just short of the soldiers and told him if they let me pass, he could pick me up in 3 hours. He nodded. I walked past the soldiers like I knew what I was doing, and they simply let me pass without acknowledgment. I looked back down the road to see the van departing.

When I entered the hotel, I saw some young Americans in the entryway and told them I was there to meet Sammy. I had been told to work with Sammy by the JPAC folks prior to making the trip. The young man I talked to left for a moment and came back and told me Sammy was in a meeting but should be back soon. I sat down among a dozen or so young American men and women and conversed with them about their participation in the JFA. None of them were in uniform, but it was easy to distinguish them from tourists by their grooming and military demeanor. I felt proud of these kids. They represented all the services and were here as volunteers to participate in finding remains of fallen MIAs. Some of these kids were sleeping on the ground in the tents out front without complaint. I was so impressed.

After a while, Sammy came forward and identified himself. I had been told he was of Laotian heritage and had been with JPAC for several years. He'd been described as one of the best experts in the recovery process. He escorted me toward the dining area where some of the Americans were seated in what appeared to be planning sessions.

"So, you know we're not going to be able to visit 1408 on this JFA?"

"Yes, I had heard that but was hoping the Lao had changed their mind. Any idea why the restrictions?"

He thought for a moment. "They haven't really given any specifics. So, what can I do for you?"

"I would like to get the latest edition of information on 1408. I was wondering what your plan of action had been."

He set up his laptop, typed in some entry stuff, and finally a Google map appeared of the area around the most likely site where Dave had died. I told him that Mr. Yui might be in Xan Noy and that I might be able to contact him on his phone, but Sammy didn't seem too interested. He outlined their plans for the investigation, and without taking notes, I did my best to take it all in. Suddenly, I became aware that he was distracted by something else in the room. Following his line of sight, I noticed some new players in the room. Some men who appeared to be official Laotian observers were watching us.

Uh-oh. It appears the Lao aren't sure they like what is going on here. It's close to noon—perhaps time for a Beer Lao?

I had already gotten a lot of information about the investigation plan, like where they were planning to do their interviews and the most likely sites they'd visit, so I decided to change the subject.

"Are you ready for a beer?" I asked.

"Too early for me, but they have a great lunch here, and you can get a beer if you like. First let me introduce you to Lt. Col. O'Brien, our Detachment 3 commander."

I had seen Lt. Col. O'Brien at previous League meetings but had never been introduced. He was busy with some other folks but stood as he saw me approach and greeted me with a smile and handshake. I expressed my admiration for the kids on his Team, and he shared similar views. He expressed his regrets about the Khoun District situation but assured me they would pick up where they left off on future JFAs. It was not what I wanted to hear but noticed he was also glancing about the room at the Lao officials, so I let him return to his other discussions.

Sammy and I hiked over to the bar attached to the dining area, and I ordered a Beer Lao and some spring rolls. I noticed that the Lao officials were still watching us but seemed more relaxed. Over lunch, I asked Sammy about making the trip on my own, and he shrugged his shoulders and said, "You can do whatever you want."

End of subject as far as I was concerned, so I changed the subject while I awaited the pickup time I had set with Vilan. We talked about some of the techniques used by the teams while they excavated for remains and how the Lao supplied much of the labor for the effort. It was all interesting stuff, but my mind was occupied by other things. Obviously, the Lao were not happy about me talking with the Team. How does my presence affect the relationship between the Lao and the Team?

After eating lunch, I told Sammy I would go to the front porch of the hotel and wait for my driver to return. I thanked him for his help and said I would keep JPAC informed of anything I discovered. Just over 2 hours had passed since I arrived, and as I sat on the porch, I pondered my next move. I didn't have a clue.

I had only been on the porch for 15 minutes or so when I saw Vilan arrive and stop outside the Laotian soldiers' barrier. As I walked to meet him, I noticed a Lao official gazing at me as I jumped in the van.

"How did it go?" asked Vilan.

"Not well, Vilan. The Team is not going to be able to investigate my friend's site because the Lao will not let them go to the Khoun District. Do you know why that is?"

"I'm not sure, but I think they're encountering some unrest from the Hmong tribesmen in the area and don't want the U.S. teams to observe any unrest or become involved."

"How serious do you think it is?" I asked.

"Not very," he said. "I think they are being overly cautious. I travel to that area frequently and know a lot of people in the Khoun District, and I don't think they would encounter any problems."

He went on to explain that he had escorted several individuals and groups into that district and had never had problems. He spent some time telling me about a movie team he had escorted to the area and said I could view the movie when I returned to the hotel. I told him I would like to see it.

"So how hard would it be for me to go back to Xan Noy this weekend?" I asked.

"We could do it very easily. How long would you like to stay?"

I told him I would probably want to be back there long enough to walk from Xan Noy to the site the village chief had described to me the previous year.

"Let me come by your hotel with a plan later this afternoon, and we can discuss it," he volunteered.

Late that afternoon, Vilan came by the hotel, and we went to my room where he presented me with his plan to visit Xan Noy and hike into the jungle to locate the hill pointed out by the Chief. The plan was unbelievably well laid out with the items necessary to make the journey and spend at least one night at the site. Tents, food, water, support personnel, and many other details were identified, with all items neatly listed and itemized. The cost for the adventure would be $6,500.

"What about permission to go to the Khoun District from the provincial authorities?" I asked.

"Do not worry, Edward, I can take care of that, and it is included in the price," Vilan answered confidently.

"I don't have that much money with me. I'll have to go to a bank tomorrow and get the cash."

We agreed to get together the next afternoon and close the deal.

I was not sure what type of transaction I would have to complete with a bank to get the cash but knew I had more than that in my account in Kansas and felt I could arrange some type of transfer.

That evening, Sivone showed me the documentary, a European film that was fairly anti-American, mostly talking about the amount of ordnance the U.S. had scattered over Laos and the resultant problems it had caused the Laotian people. I watched without emotion, but I did notice that much of the film was shot in the Khoun District and Vilan was given credit for his help with the escort responsibilities.

That night, my sleep was restless as I considered the consequences of venturing into the Khoun District even though the U.S. Team had been restricted. What kind of trouble could I get in? How will this affect the relationship between the Team and the Lao, since the Lao knew I had gotten information from the Team? As far as permission to enter the District, I would leave that to Vilan. If he went in without permission and we were stopped, I would simply plead ignorance or insanity.

About 4 in the morning, I heard a sound I recognized but surprised me. Near the hotel, someone was killing a hog—I assume to butcher it. I had heard the sound many times when I took my pigs to the slaughterhouse, but I didn't expect to hear it in the middle of a large Laotian city. The sounds went on for a while, and I estimated there would be three additional hogs in the markets of Phonsavan tomorrow. I hoped this was not a bad omen for the adventure I was about to undertake.

Phet and I met for breakfast, and I explained my arrangement with Vilan. I told him we would go to a bank to find a way to get money from my bank in the U.S. to pay Vilan for the trip to Xan Noy. Phet then told me the night before he had had a conversation with Sivone, who told him to let me know I should be careful in making deals with Vilan. Interesting.

"Will we sleep in the forest, Edward?" asked Phet.

"I'm not sure, Phet. We will make it up as we go, but I really want to hike back to the site the village chief described."

"I like sleeping in the forest, Edward. I have done it many times with my father, and I know a lot about sleeping in the forest."

"Great, Phet. You may get a chance to show off your skills."

After receiving directions from Sivone, we walked about a half mile to the closest bank only to find it closed. There was a sign on the door in Lao.

"What does it say, Phet?"

"It says the bank is closed for a national holiday, the International Day of the Woman."

"What the hell is that, Phet? I've never heard of such a holiday. I know we don't celebrate it in the U.S."

"Edward, it is a day celebrated in most Communist countries to honor women."

I asked Phet if there were other banks that might be open, and he said he did not know. He asked a local man on the street, chatting for quite a while, and then informed me there were two more banks about a mile away, but the man did not think they would be open. I decided to check it out. As the day was heating up, we found both banks were also closed and so I reassessed my plan.

It was Friday, and the banks would not open again until Monday. I was due to catch my flight back to Vientiane on Tuesday, so making this plan work would mean changing my airline schedules for my return to the U.S. Bummer! The day was becoming very hot, and I had made the mistake of wearing my cowboy boots instead of tennis shoes. My puppies were beginning to scream.

"How far is it back to the hotel?" I asked Phet.

"It is not far, Edward." My nearly 70-year-old body was beginning to rebel, and I instructed Phet to get a tuk tuk.

On the ride back, I decided to drop the idea of a journey to Xan Noy on this trip and to return to Vientiane the next day if I could. Perhaps not making the trip was for the best. It would certainly ease the possible friction with the Lao government and JPAC.

"Phet, I have decided not to go to Xan Noy on this trip. I will talk to Vilan about this and try to do it this fall or early spring and stay in the forest for a longer period. I will let Vilan know of my plans when we return to the hotel, and then we can go to the airport and change my flight to leave for Vientiane tomorrow." Phet was clearly disappointed, but after questioning me about my decision, he realized I would not be dissuaded.

We found Vilan waiting for us at the hotel. I explained the situation, and he seemed disappointed that we could not make the trip the next day. I told him I had decided to make the trip in the fall or in the spring of 2014 but for a longer period—at least 5 days at Xan Noy and the site. Although not happy, he did allow that more time would give us a better chance of success. I had told Phet to let Sivone know that we would probably check out the next day, and he returned and said she understood. I also asked Phet to reserve a room for me at the Mercure Hotel in Vientiane the next day. After searching for a bit, his reply was indicative of the way things seemed to be going: "The Mercure does not have any rooms available tomorrow, Edward."

Vilan gave us a ride to the airport, and I had him drop us off and told him we would catch a tuk tuk back to the hotel. Phet then negotiated with the Lao Airlines representative about changing our tickets back to Vientiane. Phet and the agent discussed the matter and then Phet informed me, "It will cost $90 to change the ticket." I asked him to try to get the fee reduced, but the agent was determined.

Then Phet said that because we were leaving early, he was going to go back to his village to visit his family and asked me if I would buy him a ticket to Luang Prabang. Now I was looking at an additional cost of nearly $400. "Is there another way to get to your village, Phet?" I asked.

"Yes, Edward, I can ride the bus, but the ride may take over 6 hours, and it is very rough."

I told Phet I planned on paying him several hundred dollars for his services, and if he wanted, he could use his own money to buy an airline ticket. Or he could ride the bus and keep the $400. Also, I had given him nearly $200 in kip and much of it was left. He allowed that he would catch the bus later that day so he would have more time with his family.

Phet seemed disappointed, but I knew he was about to make several months' wages for his four days with me. He also had a "new" computer. Later that afternoon, Phet left the hotel to catch his bus.

"I'll probably see you soon, Phet. I'll ask Vilan soon to have you help him plan our trip into the forest."

Phet acknowledged that he would assist in any way, and we parted ways. I felt a little bad about Phet facing the long bus ride to his village but felt like he'd had a good lesson in decision-making. He still had a wad of cash in his pocket to advance his life—or piss away. What would I have done when I was 24?

I woke the next morning feeling discouraged about leaving Xieng Khouang with so little accomplished. Maybe a follow-up trip in a few months would rectify that. Before Vilan was due to pick me up and take me to the airport, I checked out and had a nice chat with Sivone. I gave her every opportunity to repeat the message she had given Phet concerning Vilan, but she did not. As Vilan drove me to the airport, we discussed my proposed trip. He stated that this delay would help him make sure we had government approval for the adventure.

"Your English skills are very good, Vilan. How did you learn the language so well?

He hesitated before answering, "I learned English from my father. He was a guard at the POW camp the Pathet Lao maintained during the War, and he learned to speak English from the U.S. prisoners they held there."

I wasn't sure I believed his story because his English seemed too good to have been learned from a secondhand source. During the War, we had heard about suspected POW camps in far Northern Laos, but following the War, the evidence was sketchy and no POWs were ever recovered.

He then elaborated further, "There are still prisoners being held from the War."

"Are you sure? How many are being held?"

"I think there are still about 60 remaining."

I considered this story far-fetched but decided I would present it to DPMO when I returned to the States. However, this story made me reevaluate Vilan and how much of his talk was truthful. When he dropped me off at the airport, I told him I would establish contact as soon as I returned to the States and looked forward to our journey to the Khoun District.

As I waited for my flight, I watched the passengers walk toward the terminal from the incoming flight from Vientiane. Towering above most of the arrivals was Dustin Roses. No doubt he was here to assist the Team in their recovery efforts. I had only a short time to talk with him before he left the terminal and I prepared to board. I found out when he would return to Vientiane and told him I would contact him concerning my plans for a return visit to Laos. He said he would be glad to help and was on his way.

By coincidence, Sammy was on my return flight, and as we were waiting to board, I told him about my failed attempt to go to the Khoun District. He did not express any opinions, but, upon reflection, I'm sure he was glad I didn't make the journey since the Lao officials had observed him assisting me.

Upon my arrival at Vientiane, I took a taxi to the Mercure to see if they had had any rooms open up. They were fully booked. They recommended I take a taxi to the river a few blocks away where I would find several hotels.

I found a hotel that looked new and was right on the Mekong for only $24 per night. I went to dinner at an outdoor restaurant next door, but I felt like I was being watched by the Lao government. I was probably being paranoid, but I thought people at another table and also at the bar were observing me. As my comfort level diminished, I decided to try to make an early departure from Vientiane.

When I returned to the hotel, I called Mary (very early in the morning) and asked her to talk with my airline carrier and see if she could get me an earlier flight out of Bangkok. She said she would try and would email me the results. "Don't pay too big a change fee, but I can leave Bangkok tomorrow evening. Otherwise, I will fly to Bangkok and wait there." I did not mention my unease. The hotel had a small business center, and early the next morning, I retrieved an email from Mary saying she had me booked on a flight back to the U.S. later that evening for no additional fee. World's greatest wife!

I caught an early flight to Bangkok that morning and figuratively "clicked my heels" as I felt the gear come up—thunk, thunk—on my departing flight. I had a long layover in Bangkok, so I paid $60 to spend that time in a comfortable lounge. While waiting, I began working on my trip report as I had done on last year's trip to Xan Noy. It was hopeless—there was nothing to write about. I finally concluded I should chalk it up to a lesson learned on how to not get ahead of myself. I could only hope I had not adversely affected the JFA Team's efforts with respect to the Laotians.

I did, however, after returning to Kansas, send an email report to the Dinans on March 17. I described the trip and ended with the following summary:

A rather disappointing outcome to the trip. However, got some great information from the JPAC folks and plan on traveling to Hawaii to get even more. I was also able to locate a Laotian who I think can make my next trip a success. He is a professional guide who has made many trips to the Khoun

District and, best of all, his uncle is in the Laotian Foreign Ministry and can help me get the permits needed. Spent a good deal of time planning a trip late this year or early in '14. I will give you more information on this guy once I have had an opportunity to perform a little due diligence on him myself. The cost of the adventure will not be cheap—right now we are at $7,000, including bribes. Let me know if the Sons of Nutley would like a presentation. I have some patriotic friends here in Wichita who might also be willing to help. My grandchildren are beginning to wonder about their inheritance.

At any rate, today was a good day. Always think of Dave on this day and remember this day 44 years ago when he was taken from us. Like I always do on St. Patrick's Day, I planted my potatoes—a sure sign that spring is on the way. Sorry I can't report a better result from this effort, but I have confidence we will do much better on my next attempt.

Cheers, Ed (wgfp)

Chapter 20

The Luck of the Irish

It was great to be back in Kansas and resuming my normal semiretired routine. The potatoes and peas were in the ground, and the garden was ready to go. All the seeds had been ordered. I also had baby pigs due in the next month, and I would wean my calves and haul them to the sale barn soon. The days were getting longer, and late afternoon would find me on my back deck, watching the sun set, drinking a Heineken, and celebrating all the fun and interesting things I had gotten done that day. It was at this time I almost gave up my quest to recover Dave's remains.

Xieng Khouang had been a qualified disaster, and the outlook for future success did not look good. At least, this was my view as I popped the top on my first beer each day. However, the second Heineken always seemed to stir up my emotions and, despite the facts on the ground, my desire to do more. What more could I do? Right now, the best opportunities were an adventure with Vilan and identifying the Cable Guy. So I decided to see if I could get a reasonable adventure set up with Vilan and to make a trip to JPAC in Hawaii and access the Jolly Green recovery logs.

Early after my return, I, by chance, found an email from Vilan in my Junk email folder. He had written it while I was on my way home. I replied and received a response a few days later:

Hi, Edward, good to hear from u so soon. Sorry for the late reply. Been busy on the minefields. If u have any questions or inquiry, just feel free to contact me via email or mobile. Happy hunting. Vilan Phetrasy

Minefields?

Around the same time, I got a response to my report I had sent the Dinans from Charlie:

Ed, thanks for the update. I was thinking about you yesterday and figured you would be coming back around now. I'm sorry but not totally surprised that some local turmoil prevented you and the team from going to 1408. In my limited experience, turmoil is the expectation in tribal areas, and Southeast Asia is certainly no exception. My buddy from Indonesia is back in the States permanently. He had many years of dealings with Southeast Asians. His take is that they are agreeable, friendly, but cunning and very unreliable. Stay in touch, Charlie

I also got a message from Bob Maves asking about my Laos experience. He asked if I had contacted anyone at Det. 3 before my arrival. Oops, looks like I didn't cover all the bases before my trip. I sent Bob a quick reply:

Bob, thanks for the email. The trip did not go well as far as new information, but I did learn a lot. I did not talk to anyone from Det. 3 prior to this trip, and I did not contact them last July on that trip other than a short visit with a young enlisted girl while I returned some equipment given to me by Al Teel. I did get to meet Chris O'Brien while I was in Phonsavan last week as well as Jack. Sammy gave me some time when I visited their hotel, and I would like to spend some more time with him in Hawaii. When does he get back? There are some issues I would like to discuss with you using phone. Is there a best time to call you tomorrow? I'm on CDT.

Best part of the trip was a good quantity of Beer Lao—one thing the Communists seem to have gotten right. Cheers, Ed (wgfp)

We set up a time to talk the next day. When I called, I asked, "So how bad did I fuck it up, Bob?"

"Don't worry about it. There were some folks with their shorts in a wad because they didn't know you were going to be around—and I'm catching a little hell for it. No big deal."

Bob was one of the last people I wanted to piss off and was glad he didn't seem bothered by the fuss. I went on to let him know about Vilan. He said he had never heard of him but urged me to be careful. I also told him I would like to come to Hawaii and go through the SAAR logs.

"Let us know when you're coming," he replied casually. Bob has to be the most nonbureaucratic bureaucrat ever, I thought. What a great asset!

On April Fool's Day, I got an email from Vilan asking about the possibility of an exploratory excursion in the vicinity of the site. He thought a small reconnaissance trip should be carried out by his team to provide us with information and wanted to know what I thought about that.

Vilan was getting a little ahead of me, so I replied with an explanation:

Vilan, all is well and am very busy with some other matters at the time. As I look at my schedule, it seems that November will not work for our journey and Jan 2014 works best for me. How does that fit with what is going on in Laos?

I am planning on a trip to Hawaii in the near future to meet with JPAC— the folks that are in charge of U.S. site visits. Before I make another trip to the Xan Noy area, I want to get the very latest information they have available.

I am not requesting that you make a trip to the area at this time but that might be helpful later—after I meet with JPAC. I want to get some information on a previous trip they made to the area in 1994 and also try and get the name of the man who found David dead on the ground and get his description on the site.

Let me know about the Jan date. Cheers, Ed Sykes (wgfp)

A week later, I got a response from Vilan:

Dear Edward,

Jan 2014 is perfect because at the moment the situation is at its peak, three weeks ago the old city had a small incident with the Hmong. We should make plans for the future trip. So I hear you are traveling to Hawaii to meet the JPAC. It's interesting that the dates you mentioned in 1994 happened to be one of my father's bounty hunters. I believe he was a Thai informant and he probably has some good info as he worked with my father.

Anyway, give me an update when you have anything interesting.

Best Regards, Vilan

I wasn't sure what to make of the bounty hunter comments and did not ask questions. However, the comment about the Hmong problems in Khoun District probably had a lot to do with why the 1408 site visit by the JFA Team was canceled.

***

The League meeting in 2014 was scheduled for June 13–15, so we were now only a couple months away. I asked Charlie and John if they could make it, and they said they could. Charlie had been working with the AF Casualty Office, and they set up a briefing for late afternoon on June 14. They again asked that I be allowed to attend as Cousin Ed, which was approved. I was anxious to find out what the briefers would tell the Dinans about the future status of their efforts.

Prior to the meeting, the Dinans had initiated some fundraising efforts to help fund my next visit to Laos, and both Dinan brothers sent me checks. John had written an article he submitted to a local newspaper in New Jersey to raise funds and had also raised funds through the Nutley VFW and Rotary Clubs. The total donations did not even come close to what I anticipated paying for my next trip, but I was grateful. I had started talking to Vilan by phone rather than using email because I suspected his email account was being monitored by the Lao. On each conversation with Vilan, the cost seemed to go up, and by the time of the League meeting, we were at just under $10,000 for a five-day trip to the Khoun District.

It was great to see Charlie and John again at the League meeting, where I gave them the full rundown of my misadventures in Laos. On the first day, we met with Bill Habeeb, and I told him about Vilan's claim of 60 POWs remaining in Laos.

"We have been hearing those rumors for years, but we still have zero evidence to back them up. We think it is a ploy to get us to invest time and money into a search for the site."

I thought back to my conversation with Vilan and how positive he sounded about this matter. It was now nearly 40 years since the War had concluded, and it was highly unlikely such a site existed. Once again, this shed doubts on Vilan's veracity. I told Bill about my plan to return to Laos in January, and he said he would help as much as he could.

Bill also gave me a heads-up that caught my attention. He had been funded to bring on three more investigators, who would be joining the effort later this summer. He was going to give up most of his cases to them, and one of them would get Dave's case.

"Let me know when the new guy shows up, Bill. I want to get an early input," I requested. Bill promised to do that.

As the time for the briefing neared, I was anxious to see if the briefers spent any time talking about my involvement (or interference) with the JFA at Xieng Khouang and how they presented that to the Dinan brothers. As we entered the briefing room, I got a real surprise. At the briefing table was Gen. McKeague and Johnie Webb, an elderly man who I knew was the top-ranking civilian at JPAC. We were also introduced to a woman who served as Johnie's right hand.

What the hell does this mean? Are they here to ask the Dinans to shut me down? Or are they here because of the amount of pot stirring I've done and they want to reassure the Dinans? Keep your mouth shut, Ed, and see where this goes.

The briefing began with Gen. McKeague updating the Dinans on the problems JPAC had encountered with the Lao on the last JFA and reassuring them they would continue to try to move the case forward.

Charlie mentioned that I was planning another trip to Laos in January, and both McKeague and Webb seemed interested. I told them I was working with Vilan Phetrasy, who was trying to get Lao approval for my adventure. Gen. McKeague asked Johnie if he had ever heard of Phetrasy, but Johnie shook his head. "Keep us updated on what your plans are, Ed," requested Gen. McKeague.

"Yes, sir."

I expected them to give the Dinans some sort of signal that I was getting on their nerves, but it did not come. It appeared I was not a target, and my fighter pilot instincts told me it was time to join the fight. As I sensed the briefing was almost over, I raised my voice and let out my feelings. "Gen. McKeague, let me give you my perspective on this situation. As fighter pilots in this war, we all did what our country asked us to do. But if someone had told me I could be shot down and pronounced dead on the ground, and my country knew exactly where my body was and then my country had not made an honest effort to recover my body 45 years later, I would have said 'Piss on it' and found another line of work."

The silence was deafening. Finally, Gen. McKeague told us he would do as much as he could, and we should contact him with our concerns.

After the briefing, there was not much left to say. It was late afternoon on Friday and the Dinans were on their way out of town. We promised to keep in touch, and I said I would probably see them again in November.

Why November? MIT was going to dedicate the hall in Building 10 as the "Hall of Honor." Mark Rockoff had informed us about the dedication, which they were going to do around Veteran's Day. All the MIT alums that had been killed in action were inscribed on the walls of the Hall, and Dave was the last MIT alum to be so inscribed. Charlie and John had both said they would attend. Their father had graduated from MIT in the 1920s, and they felt a close attachment to the university.

The next day was my 70th birthday, so I celebrated as I always do … at a baseball game. The Washington Nationals weren't in town, so I caught the Metro up to Baltimore to watch a game and drink some beer. Camden Yards is one of my favorite ballparks, and its proximity to the Harbor makes it even more delightful. *How the hell did I get to be 70 years old? It can't be true!*

While sitting at the game enjoying a delightful day, I got a call from my fighter pilot son, Bartz. He was calling to wish me a happy birthday and give me some crap about being such an old geezer. I told him about the meeting with Gen. McKeague and recited my little "piss on it" speech. I had barely finished when Bartz said, "That's bullshit, Dad!"

"I know, but it sure sounded good."

Bartz, like almost everyone who has flown fighters, knew there was almost nothing that would have deterred a fighter pilot from experiencing the rush that comes from strapping a jet fighter to their back and giving it life. I was proud of my young son, knowing we shared similar views of this unique and special experience.

Surviving the League meeting without any complaints from the DOD folks gave me renewed energy to get on with my next trip to Laos. I began to call Vilan on a weekly basis, and he let me know he was still having some hiccups with getting approval for the visit. In mid-July, he told me about a form he had to submit along with $1,200 to get official permission to visit the Khoun District. This was the first I'd heard of this.

I sent an email to Johnie Webb and copied Gen. McKeague and Dustin Roses asking if they were aware of such a procedure:

Johnie, sorry I am late getting back to you concerning Vilan Phetrasy. As I mentioned to you and Gen. McKeague during our visit on June 14, he has given me a proposal to visit and investigate the site of the death of 1Lt. David Dinan III (1408). I had asked you if you knew of Mr. Phetrasy and had any inputs concerning his abilities/history. There are a number of hits for him on Google, but most of them are of little use for information gathering. Anything to add?

As part of his proposal he estimated the cost of obtaining a permit from the Laotian Government at $1200. I assume part of this cost is to cover his costs and some is for payoffs? Do you know of any cost associated with obtaining such a permit?

I am beginning to put together another trip to Laos in Jan 2014 and would welcome any recommendations you and your staff might have concerning my plans. Has JPAC built its schedule for next year and will it include 1408? Cheers, Ed Sykes (wgfp)

The first response I got was from Dustin:

Ed, after a bit of internal discussion within the embassy here, nobody can think of a specific permit that allows one to conduct MIA investigations. We certainly don't operate under any such permit. He could be referring to the process of paying off officials, etc., to get a letter of permission with the right stamp. If you intend to get a permit, perhaps you could ask him to provide you a blank copy, or copy of a previous permit, or something to that effect. We'd be happy to take what he gives you around to the Ministry of Foreign Affairs or Ministry of National Defense to confirm with them whether this document would actually allow you the access you are attempting to obtain.

We have no history on Vilan Phetrasy, but there are some people with the last name Phetrasy who in the past have tried to make money off the MIA issue. There was also a wartime Lao official named Phetrasy, who may be the official that Vilan is claiming relation to. Based on the history, it does sound like these guys are opportunists and not likely to deliver you any success.

I'm always happy to meet Vilan here at the embassy, so send him by sometime. Thanks, Dustin

What a great idea! I got hold of Vilan and suggested he stop by the U.S. Embassy and explain the permission process as well as his plan for the excursion into the Khoun District. He balked at the idea until I told him

I would not commit to any venture unless he talked with Dustin. He said he would do it, but my suspicion was that I had just heard the last of Vilan Phetrasy.

I sent the info along to Dustin:

Dustin, thanks for your insights. After reading the information, I called Vilan and asked about the permit process. He explained a process that was a little hard to follow. I asked him to meet with you and although he was a little hesitant, he said he would do so. He will be in Vientiane on July 21–22 and said he could do it then. I told him I would send him your contact information—what is the best way to make contact with him? I will email that information to him and will CC you so you have his email address. He knows your name and what your job is so I suppose you could give him a call any time. His English is quite good. Any suggestions? Ed (wgfp)

In late July 2013, I decided I would visit some of my POW/MIA friends in D.C. I sent emails to both Bill Habeeb and Ann Mills-Griffiths to say I would be in town and would like to drop by and see them.

Bill replied with information about when the new analysts he had told me about would start. Both would need 3 to 6 months of training, but he would assign Dave's case to one of them when they arrived so they could learn on the case. He offered to meet me on July 23, with the new analyst joining us if one had been assigned Dave's case.

The trip to D.C. went well. Bill Habeeb ended up being unable to attend, so I met with Guy Benz. I had hoped to meet one or more of the new analysts, but they were not yet on station. My meeting with Ann went well. In both meetings, I discussed my proposed trip to Laos in January but told them I might have lost the services of Vilan because of my insistence that he meet with Dustin. Without Vilan, I might not make the trip to Laos.

The following day I returned to Kansas and was greeted by an email from Dustin:

Ann and Ed,

Vilan came by my office yesterday and we talked for two hours. The conversation was surprisingly cordial, while still a bit of a poker game. I'll try to give you, as briefly as possible, my impressions, and then my bottom line.

My personal impression of Vilan is that he is like so many Lao men, trying to play big shot while possessing no real skills, compensating by running every manner of social engineering approach on people. He admits he's a "fixer" and definitely has the swagger to accompany. He lives mentally and metaphorically in that old up-country, Wild West Laos that is fortunately dying quickly. He frames things in terms of secrecy, cover-ups, plots, and schemes. He's quick to drop references to old bygone kings, CIA operatives, ravens, Hmong rebels, secret prisons, Hmong refugees in Burma being smuggled to Geneva, adventurers coming to see Long Cheng...the stories of the revolution and the postwar consolidation of power. His Laos is one in which people succeed or fail by the merits of their family background and how they leverage personal and family relationships, earned by revolutionary struggle and political alliances. His Laos does not want to cooperate on POW-MIA but appears only to cooperate for diplomatic reasons. His Laos is still fighting a Hmong insurgency in which every day, LPA [Lao People's Army] troops make contact with a strong active rebel force. He operates out of this hotel, which plays this Secret War (anti-U.S.) movie every night, and then he spends hours a day on the web talking to old veterans who trade stories of secret war events, Lima sites, old U.S. and Hmong heroes. He makes strong arguments for there being live Americans and was especially persistent that this distress signal up in Huaphan in 2003 was the real thing. While he can tell great stories to people who want to believe that things are a certain way, and make an interesting contact for tourists passing through, he can't back up virtually anything that he says with facts. He claims he learned English from his Father, who learned it from his father, who learned it directly from U.S. soldiers being held captive in the Viengxay caves. Yet he had no additional information. And he just keeps dropping story after story like that, that dead ends after the one sentence, with no additional facts. He's kind of this strange mixed-up Xieng Khouang idiot savant when it comes to Lao history and internal political dealings. I think he exists in that "dark side" of our issue, doing more harm than good, and perhaps stirring the pot a little bit. He's talked about being a "tour guide" on past adventures to go after live Americans, meet remnants of Vang Pao's forces, find out the truth of what happened in the King's prison cell, and visit Top Secret war sites. I believe he's been involved in the lot of it too, but perhaps never achieved much results. I don't understand how he's survived in Laos so long while stirring up all this type of stuff, to include his seemingly royal Lao leanings, while maintaining a strong Communist connection. I conclude he's just operating under the laws of supply and demand. He knows who has money (foreign adventurers and tourists), and he knows how to give them what they want, and he knows how to placate local authorities and do business on the shady Lao-Chinese-Vietnamese/communist side of the economy too. What he does not appear to understand is legitimate, transparent, legal commerce and upfront, honest diplomacy, as we tend to practice in the West. He rattled off so many different business ventures he's trying to get involved with, from mining in Xieng Khouang to alternative clean energy, tourism, etc. And he claims he's an LPA officer to boot. So this guy is just the full piece of work. Strangely, I've known dozens of guys just like him. They all claim they can do the same things: set up whatever business you want, sell you what you want, or whatever, and can give you a billion reasons why you just have to trust them.

Do I think he can actually get the paperwork from the people he says he can for the price he's quoting, get the logistics in place to complete the investigation and survey he's designed? Absolutely YES.

Do I think he's experienced in anthropological MIA recovery methods? NO. But maybe that doesn't matter. He claims he found a witness with a parachute. That's a good start. I don't think going after the crash wreckage is the right approach on this case, as it is quite clear Dinan bailed out. If there is some way to go after that, this is the direction you want to go. It is going to be witness dependent, and JPAC seems to have exhausted all potential leads. I'd love to try to find more though.

I raised my #1 issue with him: So let's say he can get this "permission" from the district, province, and village. So you go out there and start doing whatever (disturbing the site), what if someone who wasn't on the permission slip decides he's going to shut it down? What if it gets back to the MIA team in Vientiane and they decide that national security and foreign policy trump what a bunch of local officials say on the matter, and they come up and shut it down? To that he says, he can only arrange what he can arrange, and there is no guarantee beyond the limits of his powers. He is confident that the local authorities are all that matters and that a response from Vientiane is highly unlikely. Vientiane would not want to start a battle with provincial officials and undermine their authority, so as to risk a political backlash from a powerful province. Furthermore, they would not ruin their image of cooperation with the U.S. by targeting an old veteran who is just out on a site in the jungle not hurting anybody.

His litmus test for whether anyone is going to get in trouble is whether you are "hurting anyone" or "doing anything wrong." So every arrangement will need to be designed to meet this test. If it is set up as an old veteran wanting to get local permission and cooperation from the local villagers and military to go hike up to the site, camp out, do some looking around, as part of his personal connection to his lost brother in arms, he thinks no Lao official would have a problem with that. If you started digging deep, trading bones, funding Hmong bandits, etc., that would be a different story. I don't know where threatening the central government's monopoly on MIA investigations falls on the "doing nothing wrong" spectrum, but personally, I would expect to get shut down IF the central government found out, but I would not expect to be jailed, probably just deported. I personally do not think trouble like this is likely, but I'd prepare for the worst.

The greatest threat in my mind is failure to achieve anything of value. There are so many points of failure, and I discussed each of these with Vilan. His response continually was, "What does Ed want to achieve?" He wants to deliver that to him but needs to be clear of what he promises. I think I got Vilan straightened out to the point where he will be very specific as to what he can and cannot "guarantee." And this is a conversation you need to be having with him continually. I told him to be as up front as possible. He said basically he will offer the access, resources, and "expertise." But "success," however defined, is going to have to be a matter of Ed's own initiative, luck, research, etc. I did tell Vilan that I had the impression the idea was to get out there and just do something that would be proactive, and maybe they would find some evidence, or a witness, that then could be put back into the JPAC process and attract the "official" resources. I also said one outcome may be that everyone involved comes back with an appreciation of the difficulty JPAC faces trying to resolve these cases out in this terrain (of course, you both understand that well already). I think Vilan too understands the difficulty, and he did not intend to make a claim that it would not be terribly challenging, just that he would be willing to facilitate whatever you wanted.

In the end, I got the impression that Vilan's character was NOT worth attacking. He's definitely an odd, back-country Lao guy who speaks impeccable English (better than his Lao). He obviously comes from some kind of connected family. (The uncle is currently in Brunei working for the MFA and his grandfather was a famous Pathet Lao spokesman, and his father was raised in the caves in Viengxay as a child. He's just name dropping when he mentions his uncle, and this is not the person he plans to go to for help with this.) I'm used to nearly every older Lao guy I meet to have some odd ideas about the way the world works, and I can overlook their lack of understanding considering the circumstances. (He's actually fairly young; when I say older, I mean older than me.)

I do not think he's trying to lead on or deceive anyone (especially since we had our little talk). I don't think he's out to rip anyone off or commit some major fraud. I just think that the way he sees things is simplistic, a bit sensational, and he is overconfident. However, these are all normal traits for people in his situation. Does he want to get you excited about all this stuff, and the adventure, and get you to trust him to handle all the arrangements, etc.? Yes, of course. He's doing business. And when it comes down to it, this is a business to him. I think most the Lao officials watching this will take the money, sign the forms, and sit back not wanting to interrupt a guy doing *business. The Lao have a lot of respect for people who can get money from foreigners using whatever means possible and will not usually blow the whistle or raise some policy issue. It's considered bad form to get in the way of a man's income stream as long as he's spreading it out right and "not hurting anyone."*

So, my bottom line is: I still do not think it is a good idea to proceed with this plan because I think the potential gain does not match the risk and the financial outlay, and as an official, I have to respect the "appropriate" process until we're given permission by the proper authority (the Lao MIA team). Furthermore, if you proceed, there is nothing to protect you from whatever may happen, however unlikely. I understand your (Ed's) desire to do something for your fallen comrade. I empathize with the pain the Dinan family feels. Trying to be respectful of that, I would understand if you tried to proceed and I will not deliberately try to get in the way (however, I believe it would be smart to be up front about it with people who will likely shut you down—the U.S. Ambassador and the Lao MIA team being the examples that come to mind, but I don't intend to "rat you out" unless I feel you or the POW-MIA program are in danger). I am not as concerned about the site as I know JPAC is, but "site disturbance" is something you might want to be considerate of. I think Vilan can provide you what you need, understanding the limitations stated above. I think he now knows better than to lead you on under false pretenses or extort you for large sums of money. He is going to require money, but what he's asking is probably pretty close to what things actually cost—and the things you will need tend to be expensive. As with any salesperson with this approach, you have to be careful, verify everything in writing, hold him accountable, and be willing to walk away.

As far as my involvement is concerned, we still have time to try and figure out that piece. If I raise it through the Lao MIA team, then what they say as per my request will also determine whether your entire plan goes too. I'm always trying to find ways to get into the field, but I can't get caught sneaking around (any more than I already do). Even if you have permission all set up locally, my involvement will automatically make it a matter of central government policy.

So, with all that said, I hope I didn't just blow the lid open on something more serious or dig myself into a pit. Please understand that I am just trying to help, and this is above and beyond my official duties. If what I say when I am just trying to help you outside my normal official duties is used against me, it will be very hard to help in the future. Thanks, Dustin

Wow! The amount of thought and time put into this email was unbelievable. I was so impressed! After reading it several times, I found something new to think about each time.

The next day, I got a follow-on email from Ann Mills-Griffiths addressed to both me and Dustin:

Dustin, this excellent assessment is logical, objective, analytically sound (to the obvious limit of my ability), and forms precisely the outlines of what you, Ed, could and should anticipate. Under no circumstances should what you wish to do privately have an adverse impact on U.S.-Lao national level cooperation, nor have any adverse impact on Dustin's ability to participate, ever more fully, in the official investigation process. We've all worked too long and hard to get him where he is, and I have certainly come into contact with many over the years who fit the description Dustin provided regarding Vilan.

Thanks, Dustin, for your willingness to help Ed. Thanks, Ed, for your willingness to assist the Dinan family by personally taking on the financial and personal risk. It is most generous in both categories. If you decide to proceed, and you wish to intervene with the Lao Government, I'm willing to send such a letter to the appropriate officials, but I can't and won't make any recommendation for a course of action that only you can decide. Best to all, Ann

As I reread and reflected on these two messages, I was above all elated to have such great allies in my corner. But it was also true that they would drop me in a heartbeat if I did something stupid that endangered their efforts. On reflection, it is probably best that the International Day of the Woman caused me to abandon my recent attempt in Xieng Khouang. That effort might have resulted in such a screwup.

Thus, the decision to pursue the adventure with Vilan was mine alone. In his letter, Dustin had hit on my main objective in going—to call further attention to Dave's case in an attempt to get DOD to carry out substantial activity. I now understood that only DPMO could seal the deal. I was only a cheerleader.

So, what to do? Dustin's detailed description of Vilan paralleled what I had sensed. He had some highly questionable tactics, but Dustin and I both thought he could really pull it off. But the thing that made me decide to continue with the effort was the fact that Vilan had the cojones to visit with Dustin at the U.S. Embassy, which would have been out of his comfort zone. Therefore, the trip was still on for January, and I would make a stop in Hawaii on my way to do some more research on the elusive Cable Guy.

After many attempts, I was able to contact Vilan by phone. "How did your meeting with Dustin go, Vilan?"

"It went very well. We had a very long discussion, and we discussed a lot of issues. He is very professional and has some very good ideas."

Having passed the test by visiting Dustin, Vilan had gained personal reassurance and continued by going over the latest version of his plan with me. The only new angle was an additional requirement for some explosive ordnance disposal (EOD) equipment and personnel to ensure we would be safe from any leftover unexpended surprises from the War. He wasn't sure how much it would cost, but we would need a metal detector and one or two EOD specialists. He could take care of the details.

When I asked him about visitation clearances, he became a little less sure of himself and said he was working on it. He didn't mention the form from earlier or the $1,200 needed to purchase it. Thanks, Dustin. When I asked about the total price of the venture, he said it would be somewhere around $15,000. A lot would depend on the cost of clearances. I said I was not willing to go much higher than that, and he assured me he would do everything he could to contain costs.

Oh, what the hell. By now I had 10 grandchildren, so this worked out to $1,500 in inheritance each. They'll never miss it. A week later, my 11th grandchild showed up, so the math was getting even better.

I called Phet and asked him to contact the village chief at Xan Noy and Mr. Yui and let them know we planned on being in Xan Noy during the last two weeks in January. Phet then returned to our most perplexing problem: "What will we do about permission, Edward?"

Now mid-October, the clock was ticking. I called Dustin and was told he had left the office and would not be back for a week. However, I did get an email from him a few days later:

Ed, you caught me on the way out the door to deliver a colleague's dog to him at his new post in Uzbekistan. I'll be back next week. I did, however, want to make sure I followed up with you on the meeting I had said I was going to use to ask about your case a little bit.

I asked the officials what their policy is for old U.S. veterans to go and see a crash site they are personally connected to. He said the policy is any U.S. veteran would have to send a letter requesting his visit, with all specific details, to the MFA. Then they would set parameters and approve/disapprove. He said they do in fact approve these kinds of things and provide documents to take to the District and province. However, they also control the whole thing and make you pay for a lot of stuff, like officials, etc. He said if you just went in and coordinated locally, there is nothing they could do, but if it got back to them, then they would go out there and check out what was going on (not specific). Hope that helps. Dustin

I thought up all sorts of combinations for what "MFA" stood for, and the one I liked best was not flattering. But I finally landed on Ministry of Foreign Affairs as most likely. The guidance from the Lao was also confusing. You could go the low-dollar route (Option A) and not tell anybody and hope you didn't get caught. What was the punishment for getting caught? Or, you could go the high-dollar route (Option B) and hope your permits were approved. My initial preference was to go with Option B and see if I could get Dustin to accompany me.

The last two weeks in January became my target date, and I sent an email to Phet with these dates, asking him if he had had any luck with Mr. Yui or the village chief. I told Vilan to plan on these dates, dates he had previously suggested, and that we would go with Option A and forget about Lao government permission. I also let Bill Habeeb know that I was committed to the January mission.

Finally, I sent an email to Gen. McKeague asking about the status of the upcoming JFAs:

Gen. McKeague, I was talking with Ann Mills-Griffiths a few days ago and she mentioned the first team exercise in Laos for FY14 had been postponed because of the shutdown. I'm not sure how this will affect the rest of the year's schedule, but I would like to ask again that 1Lt. Dinan's case be part of any investigation you might conduct in central Laos.

I am still planning on another trip to Laos in early 2014. I will keep your folks posted on my plans and see if they have any new information on this case before I depart. I will also fill your folks in on any new information I might obtain. Sincerely, Ed Sykes (wgfp)

The reply from Gen. McKeague was not all that optimistic. In 2013, Congress and the president had decided to introduce a "sequester" process in the fiscal year (FY) 2014 and beyond, for budgets that required substantial drawdowns in both military and domestic spending. The Defense drawdowns were now being planned for FY 2014, which had begun in October 2013. In his email, Gen. McKeague stated that many of the early FY 14 JFAs had already been canceled, and the fate of the remainder for the year was now being decided. So, he was unsure if a JFA to northern Laos would even occur. He said his folks would keep me posted. He did not mention my planned trip to Laos. Again, this was a consistent theme in dealing with anyone involved with the POW/MIA recovery effort. They did not encourage me, but, up until now, no one had told me to knock it off. Hell, this coming year, I might be the only show in town!

Early in November, I got a phone call from Bill Habeeb. We discussed the current outlook for activity in Laos in the coming year, but he was also in the dark. He did, however, mention that the three additional analysts were now in place at DPMO in D.C. and would begin working cases at once. He would assign one of them #1408 and have him get in touch with me as soon as he had time to brush up on the case.

"Bill, I'm coming to D.C. to meet him and help him come up to speed."

Bill agreed that would be a good idea, so I told him I was going to be in Boston in another week and would fly down to D.C. to meet with the analyst. We scheduled a meeting for November 19. I was going to Boston to join the Dinan brothers for the dedication of the Memorial Hall in Building 10 at MIT, which was on the 18th. I could only hope the new analyst assigned to Dave's case was competent and trainable and would be willing to work with a crusty old fighter pilot.

A nice break in the scramble to make things work before my Boston/D.C. trip came in the form of a letter from Valerie:

Dear Ed and Mary,

Just a note to say hello. I hope you are all okay. It has been 3 months since I had my right hip replaced and all is well. I still go to rehab as I had to relearn how to walk correctly. It's very hard to unlearn bad habits. I use my cane only for stairs and have minimal pain. I never believed that I would be this far along. The second hip has been a good one as they worked me twice a day, pain and all!

I spent every sunny day at our condo pool and did my exercises in the H2O. I've met so many kind and nice people including a widower with whom I spend time. He's a mess emotionally, but we laugh a lot together and can enjoy each other's company.

I've lost a total of 65 pounds and had to buy all new clothes. I had to try on at least 50 jeans before I found 6 pairs that fit long legs and curves! My doctors are thrilled about the weight loss, which is a must for hip replacements. I've kept it off now for a year. My breast cancer has had a good outcome, and I go for my yearly checkup in a few weeks. Fall has arrived in CT and the mornings are cool.

Keep in touch. Love, Val

I shared the letter with Charlie and John, and we all were pleased with this upbeat report. It gave me hope that, perhaps, we could return Dave's remains to the U.S. while Valerie could still share in the experience.

On November 7, I received a promising email from Phet. He had heard from the Chief, who would be happy to see us again. Since the rice harvesting would be done by the end of December, he'd be available around January 22 to 26. The Chief said Mr. Yui was currently working in Phonsavan, so Phet had asked him to get Mr. Yui's current phone number and, if he spoke with him, to ask if he'd meet with us on one of those dates. The Chief had agreed, and Phet hoped to have Mr. Yui's new phone number soon to call him about joining us in January.

I love it when a plan comes together! I sent an email to Dustin with an update:

Dustin, just tried to call you at the Embassy but was told you would not be back until Tuesday. As I reread this email, I am considering asking permission as you outlined below. I would certainly be most interested in this attempt if I knew you might be able to participate. I have been working toward the last two weeks in Jan '14. Have been communicating with the Village Chief at Xan Noy and he seems pretty excited about the visit. Have also located the witness I am most interested in, Mr. Yui, and am trying to get assurances that he will be in Xan Noy during this period and be able to lead me to the site where he believes Dave landed.

Having exchanged emails with Gen. McKeague, it seems there are a lot of unknowns concerning the FY 14 schedule and was hoping this may give you some more flexibility in breaking loose during the late Jan. time frame. I am going to meet with the Dinan brothers next week, and I know they will be willing to help with letters to the Lao Government. I am also going to meet with Ann M-G next week in D.C. and see how much help she can give.

Vilan is busy putting together a business plan for the adventure and thinks the plan can be molded to fit either the high-visibility option or the low option. I expect to hear from him within a few days. If I don't hear from you, I will give you a call midmorning on Tuesday. Hope you are enjoying your time off. Cheers, Ed Sykes (wgfp)

The next day, I received an email from Dustin:

Ed, sorry I've been out. Here are my thoughts on this:

The Lao made me submit a 6-month plan covering all my work from Oct–March. Then they hacked it down to basically nothing. So based on what I am up against actually getting out and doing field work, I do not expect that I would be able to participate in anything like this. I know it's frustrating when I've already told you I might be able to, but just understand that getting kicked around like this is what you can expect in Laos. That is not to say that if the situation changes, we couldn't make some request.

Currently the mission in January is expected to go, and I'm expecting to be on it. I'll talk to Al Teel today and get a better idea from him what the investigation schedule looks like for the rest of the FY. I'll ask him if there's any way he can tailor the investigation of 1408 in such a way that you would be able to participate and get your hands dirty. If not, then you are stuck in the situation where your only options are to not come at all, or just go do it on your own. I have made it clear that I think you should try to work things officially so as to minimize your risk. However, with Vilan "backing" you up (or rather, taking the fall for you if things go bad), and the Lao team down south, January might be the time to get out there. Especially if we then head up to XK for the investigation, and you've already done the legwork, it might be a big help. I am still quite interested in seeing who the witnesses are and what they have to say. Dustin

The next day I got another email from Dustin. He had asked Al Teel about my inclusion in a JFA investigation of 1408, and it was the same answer I had gotten many times before. Although they wouldn't include me, they'd expect me to pass on any information or leads regarding location or witnesses for the Team to check out. The good news was if the March JFA to Laos did take place, 1408 would be part of the agenda.

Dustin also discussed Option A versus Option B, saying if I went with A, I'd be aware of the risks of going in with no official backing. If I chose Option B, I risked receiving a "No" from the Lao when I made a formal request, and if that happened and I went forward with Option A, it would be impossible to plead innocence.

So, I had a decision to make. Option A or B? Why not put it off until I visit D.C. and talk to Ann and Bill and the new analyst—it's only another week. I flew to Boston a few days before the ceremony at MIT and spent some quality time with my new grandson and his parents along with Mark Rockoff and his wife Beth. Charlie and John showed up the day before the ceremony, and we discussed my upcoming trip to Laos. They didn't seem to have any preferences concerning Option A or B, but we agreed it was best to forestall the decision as long as we could while waiting for DPMO to make a decision on whether they would do the JFA to North Laos in March and if Dave's case would be included. Hopefully, I would know more after visiting D.C.

On the morning of November 19, I arrived at DPMO at the appointed hour and asked for Bill Habeeb. He soon emerged from the secure area, and we got my security badge and me signed in.

"What's the latest on the JFA to northern Laos?" I asked.

"Nothing yet. Funding is still an issue."

"If it does go, what are the chances of 1408 happening?"

"Not sure, but there should be a planning session next month, and there's a lot of interest in the case. I will push to have it included," he replied.

Guess that's as good as I could expect. After entering the secure area, Bill took me to a conference room where I met a slender, dark-haired man in his mid-40s seated at a table with several documents and maps.

"Ed, this is Niall Brannigan, and I have assigned Dave's case to him. I told him you were coming, and he's been doing a little prep work on the case. I'll let you two sort it out," Bill said and left the room. *I think he's glad to get rid of my ass,* I thought.

Niall and I spent a little time going over our backgrounds. He had been born in Dublin but had spent much of his youth in northern Kentucky. He had recently been working at the Pentagon as an analyst, and this was something to do while he looked for a new assignment. He didn't say why he had left his old job but did say he was glad to be out of the "Puzzle Palace" and had no desire to go back. "I get it," I exclaimed. He then looked down at the paper assembled on the table and motioned me to sit down across from him.

He confessed, "We haven't done shit on this case."

I immediately realized why Niall might be looking for another job. An honest bureaucrat—can't have that. However, I was beginning to like Niall a whole lot.

He then began to go through the case, pointing out that much had been done in the Xan Noy area but little of it had anything to do with Dave. He then pulled out a map that had ten sites designated with markers and annotations on sticky notes, scattered across three 1:50,000 scale map sheets, and proceeded to associate each site with a report from the information accumulated from the 21 aircraft involved in the SAAR operations on March 17, 1969. As he went through this information, I could tell this guy was really good. He then narrowed it down to the three most likely sites, and two of them appeared to be in the general direction of where the village chief had taken me the year before. I told him I might be going back in January and asked if I could have his maps. "Of course."

I then told him that I most wanted to find the PJ that went down the cable and found Dave's body. I explained my trip to Maxwell, my discussions with the Jolly Green guys, and how I had run into a complete dead end. Some of the information I had discovered at Maxwell had been put in Dave's file by Bill after I'd sent it to him. However, I told Niall that the information I had gotten from the Jolly Green guys had not been helpful. I also told him of my plans to stop in Hawaii on the way to Laos and search the JPAC SAAR records for additional information there. I sensed that Niall recognized at once the importance of finding the Cable Guy. "Let me see what I can do. I'll bet we can find him if he's still alive."

At last, we finally had a DOD guy interested in digging into this situation. As I left the DPMO office, I had a surge of excitement for what this new guy might bring to the effort. At a lunch meeting with Ann Mills-Griffiths next, I told her about the new analyst and my impressions. She expressed an interest in meeting and working with him. She also listened intently as I described my efforts in putting together a January visit to Laos. She volunteered to assist with letters if I needed them.

However, I soon learned that Ann's current focus was on the budget problems at DOD as well as internal inefficiencies within the POW/MIA effort. Ann's efforts were well above my pay grade, but she was kicking over cans at the highest levels and engaged in a nasty fight. It was exciting to listen to her express her opinions in such passionate and candid terms. This woman was a real fighter pilot. She urged me to keep her up to speed with what I was trying and let me know if she could help. "Keep your powder dry, Ann."

Although I was no closer to making a decision about how to conduct my next visit to Laos, I did have a good feeling about my meeting with Niall. Having flown all those combat missions in 1969, like most fighter pilots, I had become a pretty strong fatalist. It didn't matter how good or how well prepared you were, if it was your day to encounter the "Golden BB," it was tough luck and you might just buy the farm. The attitude I developed during that part of my life, and kept beyond, was, "you never outgrow your need for luck." The coincidence of Dave being killed on St. Patrick's Day and Niall being a native Irishman was a great omen. I may have stumbled onto a shamrock!

Back in Kansas, I realized that I did not have Niall's contact information, so I sent a quick email to Bill Habeeb and he got it to me the next day. At once, I sent my first email to Niall:

Niall, got your contact info and you now have no place to hide. Please keep me up to date with any updates on #1408. Also, if you get a chance to check the schedule for the team to Xieng Khouang in March, I would appreciate it. Cheers, Ed (wgfp)

Now it was decision time for my trip to Laos: Option A or Option B? I reread the emails from Dustin and others and finally decided to go with

Option A, the low-visibility (I hoped) solution. As I studied Dustin's input, the idea of pleading insanity and blaming Vilan sounded best. The trick would be to get permission from the provincial and district officials without the Lao government figuring out what was going on. I let Vilan and Phet know.

When I talked with Vilan, he seemed a little uncomfortable with this option but said he thought we could pull it off. He also told me he had talked to some provincial officials who suggested we not travel to the Khoun District until February. They had told him there was still some unrest in the area but felt it would be resolved by then. He suggested February 11 to 22, which I agreed to. Phet was full of excitement, even though he had still been unable to contact Mr. Yui. However, he was sure he could get that done soon. He wasn't happy about the change to February but said he would be able to participate.

Great, now that I have a plan, it's time to execute it! If all went well, we would find some useful information and be able to share it for the JFA—that is, if the JFA for March in northern Laos was ever approved and if Dave's case was on the agenda.

Then on December 9, I received a shocking email from Niall:

I have tracked down the PJ who found Dave's body, and he shared an incredibly emotive recollection of that day with me. Allow me a few more days to pose more questions to him and to allow him to ponder over the maps. Leaving Dave's body behind has troubled him all these years. He wants to help in any way he can and is very willing to engage Dave's family. Cheers, Niall

Holy crap, how the hell did he do that so fast? I've been at it for over two years and never found a good clue! Maybe the Irish really are lucky.

Chapter 21

The Stream Where the American Fell

As I reread Niall's email, I was in a state of disbelief. The part about waiting a few days before I got involved with the Cable Guy was hard to digest. I want to talk to this guy! I sent Niall a short email:

> Niall, that is fantastic news! As you can imagine, I am very anxious to talk to him as soon as possible. I am getting close to finalizing my plans for my next trip to Laos and his input could be very helpful! Cheers, Ed (wgfp)

I then forwarded Niall's email to the Dinans along with an update:

> Charlie and John, this is great news! Niall is the new analyst assigned to Dave's case. I met him the day after I met you at MIT. This may lead to some great leads. I was really impressed with Niall and hope you get to meet him at the next Families Conference.
>
> My meeting with Ann Mills-Griffiths went well and she concurs with me that my best option is to go to Laos without "coordinating" with the U.S. or Laotian Governments. I will be doing backdoor coordination through Vilan with the Provincial and District Governments. Again, I would ask you not to discuss my plans with any U.S. officials, and I will not be asking you for letters to send to the Laotians.
>
> I just talked to Vilan a few nights ago, and he told me the Provincial Officials have asked me to put off my trip until Feb. due to some unrest in Khoun District. Right now, I am looking at 11–22 Feb.
>
> I got a very nice Christmas card from Valerie—she seems to be in great form and in much better spirits than when I saw her in Salt Lake. Cheers, Ed (wgfp)

I hadn't called Niall when I first got his email so I planned to abide by his wish to allow him some time to talk to the Cable Guy. However, as I savored a glass of Scotch, I couldn't contain myself and I gave him a call.

He sounded happy to hear from me and even a little giddy. He, too, was surprised at how fast he broke the code.

"How the hell did you do that?" I asked.

Niall stated that he had run into the same brick wall I'd encountered, but he used research tools I never would have thought of. It was unknown which Jolly Green unit had sought to recover Dave but merely that a rescue bird with the mystery moniker of JG-16 had lowered an unidentified PJ down to the crash site while its accompanying bird, JG-09, provided cover. The chopper crew had to be assigned to the 40th Aerospace Rescue & Recovery Squadron (ARRS), but that unit had some 10 to 12 detachments scattered all over Indochina, and no useful operational records nor nominal rolls were known to have survived for any of them. He went to the National Archives for records of men of the 40th ARRS who had been awarded campaign medals during the period that Dave was shot down. This search gave him over 200 PJ names that fit the broad criteria. He then got a mailing list of PJs from the Jolly Green Association plus scoured various internet resources and then began a campaign of emailing, letter writing, and cold-calling people on the list to determine who was aboard JG-16 or even JG-09 that day.

This was a bit of a crap shoot since not all the PJs were in contact with the Jolly Green Association. Niall first established the identity of the pilot of JG-16, but he had unfortunately died earlier that year. He then located the copilot of JG-16 and traveled to the man's home in nearby Maryland to interview him. His memory of the incident was sketchy, and he could not remember the name of the PJ on that flight. Then on December 5, Niall got an email from the Cable Guy, letting him know he was his man.

As I listened to Niall outline his efforts, I realized he had done more in less than two weeks than had previously been done in years. He must have spent almost all his working time on this search. As his first "walk-in" customer, I must have made a significant impact. Niall told me he would get me a copy of the PJ's initial notification and contact information the next day once he cleared it with the PJ.

"Roger that, Niall. What kind of Scotch do you drink?"

He gave me the name of some rotgut Irish whiskey.

"Think I'll have to start drinking that. But right now, I think I'll have another Scotch!"

Late the next afternoon, I got an email from Niall that contained the Cable Guy's story in his own words:

Niall, I was the PJ who went to the ground on 17 Mar 1969. As I recall, Col. Morse was the Aircraft commander that day. The Colonel died recently. For whatever the reason back then, I was never really told where we were exactly on that day nor any other of the missions I participated in.

I believe that the pilot was shot down in or near the PDJ. His parachute appeared to have failed as he went down into some trees on a hillside. I was lowered to the hillside below his location. As I climbed up the steep slope towards his parachute, I started to notice drops of blood here and there as if he had rolled rather violently down the slope, flinging his blood out ahead of him. I was nearly at his chute before I noticed him. He was "wrapped up" in his parachute. His helmet, which he would have undoubtedly been wearing, was missing. Many years ago, I put into words the events of that day:

The date was 17 March 1969. I was a pararescueman (PJ) stationed at Nakhon Phanom Royal Thai Air Force Base. Upon receiving notification that a pilot had bailed out of his aircraft, we took off for the area to execute a rescue. The pilot's wingman had watched him parachute into the jungle below and was keeping track of the spot. I was the PJ on the high-bird, so I would probably not be the one to go down the hoist to retrieve the man. High-bird was a backup for low-bird in case something went wrong with their helicopter (HH-3E).

The low-bird went in for the pickup. Hovering above the rugged terrain, the flight engineer began to lower the PJ to the hillside below. As the hoist cable began to unreel, the jungle-penetrator seat and its occupant began to turn slowly. As the PJ sat on the seat, he had his M-16 at his side ready to fire. Then, upon rotating into a tree, his weapon fired accidentally. Since he was the only one who really knew what had happened, he was hoisted back into the helicopter at the helicopter pilot's command. The PJ finally got it across to the pilot that the weapons fire was accidental, and the helicopter once again hovered into position to lower the PJ to the ground. He began to descend to the ground but stopped and was reeled back up again. He was then informed that a fray in the cable was discovered and was the reason for his second return to the helicopter.

At this point, it was decided to send in the high-bird. I would be descending to the ground. I began to get ready. Immediately, thoughts of a similar mission just three months before flashed through my mind. A pararescueman had descended to the ground to recover a pilot who had never made voice contact with the rescue crews. The only transmissions were from his automatic parachute beeper. It turned out to be a trap. The enemy had waited until the helicopter had lowered the PJ to the ground and was about to hoist him and the downed pilot up when they began firing. The last words received from the pararescueman were that he was hit and they were to pull up. The hoist cable snagged in the trees and snapped in two as the helicopter pulled away. Now I was to descend to the ground and recover a pilot who had been on the ground most of the day and had not made voice contact. There had been substantial activity in the air above his position, and his location could have been easily noted by anyone in the area.

We hovered into position over a clearing on the hillside below the pilot's location, and I climbed onto the jungle-penetrator seat. Slowly I descended to the ground. It is difficult to describe the feeling one has in the spinal column at times such as this; I was sure someone was out there waiting to commence firing at me. It's hard to cover your back when you are hanging from a hovering helicopter. However, I made it to the ground, and once there, I lay flat while I checked the surroundings for any sight of activity. Seeing no one, I cautiously started up the hill to where the parachute was.

As I slowly advanced up the incline, I began to notice drops of blood here and there. Then, only a few feet away, I noticed the downed pilot, lying face down with his seat kit high on his back, the broken bone in his thigh showing through his torn flight suit. His parachute and its lines were entangled in the undergrowth of the jungle. He apparently had parachuted into the trees and then went crashing through to the ground below. He had wedged between the sloping hill and some small trees and shrubs after rolling downhill.

In case this was another trap, I radioed up that the man was dead so that there would be no doubt as to HIS status at least if we didn't make it back to the helicopter. Somebody (probably a Sandy pilot) came back over the radio and said, "Then get him and get the hell out of there!" I didn't know what HE was so excited about; I was the one on the ground. Determining that it would take considerable effort (to say nothing about time) to get the man untangled from his equipment and out of the undergrowth, I decided that it was not advisable to further risk my life nor the lives of my three fellow crewmen on board the hovering helicopter. I, therefore, went back down the hill to get onto the hoist.

I climbed onto the jungle seat, strapped myself in, and gave the up signal to the flight engineer who was watching from the helicopter door. The cable began to reel in and again I hung beneath the helicopter. If this was a trap, now was the time we would find out. I was pulled into the door of the helicopter as it began to move forward for the return flight to NKP. It was quite a relief to the flight engineer as well as myself. He extended his hand for a good-to-see-you-again handshake. I also got a smile from the aircraft commander in the cockpit.

I have often regretted, since that day, that I did not take the time to retrieve the body of that downed pilot. I don't know what would have happened if I had taken the time to dig the body out of the bushes; it could be that there was not another soul within 50 miles of our location. All I know is what DID happen because I DID NOT take the time to recover the body—we returned to NKP and lived to fight another day.

Leland H Sorensen

The Cable Guy now has a name: Leland Sorensen. I now had his contact information and wanted to call him right away, but I decided to give him a heads-up before calling.

Leland, I am Ed Sykes, a retired fighter pilot, who was a close friend and roommate of Dave Dinan at Korat RTAFB. I have been attempting to assist the U.S. Government recover the remains of Dave and have made two previous trips to Laos in the past 18 months. I have gotten as far as the Village of Ban Xan Noy plus an excursion just outside the Village to a point where the Village Chief said a member of his Village, Mr. Yui, says he feels is the ejection site. I would like to send you a picture of the mountain he pointed out and see if the terrain looks at all familiar/similar to anything you might remember of that day.

I'm sure this was a day neither of us will ever forget and I wish to respect your sensitivities on this issue. However, I am now planning another trip to Laos in Feb and would like to hear inputs from you and others in the recovery effort to determine if I can use this information to give me the best possible opportunity of finding the site. If you would consent, I would like to call you as soon as you could make yourself available.

Thanks for considering this request and thanks for all of your great work in SEA. Many of my F-105 friends were saved by you and your JG crews and all of us hold you in the highest esteem.

Cheers, Ed Sykes (wgfp)

I got a reply from Leland that he had many personal issues going on and it would be better to contact him the following week. I tried to call him the next week but got no answer. I left a message on his answering machine. He did not return my call. Uh-oh. I waited several more days and finally decided to give him another call on December 23.

"Leland Sorenson," said a subdued voice on the other end. Not exactly what I thought a Cable Guy should sound like.

When I introduced myself, he apologized for not getting back with me. He was busy preparing for Christmas, and his wife had just undergone eye surgery, but he was glad I called. We began by talking about our lives during and after the War. Leland had left the Air Force and returned to his native Idaho, eventually working for a local university. He told me he and his wife, Laura, had four kids, and I told him I also had four. I then bet him I had more grandchildren than him, and he asked how many I had.

"Eleven," I stated proudly.

"You lose. I have 21," he boasted.

Without thinking, I blurted out, "You must be a Mormon."

"Yes, I am."

I had known several Mormon fighter pilots and they were all characterized by a calm demeanor, a strong family with a lot of kids, and a fierce sense of bravery and tenacity when involved in combat. I always had the highest respect when, at the gym, a pilot removed his flight suit and revealed his temple garments.

For half an hour, Leland described the events of St. Patrick's Day 1969, which matched the description he had sent to Niall in the email. However, one additional item got my attention at once. He told me that the area where he landed on the steep karst was grassy and remained grassy the entire time as he progressed up the slope. I don't recall ever seeing a grassy slope on any karst in Laos. Perhaps this would be a valuable lead. Leland said that he felt at the time that the terrain was unusual. Leland also said it bothered him to leave Dave's body and wished he had decided to retrieve it then.

I then lead up to the question I really had on my mind. "What kind of shape are you in, Leland?"

"I'm in pretty good shape and am very active for a man my age."

"Would you be willing to go back to Laos and attempt to identify the karst and the area where you saw Dave's body?"

Without hesitation, Leland answered positively, "Yes I would. This thing has bothered me all these years, and I would love to give it a try."

I explained that there might be an investigation into Dave's death conducted in northern Laos in March. If we could get started on getting his participation approved right now, we might be able to include him in the investigation.

"I would be happy to participate, Ed."

That's all I needed to hear. My "put me in, coach" efforts had been turned down by everyone I had asked, and I had tired of hearing why I didn't match the criteria. Leland matched the criteria perfectly. They couldn't possibly turn him down, but I knew we had to get working on it immediately, which I told Leland I would do as soon as I hung up.

I sent an email to the Dinans:

John and Charlie, this is the initial information I got from Niall concerning his discovery of Leland Sorensen, the PJ who attempted the rescue of Dave. I talked with him tonight and he is, without a doubt, the best witness we have to date. He still has pretty vivid memories of the rescue attempt and is very interested in assisting in this undertaking. He says he is still in good shape and would be willing to participate in the investigation. I am going to ask JPAC to include him in the March investigation and may ask you to write letters to them asking for the same.

Leland would be glad to discuss his experience with you. His phone number is below.

Niall deserves a real "attaboy" for this discovery! Cheers, Ed (wgfp)

This was late Monday and Christmas was on Wednesday, so I wasn't sure who at DPMO might be working, but I decided to try to catch either Niall or Bill. Niall answered his phone, and I asked him about the possibility of Leland going to Laos in March if the JFA Team went. He had also thought of this option but had not voiced it yet. JPAC would have to make the decision, but he would ask Bill Habeeb to call Bob Maves and ask the question.

"Great, have a great Christmas, Niall! You've already given me my present!"

Getting anything done within the government during the week between Christmas and New Year's is almost impossible, so I was surprised when I got an email from Bill Habeeb on Dec 31 saying he had called Bob Maves to tell him my idea of Leland's involvement with the JFA. He wrote that Bob seemed open to the idea and would get more information from the PJ before making a recommendation. He also informed me that Bob had told him that the northern Laos JFA-3LA (third fiscal quarter in Laos) would most likely take place beginning March 1 through April 4. *Holy crap, that's only 2 months away!* He didn't say for sure that the JFA was funded or if #1408 was on the schedule. One thing is sure—if Leland goes to Laos, Dave's case will be on the schedule.

That night, as we celebrated the passing of 2013, I reflected on the roller-coaster ride it had been. My enthusiasm and hope beginning the year after my trip to Xan Noy was followed by my disaster at Xieng Khouang. However, Niall's discovery of the Cable Guy and now the possibility of sending him on an investigation had turned the year around. I was also hopeful my trip to Laos in February would yield some new information. *I'll drink to that*—and I did. *Happy New Year!*

***

Two days later, I decided to make a trip to D.C. in mid-January to work with Niall and Leland (by telephone) to see if we could pin the site down more closely using some of the maps we had been discussing. I also wanted to visit Bill Habeeb and Ann to push again for the inclusion of Leland in JFA-3LA.

By January 6, I had heard nothing, so I called Bob Maves to see if he had pursued the idea of using Leland any further. I got his answering machine, so I left a message and sent him an email:

Bob, an investigator at DPMO (Niall Brannigan) has located Mr. Leland Sorensen, formally a PJ with the USAF. He was the PJ that went down the cable as part of the crew of JG-16 on March 17, 1969. He discovered the body of 1Lt. David Dinan III (1408) entangled in rocks and trees. He determined he was deceased and because of the difficulty of removing the body and the possibility of hostile forces in the area, JG-16 departed the scene without Lt. Dinan's remains.

I am hopeful that Mr. Sorensen can be included in the team that investigates the recovery of the remains of Lt. Dinan during your upcoming [JFA-3LA] in March. I have interviewed Mr. Sorensen and he has very vivid memories of the recovery effort and the terrain he encountered and the surrounding landscape. We have looked at maps supplied to me by DPMO, and Mr. Sorensen feels there is a strong possibility that this is the correct site. Based on my interview, I feel he can pinpoint the exact location where Lt. Dinan's body was last observed by any U.S. witness.

I request you contact Mr. Sorensen and interview him yourself. Mr. Sorensen is still quite fit and actively hikes in his home state of Idaho. I feel we should take advantage of his knowledge and current good health and involve him in [JFA-3LA]. Not sure but sounds like the kind of guy that might become a real fan of Beer Lao. Cheers, Ed Sykes (wgfp)

OK, so I embellished the possible success of Leland a little, and being a Mormon, I was fairly sure he didn't drink beer. However, I wanted this bad. Even if Leland did not achieve success, it would mean that the case would get some attention after 45 years. That would be a success. "We're coming to get you, Dave!"

I then sent the Bob Maves email to several others. The first went to John and Charlie Dinan along with Mark Rockoff:

John and Charlie, this is an email I sent to Bob Maves, the SEA lead at JPAC. I am hopeful JPAC will include Leland in their upcoming 3LA investigation in Laos. If I encounter any resistance to this proposal, I will ask you to roll in on some heavier hitters at JPAC. I am hopeful this is a "no-brainer," BUT the history of neglect in this case makes me take nothing for granted.

Mark, if it really gets rough, the MIT connection might be useful. Hope all had a great New Year. Cheers, Ed (wgfp)

I also sent a copy to Ann and asked for her help. I let her know I would be in D.C. the following week and requested a meeting.

Meanwhile, my Laos trip was still a goat rope. Neither Phet nor Vilan were able to nail down solid assurances from the provincial officials that we would be able to travel to Xan Noy and spend several days there. I knew Vilan was going to be in Vientiane in early January, and I asked the two of them to get together and come up with a strategy to get permission. I recommended they decide on the best approach, and then I would fly them to Xieng Khouang and they could meet with the officials together.

On the morning of January 8, I awoke to find an email from Phet:

Edward,

This evening I met Mr. Vilan in Vientiane. We had dinner together and long discussed about our trip to Phonsavan. I told us many things that [would make it] possible [for us] to get [to] Xan Noy village. I told him to call you and explain all those things … for example, we have to do permission agreement with Lao authority, rent 2 4x4 ranger cars with F4 to be able to go up the mountain, hire 6 Lao soldiers to protect us because that area is unsafe site to visit; as Vilan told me that, we will spend big amount of money to be safe and we can do whatever we want to… Vilan will go back to Phonsavan on this Friday, so I think [I] will go to Phonsavan by this Friday to meet Mr. Yui (most important guy to work with if without him we cannot work well at the forest) and also will meet some highest Lao authority in Phonsavan. I think we can meet Lao soldier at his home on Friday or Saturday will be no problem because they are Vilan's relatives. Tomorrow I will call Mr. Yui that I am going to see him in Phonsavan soon. Have you sent my transport costs? if I can get by tomorrow or on Friday would be great …

I started working on a MoneyGram to get Phet the money he needed to travel to Xieng Khouang, but before I could get it sent, I got a second email from Phet:

Edward, we can try another way if we think together might [be] possible for us to go. Let's make a very clear of our trip plan. Think about the amount as Vilan told us is quite high to pay, we know that he has everything in hand, all the important tools with him, what we really need are equipment (GPS, so on) and someone to take care [of] us during visiting the site at Xan Noy. So, I want you to think this plan carefully before making decision. I worry that we will waste money freely, nothing gained.

I always happy to work for you with my heart… I will call Mr. Yui tomorrow morning. I called the chief of Village, Mr. Khammerng, he's waiting to see us, drink and eat together with us soon. I just gave him a call last 7 hrs, he thought we will be at Xan Noy this month, and I told him again we will be there next month due to our plan changed again. I will inform him again. Phet

By now, it was quite late at night in Laos, so I decided not to send the money to Phet until I had a chance to talk with him the next day. I also was concerned by his reference to the amount it was going to cost to make the trip after discussing the figures with Vilan. Phet must have stewed over the whole thing during the night because I got a third email from him about 10 P.M. in Kansas:

> Edward, I have some good points to share with you: Let's do what you want to do in Laos. Forget about Vilan's plan, we make our own plan like we did last time at Tourism authority office. Here below are my points:

> 1. We visit Tourism police for ask them to give us a permission agreement paper to visit Xan Noy village, we can stay overnight in Khoun District, day time we go to work in Xan Noy and back stay overnight at Khoun district for our security. We don't need to stay overnight in the forest.

> 2. We rent one car and hire 2 police to come with us.

> I think this plan will work easier. Forget about Vilan. I know he's very busy these days and of course we have pay him a lot.

> Keep this message, don't let Vilan know our plan.

> For the agreement paper from Police office, I will ask them and negotiate how much money we have to throw to them. We just tell them that we are going to visit Xan Noy for seeing nature, people's life and collecting data raising fund to help them in the future; in order to get permission to reach Xan Noy, 2 police, we can feed them with Lao wine, and small tips like we did last time. They will be very happy and can tell them to keep word in their mouth.

> If you agree with me to do that, I will ask the police number and talk with them about that soon.

> I want to see our successful plan, but you have to prepare your technologies that [will be] needed. Phet

I felt like I was entering the cone of confusion. I had always sensed that Phet did not care much for Vilan, and now he wanted to make a break. It was late at night in Kansas, but I knew it was early morning in Laos, so I decided to call Vilan and get his take on the situation. After a couple of attempts, I was able to get him and asked him how the meeting went. Right away, I could sense he was not his usual optimistic self.

"It went OK, Edward, but I am beginning to think that getting permission to do what we want may be difficult. The officials are all very nervous about our proposal. I think to do what you want may be very expensive."

"How much are we talking about?" I figured that was a fair question to ask.

"I think it will cost about 24 to 25 thousand dollars," he replied.

I explained that I was not interested in paying that much, and I fully expected him to make a counteroffer or whine about all the time and effort he had put into this venture, but he did not. He simply stated that he understood, and I thanked him and that was the end of the conversation. As I lay in bed that night, piecing together the conversation, the only conclusion I could come up with was that Vilan had concluded that there was too much political risk in going through with the venture and he would price himself out of the market. Only a guess. Phet had said he was busy and perhaps that was the reason. At any rate, I would never see or hear from Vilan again. On reflection, he was one of the most interesting people I have ever known, and I wish him no ill will.

The next morning, I called Phet and reached him on the first attempt. I repeated my conversation with Vilan and when I had finished, he replied, "That is good, Edward." He went on to say he would begin working with the provincial officials and see what kind of arrangements he could work out. He would also contact Mr. Yui and the village chief once we had gotten permission to visit Khoun District.

"Great, Phet, keep me posted." I then told him that the man who had been sent to retrieve Dave's body after his ejection had been located and might be sent to Laos in March in an attempt to locate the site. We would make our trip in February, and if we found any new information, I might stay over to assist the U.S. Team with our findings or have him visit the U.S. Team and give them our findings. He was excited about the prospect of being further involved.

Shortly after talking to him on the phone, I sent him an email concerning Vilan:

Phet, call the police and see what they say. I'm not sure we can keep this a secret from Vilan. Do you think he would cause us difficulties if he knew? Your idea is what I have wanted to do all along, and I think it might work. Tell them we would like to visit our friend the Village Chief who has invited us to stay in his village. I think we should let them know that I lost a friend in Laos and would like to visit the area where he died. Let them know that we will not disturb anything or take anything from the forest. Let me know what they say. Cheers, Ed (wgfp)

A short time later, I got an email from Phet:

Edward, No no, Vilan can't cause us problem. We can go there with 2 police, 1 driver, 2 of us and chief of village. Just do not tell Vilan that we are going there. I will call the police and see what they will say to us. Take a good care, Phet

An adventure involving Vilan was no longer an option, but the possibility of a "cowboy" excursion with Phet was still in the cards. However, I had remaining doubts about our ability to pull it off without getting in trouble with the Laotians. The JPAC investigation into Dave's case (3LA) was not a certainty, and the possibility of including Leland in any investigation was even more of a long shot. Also weighing on my mind was the bad taste I still had in my mouth from my Xieng Khouang adventure of the previous year. I wasn't sure I wanted to barge into a JPAC investigation without an invitation as I had done before. I was running out of time as JFA-3LA was scheduled to begin in March, seven weeks away—if it even happened. Perhaps my trip to D.C. next week will yield some answers.

Before leaving for D.C., I sent an email to Leland asking if he had heard from Bob Maves yet. Not yet, he replied. A week had passed since I had asked Bob about the inclusion of Leland. This was not reassuring. I set up a time to visit Ann and asked her to see if she could do some gentle nudging, and she said she would do what she could.

Traveling from Kansas to D.C. in mid-January is not without weather risk, but it turned out to be rather nice. Indicator of good luck? I visited with Bill Habeeb and Niall Brannigan at DPMO, and they had some reassuring news concerning 3LA. It looked like it was a go, but they weren't 100 percent sure Dave's case would be included. Niall told me he was doing a predeployment interview with Leland, which he would forward to Bob Maves to determine if Leland was eligible to be included in the JFA. I was certain Niall would push for Leland's inclusion as hard as he could. I also let them know I had canceled my effort with Vilan but may try to conduct a smaller-scale investigation with Phet.

At lunch with Ann the next day, she also promised to do what she could to have Dave's case included and would discuss it with Gen. McKeague. She was not surprised with my problems with Vilan. She then launched off on her quest to reform and reorganize the POW/MIA recovery effort. She told me she had met with the Secretary of Defense, Chuck Hagel, and expressed her dissatisfaction with the entire effort—especially its lack of capable leadership and management. She described her meeting with the secretary in detail. It sounded like a doozy! As she told me all the things that needed to be fixed and what had to be done to fix them, I was again impressed by the drive and dedication of this woman. She certainly had earned admission into my personal Hall of Heroes!

The next day, I returned to Kansas and was greeted with some interesting emails. The first one was from Phet:

Edward,

I have called tourism authority this morning. I explained them all of our plans to visit Xan Noy village, but I would like to know when would you [be] available to come to Laos, to Phonsavan? Could you tell me how many days again would you like to stay overnight at Xan Noy as the chief of tourism police told me that if we might stay overnight at the village, we have to pay more for the permission agreement letter.

He told me it will cost around $2000–$3000. Now they are on discussing about that and their staff overnight allowances with us, and I ask them to put all things in the paper that we need to be paid for. Tomorrow I will call and see what they will say.

Edward, could you explain me secretly that, when we arrive at Xan Noy village, what will you do in mind, when we go to crashed site and found a place you want then what will you do with it, when you can see the born [bones], will you take it along with you or just keep it there?

I would like to know this, in my mind, and expect that we will have to pay much money for this trip, for example, small tip for chief of village, Mr. Yui, Mr. A.B.C.D., for police, car rent, and food for everyone. My estimation for this trip budget is around: $6000–$7000. If this is true, do you able to pay for it? What do you think? Take care, see you soon. Phet

Holy crap! This was looking like a high-priced shot in the dark. No telling how much more the Lao will make me pay. By the time it is all said and done, this will be a 10-grand adventure. But I had been prepared to spend 16 thousand with Vilan. Brace yourselves, grandkids.

I also had an email from Bob Maves addressed to Niall and me that requested Niall send them his interview paperwork today so they could determine if Leland was physically capable to deploy on the JFA. A third email from Ann confirmed that Dave's case was on the schedule, per Gen. McKeague, and JPAC was awaiting the reports from Brannigan. So good news about 3LA, and I was especially happy to note that Ann had discussed Leland with Gen. McKeague and he was receptive to the idea. Real progress! The problem is timing—there is little time left.

The next day I received an interesting email from Bob Maves that indicated that Bob was aware I had called off my adventure with Vilan:

Mr. Sykes, Request to know the price of your guide. Request to know price of permit. How much did they raise the price that you now are not going to go to the site?

My reply:

Bob, my Commie friend raised his cost from $7K to $25K. I don't know the breakdown of costs because I rejected the idea at once.

I am now working with my original guide (Phet) who is working with the Province Tourism Office and their fee for a permit is $2–3K. Having worked with these guys before, however, I think all of this stuff is negotiable. I am going to wait and see what you guys do in March before I commit to another trip. However, if you guys were to take Leland Sorensen on your investigation, I would like to be included in the brief and debrief in Phonsavan. Good enough excuse to travel in search of a Beer Lao.

Phet has been in contact with Mr. Yui on many occasions recently, and I feel you should include Yui in your investigation. He is currently working in Phonsavan. I'm not sure the site he is describing is Dave's, but I'm guessing it may be part of one of your cases. I will try to give you a call next week and discuss some of these issues with you. Cheers, Ed Sykes (wgfp)

I didn't have to wait for the next week. I got a reply from Bob later that day:

We are going through the process and requested the funding for Mr. Sorensen. Of course, a lot of complaining already since less than 60 days from deployment. God forbid finding a new witness.

Please provide Mr. Yui's full name and address. Better to let our Lao Officials know we want to interview him than for him to magically appear at the hotel. Also, do we need to interview your Phet too? VR, Bob

More good news. I replied to Bob at once:

Bob, agreed. I will ask Phet to obtain that information for me. Glad to know the funding request has been made—hope you can make this happen! I'm sure Phet would be willing to talk to you. His experience on this is about the same as mine, although he may have better insights because of his language advantages and his discussions with the Village Chief and Mr. Yui.

I know there was discomfort when I showed up at the hotel last year. Would it be better for me to not attend any discussions or is there any "legal" way for me to be an acceptable spectator?

Are you in the office this morning? Cheers, Ed Sykes (wgfp)

Bob emailed to answer he was leaving the office and I should call him the next day. He also stated that my participation in any debrief after Leland's investigation would be up to the interviewer.

I sent an email to Phet asking if he'd talked to the tourism officials. I also let him know that Sammy, a Laotian who works for the U.S. government investigation team, wanted to ask him questions about our trip to Xan Noy and his discussions with Mr. Yui and the village chief. I asked him to cooperate so they would have the information before the team's own trip to Xan Noy in March.

A reply from Phet came quickly:

Dear Edward, I have read all the message that you sent to me, it means U.S. team trip March, you won't come with them, right? Last week I could call Mr. Yui and he was working outside of Phonsavan district, sometimes there is no signal available there, he said. Here are his mobile: 021xxxxxxxx or 020xxxxxxxx. Ask Sammy to keep trying call him.

For Tourism Folks, I am waiting the answer as I have asked them from last week, they told me that they will have a meeting to discuss about that because, as we knew that Xan Noy is not tourist site, closed not open for tourist to visit. Still danger with the unexploded bomb.

I would be so happy to cooperate with U.S. team, and openly to give all the answers we have had from our last trip. How many days are they going to visit Xan Noy? Will they stay there or in Phonsavan like they did last time? Phet

So many questions I was unable to answer. The next afternoon, I called Bob. I first thanked him for his support of including Leland on the JFA.

"It's a no-brainer. I just hope we can get all the necessary hurdles taken care of," he replied.

He also told me that the JFA would be smaller than a normal effort to save money and thus would probably be staged out of a location closer to the site of Dave's ejection (Ban Ban village). It would save money, mostly on "blade time" (helicopter time). All the helicopters were leased from the Lao at their insistence. He gave me some information about getting to the Ban Ban area but was noncommittal on my participation.

"It's up to you, Mr. Sykes."

I had heard that before! The amount of optimism expressed by Bob concerning including Leland on the JFA caused me to rethink my plan. I was running out of time on my February trip and still didn't have a good feeling about the Lao giving me permission to travel to Khoun District. It seemed wiser to let the Team complete their investigation with Leland, and then, if they found no hard evidence of where Dave died, I could make my own journey later that year. As far as visiting the Team during the JFA, I was torn. That night when I was fairly certain Phet would be awake, I tried to call him but got no answer.

I sent him an email:

Phet,

… This year the team will not be staying in Phonsavan—they will stay in Ban Ban instead. Do you know where that is at—it is very close to Vietnam. I don't think you can fly there on Air Lao. JPAC guy said it is a 3-hour drive from Phonsavan. Are you ready for another adventure? Cheers, Ed (wgfp)

When I awoke the next morning, I had Phet's response. He knew of Ban Ban, which was the village after Xan Noy and on the way to Vietnam's border, an important site from the War. He reiterated his desire to work for me then asked how long the U.S. Team would be at Ban Ban.

369

He also asked for the exact dates in March and if he would need to deal with the Lao authorities or if the Team would manage by themselves.

I called and managed to catch Phet on my first attempt. "Phet, I have decided not to make a trip in February. Time is running out, and I'm not sure we'll get permission from the Lao officials. The U.S. Team will be in the Ban Ban area from March 1 to early April, and I may decide to join them there. I would like you to go with me if I make the trip. It is beginning to look like the Team may take the soldier who found Dave's body with them to help locate the site. If they don't locate the site, I will plan another trip to the area in June or July and we will camp in the forest like we planned before. I will let you know soon if I will join the Team in Ban Ban."

As we talked, I could tell Phet wasn't disappointed about canceling the February trip. I suspect he had been unhappy with his role as negotiator on my behalf with the Communist officials. I asked him to find out as much as he could about the Ban Ban area. I also mentioned that I would like to avoid further contact with Vilan.

Within a few hours, Phet emailed me with information about Ban Ban, a village in the city of Muang Kham. Although there were no hotels in Ban Ban, there were many villages nearby where hotels were available. He asked for more details of our trip, like which dates and where, so he could manage his schedule. Phet also asked which company the U.S. Team works with in Laos to handle their trip. He mentioned that if it's a company Vilan works with, he would probably come with the Team (but we shouldn't be afraid of Vilan); otherwise, we wouldn't see him at all.

I replied to Phet that I would give him the information he desired as soon as I had it. As I had discovered, the schedule for JFAs was flexible, and if I was going to join the JFA Team, I would not have a lot of lead time.

About this time, it dawned on me that I had another factor to consider. My little "hobby" farm in Kansas brought me a great deal of delight. And one day as I was feeding my three sows (the "Spice Girls": Ginger, Cocoa, and Cinnamon), I noticed they were beginning to get "piggy." I checked my records and found they were due to have baby pigs in mid-March. This happened only twice a year and was one of my favorite events. As I edged past 70, I was learning to look for and appreciate the simple pleasures of life, and this was one. Without a compelling reason to venture to Laos in March, I would just as soon stay home, especially if Leland was included on the JFA. If they found nothing, I could still get together with Phet in mid-summer.

As I waited to hear about the progress of including Leland, I had an interesting exchange of emails with Niall. Niall was busy working with Leland, sending him maps and keeping him updated with his research. Niall had located the Air America Huey helicopter pilot who was in the area the day Dave died and was the guy who found and guided the Sandys to Dave's parachute. That pilot had had his own mystery moniker of "FG," but the Air America veterans are a tight-knit group, and Niall was soon put in touch with Nikki Phillipi, who was then the federal government's longest-serving employee in the Pacific Northwest. Niall included me in most of his emails to Leland, keeping me in the loop regarding his research.

One email from January 21 said he'd made contact with the Air America pilot, Nikki Phillipi, who had saved his logbooks from the period, which could give us a lucky break. He also mentioned that he had botched the location for the site on Map 3 on the hardcopy maps he'd sent to me and Leland, apologizing for the brain fart. I replied, asking if he'd had a chance to talk to Nikki and wondering where he lived. I also informed him that Bob Maves was pushing hard for Leland to be included in the spring JFA.

The next day, Niall got back with me:

I did indeed speak with Nikki, Ed, and he's a real live wire. He had a tour in Vietnam with the Army early in the war, went home for a spell then went out to Laos in '67 with Air America and remained in SEA until the fall of Saigon when he was one of those pilots extracting folks from the U.S. Embassy. ...

His logbook records him operating out of LS20 on St Paddy's Day '69 but makes no mention of a SAR mission... He indicated that that wasn't unusual for his logbook in cases when Jolly Greens showed up to do the actual rescue, and he was a "bystander." ...

The case resonated somewhat with him, but he recalled it being an F4 back-seater, not an F105 pilot. He will eagerly examine the maps, and I will get those posted to him tomorrow, unless the Govt shuts down once again. ...

Nikki is living in Seattle and would love to speak with you. Cheers, Niall

I replied at once:

Niall, great work. I will give him a call and swap some lies and pretend we are still relevant. Would be great if he could come up with some additional information. Cheers, Ed (wgfp)

The reply I got from Niall reaffirmed why I liked him so much:

Ha, Gen-Xers and Millennial babies have far more of a challenge demonstrating their relevance in my eyes, Ed! And as for lies, and those who tell them, it'd be a dreadfully boring planet without them. Niall

He must have been sipping some of his Irish whiskey to have come up with that. Such wisdom in one so young.

Niall later recalled that Nikki was subsequently unable to contribute anything not already known about the case, but in noting that the loss location he radioed in for Dave's crash site was isolated, to the south of the nine other reported loss locations, Nikki was adamant that any coordinates he reported would have been accurate.

***

I didn't hear much for several days, so I assumed things were going well and Leland was going to be leaving for Laos in another month.

Then Leland dropped an email to me that gave me a real start:

Ed, this is to update you on my Laos situation. The immunizations required have become a bit of a problem. Leland

He had attached a copy of an email he had sent to the JPAC medical staff:

JPAC Medical,

I am to accompany a JPAC team into Laos at the end of February (2014). I have received a list of immunizations that are required (SUBJECT: Specific immunizations for JPAC areas of operation – 10 May 2013). I am encountering difficulties finding availability of some of the shots.

Also, I find they are rather pricey (e.g., Rabies = $320/dose, JEV = $267/dose). All of the shots will be well over $1000, if I can even work them into my schedule. Scheduling the shots is turning out to be a bit tricky, also. There is no way the Hepatitis Series can be COMPLETED in 1 month. I think that takes months to do.

If this immunization conundrum turns out to be a "deal breaker" for this trip, I need to know immediately so that I do not start something that will end up being pointless to have started in the first place. Leland Sorensen

Really?! I had made two trips to Laos and never even considered that I might need some shots. Nor had I had so much as a sniffle. I immediately got on the phone to Leland.

"How many shots do you need, Leland?"

"It looks like I need 22, and I can't get some of them done before I get there because they are a series," he replied.

"Do you still want to go, Leland?"

"Yes, of course. This has bothered me for years, and I want to try and make this right," he stated in a positive tone.

"Great, let me call Bob Maves and see what he can do to help," I stated and promptly called Bob. I explained the situation to Bob, and he went into an obscene rant about some folks I didn't know. "Let me fix this, Mr. Sykes."

Later that day, Leland got an email from Al Teel explaining that there was some confusion about Leland's status. The 22 shots would normally be required of contractors JPAC hired to stay for the entire JFA and didn't apply to him. He would still be required to get a few shots, however. He also let Leland know he would get all the shots he needed at Mountain Home AFB in Idaho, and he wouldn't have to pay for any of them. I could only imagine the conversations Bob had had with some of the fine folks at JPAC to get this straightened out this fast. Leland also got an email from the immunization office at Mountain Home letting him know they were waiting for him. Thank you, Bob Maves!

As part of these discussions with Bob and Leland, I was told that Leland's participation was fully approved and he would depart the U.S. on February 27 for Hawaii and fly military air with the Team to Laos later that week. He was then scheduled for a week of "vacation" at Vientiane before proceeding to northern Laos. The search for Dave's site would take place between March 14 and 16. *Glory.* So, if I wanted to be part of a debrief, I would have to be in Laos when the Spice Girls were due to pop. What to do?

I sent an email to the Dinans and Mark Rockoff:

Charlie and John, Leland has been approved for travel to Laos from Mar 1–17. The only holdup right now is getting his immunizations. I am thinking of joining up with him at Ban Ban (on the Ho Chi Minh Trail) right after he returns from his investigation journey (I will try to go along with him but don't think JPAC will approve that). I think it is rather ironic that this investigation may finally make its first trip to the site of Dave's death exactly 45 years later. Cross your fingers, this may be the break we've been waiting for! …

Cheers, Ed (wgfp)

Within a day, I got an email from Phet concerning a possible trip to join the Team in March:

Edward, have you gotten the schedule plan with U.S. team or not yet? Yesterday I just received a phone call from Tourism police, they told me that, during end of March is coming of Lao new year in April 14, 15 [etc.], so Lao authority will watch closely with all visitors. They told me that for the documents are no problem, they can provide us if we need it for security, but we cannot stay overnight at the village for more than a week. This time we have to cooperate and work with Lao authority smoothly in order to work well. I would like to let you know that in March between 10–14 I may go to visit my family in northern Laos. Anyway, when you have the plan, do not forget to let me know. Have a great day, Phet

Sounds like the Communist bureaucrats have laid down the law with Phet; "we have to cooperate…" is not very subtle. The proposed visit to his village on March 10 to 14 concerned me since Leland's investigation was scheduled from March 14 through 16, but I was sure I could find my way around without his help. I sent him a response:

Phet, the schedule is beginning to come together. Right now, I will come to Vientiane on March 10 and fly to Xieng Khouang on the 12th. Would then travel by taxi to Ban Ban and stay there until the 16th. I would then go to Vientiane and return to the U.S. I would understand if you went to see your family during this time, but I would really like to have you join me on the trip to Ban Ban. The days I really want to be in Ban Ban are 13–15 March.

I am glad to hear you are working with the tourism people. If this trip goes well, I will probably try to return in July or August and visit Xan Noy. Cheers, Ed (wgfp)

The next day, February 1, I got a response from Phet:

Edward, That's cool but could you ask them to delay little bit, from 1st– 7th or 15th until end of March. The problem, on 10–14 will be my younger sister wedding party at my home, and my parents want me to be a part of this party. Thank you so much, Phet

At this point, I realized it was becoming more likely than not that I would be dealing with the Spice Girls in mid-March. Going on a trip without Phet would not be as much fun; besides, I wasn't sure I wanted to take care of all that kip. I responded to his email:

Phet, you must go to your sister's wedding!

I find there is no way I can have them change the schedule. I'm sure I can get to Ban Ban by myself and this trip does not involve going to Xan Noy. I will probably do that later this summer. Where in N. Laos does your family live? Is it easy to get to? I would like to meet your family if it is not too hard to get there.

You must go to your sister's wedding! Cheers, Ed (wgfp)

Even as I continued to look at airline schedules and look for hotels in the Ban Ban area, I was having second thoughts about going since, based on my previous experience at Xieng Khouang, I might just complicate things. Since it looked like Leland would probably make the trip, that would count as a win.

Around 8 in the morning on February 10, my phone rang; it was Leland. He was sitting outside the gate at Mountain Home AFB where they had refused to let him on the base. Supposedly JPAC had informed the base of Leland's needs, but somehow, the order hadn't gotten to the main gate. Leland was clearly exasperated. He had driven a long way to get the shots and now was sitting in his car on a cold Idaho morning. I told him, "Don't leave! I'll figure out something."

I first decided to call Bob Maves and leave a message on his phone (it was 3 A.M. in Hawaii). I would then call DPMO in D.C. and see if they could straighten it out. After dialing Bob's number, I was shocked to hear, "Bob Maves."

"Bob, what the hell are you doing at work at 3 in the morning?" I asked.

"Oh, I come in early a lot and work in the lab," he stated and then went into an explanation about getting more done when the scientists weren't around. I explained why I was calling, and Bob grumbled about incompetent people.

"I'll take care of it, Mr. Sykes," he assured me. Bob's dedication continued to astound me. I called Leland back to tell him help was on the way. He seemed relieved but not happy. A few hours later, I got confirmation from Leland that he had gotten his shots. He would have to return to the base later to get a few more, but he was now committed. Shit hot!

Three days later, I got an email from Leland that sealed the deal as far as his intentions to participate in the JFA. Leland wrote, "You know you're from Idaho when…you are the Top Story on the 6:00 News—two nights in a row." Then he included links to lengthy segments he had done on a local TV station. Watching these interviews, I saw that Leland was really fired up about his upcoming adventure. In the process, he had become somewhat of a local legend.

I quickly sent copies of this email to everyone I could think of concerned with the case, including the players at DPMO and JPAC and Gen. McKeague. Might as well let them know there was no turning back on including Leland. I soon got a reply from Gen. McKeague saying, "We're looking forward to executing this mission with him."

I also sent the videos to the Dinans and to my family with instructions to my own kids:

Please show my grandchildren what a hero looks like (wgfp)

***

Decision time: After assessing the situation, I decided not to make the trip to Laos and wait for the results of Leland's investigation. Although this trip didn't work out, I knew there would be other adventures. I called Phet to let him know, and although he was disappointed, he was also relieved that he didn't have to worry about my trip interfering with his sister's wedding. He had previously asked me to bring a Kindle and a battery for his computer, which I had agreed to do. Since I was not coming, I told him I would wire him the money to buy the equipment. We both fully expected that I would be coming to Laos in midsummer, so we discussed that upcoming trip with optimism.

On February 24, I got an email from Leland:

Ed, I don't think that I am going to be "vacationing" in Vientiane as originally planned. Al Teel said something about me moving north with the team so I could "help out" instead of sitting around. (I told him that I was more of a worker than a vacationer.) I am going to the Idaho Falls area (Ammon) on Wednesday evening so that I can leave Idaho Falls on Thursday morning (0600 hrs). I guess this is REALLY HAPPENING! Leland

Yes, it was really happening. I sent an email to Bartz to let him know my decision to not go to Laos in March since the PJ (Leland) who had found Dave's body was going. I was going to stay out of the way and wait to hear the results of the JFA. My fighter pilot son replied, "I actually think that is a good decision to stand by on the trip…kind of a mature decision for you. I'm afraid you may be all grown up."

Can you put a 45-year-old son up for adoption? I realized later that day I could even find a replacement if needed. I had sent Phet some money to buy the electronics he wanted and received this email from him:

Dear Edward, thank you so much for your kindness! I regard you as my father… I am waiting to see you soon!

Have a good health and take care! Phet

I explained to Mary that we had a new child. She was not thrilled.

With my trip canceled and Leland on his way to Laos, I took a few weeks off and concentrated on getting my garden ready for spring planting and keeping an eye on the Spice Girls for their impending hatch. I heard nothing from Leland or anyone else concerned with the search for two weeks.

Then on March 13, I got an update from Niall that made me amazed that Leland, who we had located less than 3 months ago, had been included in the investigation:

Hi Ed,

... JPAC's schedule had [Leland] idle out there for more than a week, but they finally began the site surveys on Tuesday, and he's due to fly home on Saturday. I was at DIA yesterday for some training, and hoped that there was some special serendipity in the air when I saw that the historic aircraft they have on a plinth in front of the HQ building is an F-105D! The "Ohio Express"...

As soon as I get some feedback, I'll let you know. Cheers, Niall

As I prepared to spend the night in my barn helping Cocoa deliver a litter of pigs, halfway around the world, Leland was helping fulfill the promise I had made to Dave two years earlier in Xan Noy: "We're coming to get you, Dave."

Several years before, I had built sleeping quarters in my barn for nights when I had to stay out there for pig deliveries. Cocoa didn't let me down, and over a period of 5 hours, she delivered a nice litter of 12 babies. By the time the sun came up, she was nursing them comfortably. The birth and discovery of their mother's source of nourishment is fun to observe, and watching little piglets fighting for position at their mother's teats is one of nature's most glorious sights. About 7 in the morning, all was going well so I decided to head to the house for breakfast.

The morning was rather warm for mid-March, and the sun was creating one of those great sunrises that few places do better than Kansas. As I approached the house, I heard Mary yelling my name. I looked up to spot her standing on our back deck.

"Ed, you have a phone call—it's Leland."

I hadn't expected that. I hurried to the phone. "Hello, Leland."

I had never heard Leland this enlivened. "We found the site!" he exclaimed and went on to excitedly describe the events of the day. It was the final day of the investigation, and they had found nothing up until then. In midafternoon, they had landed at a village and found a man who said he knew a place that might be the spot they were looking for. The villagers called the spot "The Stream Where the American Fell." They decided to check it out and used the helicopter to get close to a foot trail. They walked along it for a while, and then the man led them up the steep hill a ways and said he thought this was the spot. No evidence presented itself to indicate a pilot might have come to rest there. After looking around the area for a bit, they realized they were running out of time and headed back down the hill.

As they were returning to the helicopter, the man suddenly stopped and said he had shown them the wrong spot. He pointed up the steep hill, and again, they began a difficult climb. Eventually the terrain leveled off, and there appeared to be a stream bed. Almost at once, they found evidence that something had happened in this spot. A few small pieces of nylon parachute material and some metal bits that appeared to be parts of a parachute harness were scattered about. One of the Team members found a piece of plastic imbedded in the sand and picked it up, trying to determine what it was. He studied it for a minute and said he wasn't sure what he had found.

"Let me see that," Leland requested, and after he studied it, he recognized it as a pretty well-preserved military identification card. He was even able to make out part of the first name and most of the last name.

"This is the guy we're looking for," he had told them in disbelief.

Now I was in disbelief, and I had him repeat the details. I wasn't sure how to express my feelings, but after gathering my wits, I asked, "Do the Dinans know?"

"Yes, Ed, I called Charlie before I called you, and he was very surprised."

We babbled on, exclaiming how crazy this was. Dave's military ID card had been laying in plain sight for 45 years. Leland promised to give me a call when he got back to the U.S. and fill me in on the details. He would start back to Idaho the next day, and I sensed he would travel home a much "lighter" man, having lifted this burden from his shoulders.

Figure 10 - The "Cable Guy," Leland Sorenson, found Dave's military ID nearly 45 years after he'd found Dave's body near "The Stream Where the American Fell," 2014. Photo courtesy of Leland

I sat on the back deck for a few minutes thinking about the situation. This was too easy. Was it possible a villager had had this card in his possession for years and had planted it at this spot to attract future U.S. teams and the money that came with them? I had heard stories of this occurring in the past, so it left me with a hint of doubt, but the other evidence made me think it was real.

Once inside, I described the phone call to Mary, who also expressed her doubts. Despite those lingering doubts, I wanted to share this unbelievable news with family and a few others.

The email went out:

Got a phone call from Leland Sorensen in Laos this morning. Says they have good reason to conclude they have found the site where Dave died. Biggest find was Dave's Government ID card. I think they also found a few of the metal parts of his parachute/survival gear. This is great personal news for me, and I now know there is a good chance I will get an opportunity to see Dave buried on his Country's soil. It's not even 8:00 AM here in KS but I think I'll have a beer—it's past five o'clock somewhere—right now that would be Laos.

This smile on my face has been 45 years in the making and will stay with me until I get to have a beer with Dave. Cheers, Ed (wgfp)

Chapter 22

Making the Team

The response was immediate. The first came within a few minutes from Bartz:

Shit Hot!

The rest of my children jumped on board, and I crowned myself Father of the Year.

An email from Ann Mills-Griffiths made me think I had not been too big a pain in the ass throughout this process:

I am, of course, thrilled and relieved for all of you and will hope for remains recovery to top this off. You have made it very easy to work with you, and I could only hope that the others for whom we are still searching have at least one friend, in addition to family, who are as determined AND responsible as you have been and are. Best, and enjoy that beer. I'll raise a glass at lunch, scheduled with our U.S. Ambassador to Laos, Dan Clune (a GREAT supporter), and tell him how pleased you are. Best, and stay in touch, Ed. Ann

I called Charlie that day, and he was, of course, excited. This was more than we had expected. I was sure the Air Force would give him more information soon and asked him to let me know what they had to say. We also discussed Valerie, and it was decided that he would call her and let her know of the discovery.

A few days later, he sent me an email about their conversation:

Ed, I just got off the phone with Valerie. She sounded pretty upbeat and said that her hip replacement went very well, her breast cancer was apparently one of the less aggressive types, and the treatment was a lumpectomy and hormone treatments. She's been clear for almost a year and a half and is happy to be back in Connecticut (and out of Utah). She was genuinely surprised at the news and asked when they were going to start excavating, when we could expect results, etc. I told her that nothing is certain, but we're much better positioned now than we were prior to the find. I also told her that the end game is a burial at Arlington, and she said that she would definitely attend. I told her I'd keep her in the loop, and she gave me her e-mail…

I think things are looking pretty good for her, and she certainly was happy to hear from me and excited about the news. I'll keep you posted. Charlie

I had recently found out that Valerie had left Utah and moved in with her sister in Connecticut. I was glad to hear that was working out. My visit with her a few years earlier in Utah made me think she was not happy there. I was also excited to hear that she planned on attending any ceremony we might have for Dave. Perfect. I had started to build an outline for a book about this adventure, and it appeared it could still have a fairy-tale ending.

Somehow, I had forgotten to keep Phet up to speed with this development, but an email from him jogged my memory. He told me how well his sister's wedding had gone and asked how my life was. I sent him an email at once:

On Mar 15, the U.S. team along with Mr. Sorensen found the site where Dave was killed. They found his ID card and other parts of his flight equipment. However, they did not find any remains, and they will come back later to do that. Because of this, I am not sure I will need to return to Laos in search of Dave's site. However, I would like to return some time and celebrate with you and some of our other friends. I will let you know when I am coming back. Cheers, Ed (wgfp)

Phet responded that he was happy about the news and thought a celebration in Laos was a good idea. I next let Phet know that there was a good chance I would return to Laos when an attempt was made to recover Dave's remains. He expressed excitement about the prospect. He was now attempting to obtain a scholarship to complete his master's degree in the U.S., and I assured him I would do what I could to help.

A week after returning from Laos, Leland sent me an email saying he was once again a "hometown hero" on the six o'clock news. He included the links to the broadcasts. Impressed by the extensive coverage and the good interview, I forwarded the link to everyone I could think of.

I had not talked to Leland since the phone call a week earlier, but I figured he had settled in and gotten over the jet lag, so it was time to get the full scoop.

Leland was still bubbly about his adventure. He first made an admission. "I didn't recognize anything, Ed. We flew all over the area where it was likely I might have landed, and nothing clicked in my mind. We then began to stop at villages in the area and ask for witnesses that might have information. On the very last day, we found a native who said he had visited a place many years ago where he had been shown a site where there were still signs of a downed American airman. In fact, the natives referred to it as 'The Stream Where the American Fell.' After leading us to a site that produced no evidence of a downed airman, he remembered the eventual site where we found the ID card and lots of other evidence."

I asked Leland about the grassy meadow he had described in his first account of the incident. He told me the villagers told him the meadow was probably a result of burning the natives did to clear the jungle for grazing. After a few years of grazing, the nutrients in the soil were used up and it was allowed to return to jungle.

The riddle of how Dave's body got to the streambed after being wedged between a karst and a tree further up the hill was unsolved. In talking with those at the scene, Leland was told it was most likely that some natives had pulled the body out and stripped it of anything they thought might be valuable, and then it was released to roll down the hill, ending up in the streambed.

At any rate, Leland's confidence that they had found the correct site was quite high, calling it a "miracle." It had happened at the end of what had seemed like a hopeless search. We discussed the possibility of a special dig to immediately recover remains and close the site. It would be nice, but that's not the way things work in this large bureaucratic organization.

We agreed to stay in touch, and I told him I would keep him up to date with the proceedings. Finally, Leland confessed that his mind was now more at ease with the events of St. Patrick's Day 1969. *And well it should be!*

Time to check in with Bob Maves at JPAC.

"This is Bob."

"Bob, Ed Sykes here. What do you think?"

"Dig, baby, dig!" he replied at once.

When I asked if there would be a special dig, he said he doubted it. He then went into a tirade about how decisions were made above his level. His expectation was that the dig would be scheduled on the March 2015 JFA to northern Laos.

"Do you suppose anyone will tamper with the site before we dig?" I asked.

He told me that was unlikely as there were safeguards against such activities, but he did not elaborate.

Around this time, I got an email from Dustin Roses asking for contact information for Vilan Phetrasy. He said he wanted to make sure that Vilan did not attempt to disturb the site. Thinking this was a good idea, I got him the information. I never heard any more about this concern, but I assume Dustin did what was necessary. Dustin also let me know that his tour in Laos would soon be over, and Stony Beach was in the process of choosing his replacement. I hoped his replacement would be established by the time of the spring dig. I told Dustin I was sorry to see him leave his post and hoped the new guy would be as helpful.

So, now that we had a pretty good idea that the actual site of Dave's death had been found, I became curious about which of the 10 locations that Niall had presented me with was the real site. It did not take long to determine that none of the coordinates we had been given were correct, and most weren't even close. The closest of all the sites we had considered was the one that the Air America pilot, Nikki Phillipi, had recorded in his logbook. However, it was almost two miles from the site where Leland had found the ID card. As I reviewed my trip to Xan Noy, it appeared that the village chief was also in the ballpark. However, finding the villager who took the Team to "The Stream Where the American Fell" made all the difference.

***

In the summer of 2014, Charlie, John, and I met up for the League meeting knowing that there was not much more to do until the site was dug. The meeting with JPAC was upbeat. We were all impatient for the dig but couldn't do anything to speed up the process.

In September 2014, Charlie invited me to attend a local family briefing in Pittsburgh that was being conducted by DPMO. I didn't even know they did local briefings, but I would certainly attend. Charlie picked me up at the airport, hosted me in his home for a couple of nights, and introduced me to some good varieties of Scotch. We rambled on for several hours the night of my arrival, and I'm sure the Scotch only added to our optimism. We marveled that we might actually pull this thing off.

I awoke the next morning to the wonderful smell of coffee and bacon and followed my nose to the kitchen where I found Charlie and his delightful wife, Joan, preparing a country boy's breakfast. Joan, the athlete in the family, participated in several running events, giving her a fit, trim appearance. We talked about our families and the prospects for Dave's return.

"Ready to check out the meeting?" Charlie asked, checking his watch.

"Let's do it!"

The DPMO meeting was in a hotel in downtown Pittsburgh, not far from Charlie's home. As we entered the conference area of the hotel, I was surprised by the size of the crowd. Several hundred people were there, all interested in some member of their family that didn't make it back from past conflicts and was still unaccounted for. In the meeting area, I spotted Niall coming toward us with a big smile on his face.

Niall exclaimed, "Come with me! I have someone I want you to meet."

He led us to a front-row table where several men and women in business attire were seated. I recognized a few of them from my visits to DPMO. Niall introduced us to a man I didn't recognize, and with little ceremony, the man looked at Charlie and said, "I have something for you, Mr. Dinan." He picked up an object in a plastic case from the table, and I immediately recognized its contents—it was Dave's military ID. I thought I saw Charlie's hand shake as he accepted it, then he stared at it for a long moment. He thanked the folks at the table and then went silent. We found a seat near the back of the room, and Charlie let me examine the ID. I was amazed at the quality of the card after 45 years on the jungle floor. I don't think either Charlie or I paid much attention to the round of briefings that followed, but after the briefings, a number of television and newspaper reporters wanted to talk to Charlie. He did great! One TV station also interviewed me, but I'm not sure I did as well.

***

As 2014 came to a close, we were comfortable there would be a dig in early 2015. I had let Phet know that I would return to Laos to monitor the progress of the dig, and he was eager to participate. We had not been given exact dates for the JFA but knew it would probably take place in the March–April time frame. Our excitement level was high. What could go wrong?

Then in late January 2015, we got an email from Niall that burst our bubble:

Charlie, John & Ed – I've just returned from a biannual coordination conference with JPAC in Honolulu—yes, I was TOILING away there—where I learned of an unpleasant development regarding Dave's site excavation. All 2015 activities for the northern half of Laos have been bumped to 2016, due to a decision to close the southern base camp at Ta Oy (Saravane) by the end of this year, hence, only southern sites will be addressed for the remainder of this year. Ann Mills-Griffiths is in SEA right now with a delegation, so you may want to touch base with her.

As for the Stony Beach rep in Vientiane, the new fellow, Duffy Spivey, will not arrive until March.

I loathe being the bearer of this news, Niall

I read it again and then a third time, but it didn't come out any different. Unbelievable! My first instinct was to begin making phone calls and raising hell. However, having been in leadership positions in the military, I knew that I didn't hold the biggest hand grenade, and all I would do was lose some credibility with the folks we needed most to get the job done. I deduced it would be somewhat akin to strafing a gun site, and we know how that turned out for Dave.

Once I had calmed down, I sent a reply to the group:

Niall, thanks for the update and your efforts in this case. The news is very disappointing and my son's admonition regarding JPAC—"they wait until you die and then they close the case"—may have some merit. However, on the other hand, we're certainly a lot further ahead than we were twelve months ago. Hopefully, we'll get together at the June conference. Ed (wgfp)

The June League meeting was probably our best chance to roll in on the DOD folks with guns blazing. Charlie and John both agreed to be there, and we needed some strong assurances from the top that we would get a dig in 2016.

A reply to my email from Niall was also discouraging:

All, I hope to see you all at the June conference too, if I'm still here, for I'm due to turn into a pumpkin on 1 July, whereby I return to my cellblock in the Pentagon. The reorg has yet to indicate any opportunities for me to stay here permanently, but with a new interim Director…maybe the logjam will finally break…

Things were turning to cow manure in a big hurry. No dig until '16, and now they're going to let one of their finest investigators go to some meaningless job somewhere in the Swamp.

I replied to Niall, offering my assistance:

What can we do to help you secure a permanent position with the POW/MIA effort? I know Ann may be able to help. Would a letter to the right folks help? This is a job you are clearly well suited for and hope we can keep you on board. Ed (wgfp)

Niall replied to tell me Ann had already offered to help, and he was hopeful things would work out in the end.

I also sent an email to Phet, letting him know I would not be coming to Laos in 2015. He was clearly disappointed but assured me he would be ready to help if I did return for a dig in '16.

***

As usual, I sent an email to the Dinans on St. Patrick's Day:

Charlie and John, as I do every year, I plant potatoes on St. Patrick's Day and always think of Dave and his departure from this earth 46 years ago. I'm sure you also take time to remember him on this day. I feel especially bad that there is not an excavation team at work in Laos as we all expected might happen this month.

I spoke with Ann Mills-Griffiths a few weeks ago and she is hopeful that the reorganization that is going on will make the POW/MIA efforts more efficient. She asked me to pass along her best wishes to you both.

I hope one or both of you are planning on attending the conference in D.C. in June. I plan on being there. If you do attend, please list your long-lost Cousin Ed as one of the family members so I can attend the briefing.

The Dinans did include Cousin Ed, and during the family briefing, we were told that Dave's excavation was on the schedule for 2016 and should take place in the Spring JFA. Having been shut out of the 2015 schedule during the midyear review, we strongly urged them to not dash our hopes again. They indicated it was pretty much a done deal.

During an afternoon briefing, we had a chance to hear from the commander of Detachment 3, Army Lt. Col. Marcus Ferrera. In Laos, he would be in charge of the Team that would excavate Dave's site. His briefing was good, taking lots of shots at the Lao government for making their mission so difficult. He was plainspoken and nonpolitical, and I liked him at once. During the break, I found him at the back of the room and noted that he was built like and moved like an F-16 fighter pilot. Along with the Dinans, I approached him and we began to introduce ourselves.

"Col. Ferrara, my name is Ed Sykes, and these are the—"

He interrupted, "I know who you are, Ed Sykes. I was in the room at Xieng Khouang in 2013 when you were there looking for information about your friend. You be there when we do the dig next year, and I'm going to put your ass on a chopper and take you to the site!"

He hesitated slightly then continued, "You're going to have to work hard, and it's not easy work. Are you up for that?"

"Hell yes!" I responded, not believing my ears. I immediately had doubts about whether he would come through. He would face substantial pushback from his superiors that could adversely affect his career. I played along. The ball was in his court, and this was the first time a "coach" had seriously considered putting me in the game. The conversation was short as several people were waiting to talk with him, but I got his contact information and told him I would be in touch.

A second speaker that afternoon was a Navy captain, Ed Reedy, who I deduced was in charge of the JPAC forensic lab. (For those unfamiliar with the strange rank structure in the Navy, a captain is comparable to an AF/Army colonel.) He pledged to the crowd that he would find ways to speed up the identification of remains at the lab.

Following the briefing, the Dinans and I went to the bar for a celebratory beer. I excused myself for a minute to go to the boy's room, and when I returned, I found Capt. Reedy standing at the Dinan table talking to them about Dave's case. He was in Navy dress uniform with stripes on his sleeve, so I decided to go into the standard fighter pilot trash talk with our Navy brethren. The Navy dress uniform looks similar to that worn by commercial pilots. He looked at me as I approached, and I studied him for a moment and then asked, "Which airline are you with? Which airplane do you fly?"

Without hesitation, he got right in my face and yelled, "You just pushed my last button!" He went on to berate me for asking such a silly question. He gave a good performance. When he hesitated a moment, I asked the Dinans if they would like to join me in singing "Row, Row, Row Your Boat" in rounds, and I began to sing. Now Capt. Reedy, realizing I wouldn't be tethered, smiled and began a somewhat civil discussion with me about his job and Dave's case. After a few moments, he excused himself and left without a hint that I had pissed him off.

Note to self: Make sure he's not holding a grudge. I had long ago adopted a "conservation of enemies" theory that you should choose your enemies carefully and always limit the number of enemies to three or less at any one time.

So, following the meeting, I sent Capt. Reedy an email:

Ed, glad I had an opportunity to "push your last button" Friday. I haven't been attacked by a Navy puke that well since the pre-Tailhook days. I was very impressed and it brought back a lot of great memories.

Short introduction—I am a retired AF Col. who flew Thuds in SEA and was a friend and roommate of 1Lt. David Dinan. We left his body on the side of a hill in Laos in 1969, and I have been attempting to return his remains to his Country.

I was impressed by your speech and hope you are successful in accelerating the identification process—time is running out for many in that room. One of my last buttons is that my Country has not yet returned Dave to his home, and I would like to see that done before some Navy puke puts a hole in me for one of my obvious butthead remarks. Ed (wgfp)

The next day, I got a response back from him:

Thanks for your note, and appreciate you taking jibe in the spirit in which it was given. A little good-natured interservice rivalry is always good to keep us on our toes. As I mentioned then, I've gotten that comment from drunks in airports while traveling, you were the first non-child who asked the question. That's how I knew you were messing with me.

I have little to nothing to do with the research of sites, but if you need me to point you toward the right people, please let me know. I assume that after all this time, you gave that information long ago. I am standing by to assist you, however I may.

Next year, I'll wear my whites so that you can ask me for some ice cream. Make sure that you dress up special as well so that I can return the favor.

With deepest respect, and thank you for your sacrifice, Ed

I replied to Capt. Reedy:

Ed, thanks for your kind offer. I doubt that I will need your help in the short run, but I have added you to my "good guys" list in my quest to locate Dave's remains.

My favorite flavor is chocolate almond. Don't think the AF has any comparable outfits that make us the subject of so much abuse, but perhaps I could wear a muscle shirt and speedo. About the same. Cheers, Ed (wgfp)

I felt like I had taken care of my conservation of enemies. About this time, Ann's efforts came to fruition, and DPMO and JPAC were consolidated into one organization, the Defense POW/MIA Accounting Agency (DPAA), with headquarters in D.C. and operational headquarters in Hawaii. The bureaucrats were all scrambling for cover as they always do when change occurs. One good outcome from my point of view was that Niall Brannigan, although unfortunately not involved with the accounting of Vietnam War remains, was permanently assigned to DPAA as the civilian Director of the Detachment in Europe that was responsible for the recovery of WWII remains.

By the fall of 2015, I was chomping at the bit to get back to Laos and participate in the dig for Dave's remains. Wanting to find out the status of the Spring JFA, I sent an email to Marcus Ferrara:

Marcus, Ed Sykes here. We met at the POW/MIA Families meeting in June and discussed this case. We are a little more than 6 months away from your spring visit to North Laos and was hopeful that everything was still on track. As I recall, the plan was to double up on this adventure since last year's visit was canceled, and you thought May would be a good target date for an attempt to excavate Dave's site. Any new information?

Hope all is going well with your mission. I recently had lunch with Ann Mills-Griffiths and she is certainly one of your fans.

Hope to see you next spring and look forward to downing a "Beer Lao" or two if you are up to it. Cheers, Ed (wgfp)

I was careful not to mention his assurance that I would be included in a site visit because I was not sure this information should be shared with his boss. It was up to him to share this information up the chain, and an insecure email was not the best way to possibly elevate his offer.

A few days later, I got a short message from Marcus:

Ed, Everything is going according to plan. We have two open sites we have to finish before we get to Dave Dinan's site, and I expect those will be complete in April. We kick off our next mission next week and I have great expectations! Marcus

Shortly after that, I got a "P.S." email from Marcus:

By the way, the plan is to triple-up on the sites around Muang Kham. That's to protect against "Murphy." :-)

I wasn't sure which "Murphy" he was talking about, but the Lao government's messing with Dave's investigation in 2013 at Xieng Khouang would certainly qualify.

I replied to his email:

Marcus, God I hate Murphy! Have had some really fun times at Muang Kham—but never on the ground. Back in the day, that whole area was known as Ban Ban Valley and it was heavily fortified—as evidenced by all of the sites in the local area. Just by coincidence, I may be in the area next year in early May. Wishing you the best on your ongoing challenges!

I got a quick reply from Marcus: "Coincidence would be better in late May ;-)"

My response to him was short: "Wilco."

Through this short exchange of emails, Marcus had subtly let me know that I should plan on being in Muang Kham in late May. He gave me no other offers for assistance and did not speak of my role in the exercise. Perfect!

I also sent an email to Duffy Spivey, the Stony Beach person assigned to the U.S. Embassy in Laos. Duffy had replaced Dustin Roses, and I had downed a few beers with him at the League meeting.

Duffy (think that is the way I was introduced), we met at the POW/MIA gathering in D.C. in June over a few beers with the Dinan brothers.

Last time I picked up any information on your travels, I heard you were still in Bangkok awaiting your move to Vientiane. Was curious if you had made it to Laos yet. I used to call Dustin from time to time and find out what was going on in Laos and it was always interesting to get his perspective on things. The phone # I used was 856-XXX-XXXX. Is that still a good #?

If you have made it to Laos, I hope all is working out and you are settling into a good relationship with the Lao. I look forward to visiting with you over the next several months—possibly in Laos. Cheers, Ed Sykes (wgfp)

Based on my earlier meeting with Duffy, I had surmised that he was not much like Dustin Roses. Dustin was detail oriented and liked deep, well-thought-out explanations of his views. Duffy, on the other hand, jumped right in with the first thing on his mind and put it in quick, clear terms. The differences didn't matter much to me as long as I could use him to keep me out of trouble.

As 2016 rolled around, we had not heard any news about the upcoming dig. I called Marcus at Det. 3 in Vientiane and held a brief conversation without any specifics concerning my role in the upcoming JFA, but I forwarded the highlights of the conversation to the Dinans in late January:

Charlie and John, just talked with Marcus Ferrera, Team Chief at Det. 3 in Laos, and he says everything is still on track for a dig in late May.

He was very upbeat as he had just returned from a very successful field trip. This guy is great, and if there is anything to be found, he is the guy I want in charge. Let's cross our fingers for a successful 2016.

In early February, I started work on my book and made a trip to the Air Force Museum in Dayton, Ohio, to review some historical records. I also gave Valerie a call to let her know I was beginning work on the book, and she agreed to give me some help. My confidence about being able to tell a fairy tale was growing as I prepared for the dig.

In early March, I sent an email to Marcus to try to nail down my travel dates:

Marcus, hope you are having continuing success. By coincidence, I may be in the area from 15–22 May. Hope you will be there and I get a chance to buy you a cold one. Ed (wgfp)

I got a somewhat discouraging reply:

Currently sitting in Muang Kham, waiting for the weather to clear. It's been particularly bad this mission. I'll be here again next mission but will go back to do an HHG shipment to my next assignment around 21 or 22 May…

HHG refers to his household goods as he was being transferred to a new assignment. The tone of this message was not all that positive, but I now set my travel dates to ensure that I would be in Muang Kham from May 14 to May 24. I called Leland to get information on the hotel they used during his stay, and he gave me a great deal of information about the hotel, including a picture and features of the area. I then sent Phet an email asking him to see if he could secure rooms in the JFA hotel.

In late April, I sent the Dinans an email (and copied it to Valerie) with my plans:

Charlie and John, I think my plans for my trip to Laos are in the oven. I leave the U.S. on May 12th and return on May 24th. After arriving in Vientiane, Phet (my Lao friend/guide) and I will proceed to the village of Muang Kham, which is right on the Vietnam border and very close to the site where Dave's aircraft was hit by anti-aircraft fire. I suspect the actual site was just north of Muang Kham near the village of Ban Ban, which was a very hotly contested junction on the Ho Chi Minh Trail. The site of Dave's ejection is a few miles SE of Ban Ban/Muang Kham.

I have been in touch with LTC Marcus Ferrara and Duffy Spivey in the past few months, and the work in N. Laos is proceeding as planned but they say the weather (low clouds and rain) has not been cooperating. I am optimistic that there will be a dig at Dave's site this spring, but I won't be betting any $$ at Vegas on it. I will not be attending the POW/MIA Families Conference this year as it conflicts with our 50th anniversary celebration we are planning in Colorado, and Mary insists that I be there.

I will let you know how things are going—cross your fingers. Ed (wgfp)

Near the end of April, I received an unsolicited message from Marcus:

Weather's been brutal, Ed. We're starting back up next week after taking a break for Lao New Year. We're behind where we thought we would be.

Well, crap. This is sounding like 2013 all over again. Last time it was the Lao government, and now the weather was emerging as the enemy. Nothing to do but push on and do all I could to make it work.

Phet sent me a hotel update a day later, and the plan was complete:

I have called to Sengdeun hotel this morning, they said that it is full already because U.S. team is staying there around 40 rooms, and some of U.S. team will also come more to join the team from 3 of May, so I asked other hotel name is: Phoutthachan hotel (Mae pheng), which is near by Sengdeun hotel and we can walk. This hotel mostly Lao team (aircrew, staff) stay and some of U.S. team also stay there…

I have booked 2 rooms with Air condition (single room, the best one) for two of us. Do not worry!! :) One more thing: I have asked the owner of the hotel that she does not accept US dollar. So, we have to be sure for the Lao kip in Vientiane before flight to Phonsavan and Muang Kham in order to avoid problems of withdrawal Lao kip from local ATM. Phet

Just before leaving for Laos, I sent an email to Ann:

Ann, … I know that this spring is Marcus Ferrara's swan song—I think he leaves Muang Kham about the same day I do. There will be a replacement in place, but he did not give me a name. Apparently the weather in the area has been very bad and the team is running behind schedule. I'm hoping my presence might increase their interest in Dave's case, but I'm not sure that's the way it works. At any rate, I want them to know an old fighter pilot brother is still interested.

I regret that I will miss this year's Conference. Mary and I celebrate our 50th this summer in Colorado with all the kids/grandkids and she has not given me permission to break away for a day or two, and I'm way too old to be looking for a new wife. However, were we to have outstanding results in Muang Kham, I might risk a divorce and make a quick trip to D.C. and spend some time with the Dinans (they will be there).

Thanks for being such a great soldier—you remain one of my life's biggest heroes.

Cheers, Ed (wgfp)

Then I received an email from Valerie. I had forgotten that I had copied her on my email to the Dinans a week earlier.

395

Dear Ed, I read your email & hope you get to go to Laos, weather permitting. It has been too many years & soon all of us will be gone or senile. This is a travesty that has dragged on for too long! I survived pneumonia & was released 1 mo. ago to do whatever I was doing before I got sick, which was nothing when you can't breathe. I started doing cardio exercises with the old ladies at our senior center 3 days a week. Thank goodness for the chair we use for support. I can breathe now but my stamina is limited to the half hr., can't complain, at least I can do that. I have a 53 yr. high school reunion coming up in Sept. I am going with two of my former friends from high school. I will see my old beau who was married at my 20th reunion & wanted to hook up for a week. I told him I was not a homewrecker. Four of my male pals have passed away, including the former capt. of the football team. Of course, I was looking skinny & hot then! Maybe I can pull off elegant for an old lady this time! We are all awaiting the opening of our pool the end of May & can come out of hibernation! If you were a single guy here, the ladies outnumber the men 8 to 1 because they are all dead. They throw themselves at the first new available guy. My married male friend & I watch all the shenanigans. Give my regards to Mary & Scooter. Love, Val.

Reading this made me more determined than ever to make this trip a success. And maybe, if the "coach" puts me in, I'll be at the site when it occurs.

Chapter 23

Muang Kham

Once again, the realization of how big Earth is hit me as I headed west toward Laos. Twenty-six hours after departing Kansas, I made it to Bangkok. My prospect for success was quite elevated. The highlight of the flight over was a delightful shower I took at the Cathay Pacific Lounge in Hong Kong. The shower had the largest head I had ever seen, and the water felt like a delightful rain. The light sound of Asian music added to the allure. As I started the next leg of my flight to Bangkok, I was in an upbeat mood. I checked into the Novotel at the Bangkok airport, enthusiastic for a successful conclusion to this adventure. A few Singha beers at the poolside bar added to my euphoria—what could possibly go wrong?

With the help of a friendly tourist, I was able to fill out my visa form and enter the country. Exiting security, I found a smiling Phet. Although it was nearly three years since our last meeting, he looked much the same and was ready for a new adventure. Soon we had purchased tickets for our journey to Xieng Khouang the next day. Once again, I stayed at the Mercure, and Phet and I celebrated with a Beer Lao at the swimming pool. He did a little better at swallowing this one. I told him that our trip was uncertain because of the circumstances with both the U.S. and Lao governments not knowing which way was up. As always, he was inquisitive, but I did not have many answers for him.

Over the past few years, Mary and Phet had begun to correspond as she helped him with his English. Unbeknownst to me, he sent Mary an email shortly after my arrival:

I have met Edward at Mercure hotel in Vientiane capital this afternoon already. Edward is doing well and we have talked and made our plans for our trip to Muang Kham. We will leave from Vientiane to Xieng Khouang province at 9:30 a.m. and then heading directly to Muang Kham.

Edward now he is enjoying with swimming pool and relaxing himself with some Beers Lao.

We will look for your gifts during our mission in Muang Kham and in Vientiane. We will try to get things you want. Sometimes, Edward will try to call you through my cell phone :). Have a great day and take a great care, Phet

Phet returned to his apartment in Vientiane, and I caught up with some much-needed sleep. The next morning, he arrived on the back of a motorbike with a friend. We got a taxi to the airport and soon were on our way to Xieng Khouang.

Phet and I both wondered if Vilan might be at the airport in Xieng Khouang picking up guests for his stepmother's hotel, but he was not to be seen. I was sure he would have had many questions as I had not told him I was returning to Laos. Phet seemed even more relieved than me.

Phet had arranged for a taxi to Muang Kham, and soon we were on the road for the hour and a half trip. It was along this route that the North Vietnamese had sent supplies to support the Laotian Communists. As we traveled, I attempted to recognize target areas I might have bombed, but a half century had wiped those locations from my memory. The road was mostly paved, but in some places, it was probably not in much better shape than it had been when we looked for trucks carrying war supplies.

We arrived at the Phoutthachan Hotel and checked in. As we entered the hotel, I noticed a small outdoor canopied structure with picnic tables on one end and a kitchen on the other. This must be the hotel bar and restaurant. Beside the kitchen door was a wooden crate with four cute ducks slowly walking around their cage.

"What are the ducks for, Phet?" I asked.

"They are tonight's main course at the restaurant."

Phet suggested we go to lunch and I agreed. We walked up the hill to an open-air restaurant directly across from the Sengdeun Hotel where the JFA Team was staying. Phet attempted to get me to order a delicacy of the area, a "cave creature" that was delicious. I asked him what it looked like, and he took me over to a refrigerator and pointed one out to me through the glass door. I know a rat when I see one, so I opted for the always safe Lao noodles. As we ate lunch, I noticed four middle-aged men with Aussie accents and guessed they were the helicopter pilots waiting to make their afternoon pickups. I saw no need to go to the JFA hotel until after the teams returned later that afternoon.

After lunch, we wandered back to our hotel and I afforded myself the luxury of a nap. I wanted to get out of the heat for a bit. For the first time in my three trips to Laos, I felt like I was overheating. Phet had a room across the hall, and we agreed to get together in a couple hours and go find Marcus. I drifted off easily and awoke to the sound of helicopters operating in the direction of the hotel up the road. I listened for a while, wondering if I would get a chance to be part of one of those teams that were out there conducting the search.

Around 4 P.M., I knocked on Phet's door, and it took some time before he answered, the shadow of sleep still over his face. He was obviously enjoying the accommodations of our hotel. I gave him time to get ready, and he eventually knocked on my door, and we headed up the hill for dinner and, hopefully, a meeting with Marcus.

The hotel appeared to be a long, narrow building, and from the description I had received from Leland, the restaurant was in the back. A small military tent and a few Laotian guards stood in front of the hotel, whom I ignored, passing along the side of the building to where I supposed I would find a restaurant. The guards paid no attention to me as Phet and I proceeded to the back of the hotel, where we found a large elevated wooden deck with a metal and canvas cover that served as the restaurant.

As we climbed the deck stairs, I spotted Marcus seated with two others who had the appearance of fellow Team members. A laptop in front of him, Marcus was busy discussing what seemed like official business. Wanting him to know I was there, I headed toward his table. When he noticed me approaching, his face remained expressionless.

"Marcus, Ed Sykes here. I'm ready to go to work," I stated.

I got a subdued response. "Hi, Ed. How was your trip?"

"The trip was fine. How is it looking for digging Dave's site?"

Marcus looked away for a second and then looked back and said, "We're not going to get to Dave's site on this JFA. We have had all kinds of problems with weather, and we're way behind on all our digs. It's unlikely we'll be able to close any of our sites this JFA."

I was stunned, to say the least, but this wasn't the first time I'd been pushed sideways. Perhaps they would make Dave's site a primary dig on the next JFA.

"So, when is the next JFA supposed to be coming to northern Laos?" I asked.

Without the slightest hesitation, he gave me a blank stare and put the knife in my heart. "It doesn't look like we'll be back in northern Laos for another 5 years."

I was too shocked to respond. An outburst would be unproductive, especially since I had no preplanned response to that kind of news. Marcus looked at me, anticipating some type of outburst, but when it didn't happen, he returned to talking with his cohorts.

I turned toward Phet, who stared at me sadly and then asked, "What do we do now, Edward?"

I motioned him toward an empty table, and we sat down.

"What will we do now, Edward?" he asked again, and I replied, "I'm not sure Phet—go find a waitress and get me a Beer Lao."

He got up and approached some waitresses huddled nearby; he spoke to them and soon a Beer Lao appeared in front of me. We ordered food, and while we were waiting, I could hear Marcus visiting with his wife on his computer. I didn't want to interrupt that, but I also wanted to make one more run at Marcus to change these circumstances.

As our food was delivered and we began to eat, I heard Marcus finish his call. I looked over my shoulder and noticed him approaching our table.

"Going to knock off for the day," he said. "Where are you staying?"

"Down the road at the Phoutthachan."

"Good choice," he replied. "They have air conditioning, and I've been able to sleep better since I moved down there. Later," and he departed the deck and disappeared down the alleyway.

"What will we do now, Edward?"

"I don't know, Phet."

"Will we still be going touring, Edward?"

400

Touring was the last thing on my mind, but I know Phet had his heart set on it.

"Yes, Phet, but not tomorrow. I want to try and spend some more time with Marcus tomorrow. Maybe the next day."

He nodded, but the disappointment was written on his face. Several members of the JFA teams were gathering at the restaurant, and the feeling of pride in our military momentarily overcame the disappointment of my current situation. I remembered my encounter with some of the young soldiers in 2013 in Phonsavan, and it was satisfying to see these young, energetic kids enjoying themselves even though they'd been toiling in difficult conditions with only one day off every 14 days. How un-American can you get.

Phet and I made the short walk down the hill, and as we turned into our hotel, I noticed some Americans at the yellow shack that served as the restaurant/bar. I scanned the group to see if Marcus was among them, but he was nowhere to be seen.

"Let's have a beer, Phet."

"OK, Edward."

We grabbed a seat at the end of one of the picnic tables, and as the waitress brought me a Beer Lao, the American team members got up and left. I had wanted to talk with them and now they were gone. I slowly finished my beer in hopes that Marcus would appear, but it was not to be. Phet was clearly not enjoying himself, so we decided to call it a night. All the optimism that I had brought with me to Laos was now a smoking hole.

I have been a sound sleeper my whole life, but that night was the exception I will never forget. The disappointment of Marcus's revelations was too much to bear. I lay awake for hours trying to develop a plan that might convince Marcus to do some kind of a dig, and in the end, I could only think of one angle that might work. I rehearsed it over and over.

Early the next morning, I stationed myself in the lobby of the hotel, hoping to catch Marcus on his way to work. There were a few Team members gathering and exiting a side entrance to the hotel, and I asked one of them if they had seen Marcus. He told me that Marcus had already left and was probably in the van outside waiting for the rest of the group.

So I went outside and found the van with several members already inside; Marcus was sitting in one of the rear seats. Realizing this would not be a good time to single him out for a discussion, I asked, "Marcus, any chance I could discuss some issues with you today?"

He appeared to be a little put off and, after glancing at his watch, replied, "Let's meet at the hotel lobby at 1700."

"Roger that," I replied as the last members of the Team assembled and the van took off up the hill.

So, I now had 9 hours to wait for Marcus, and based on his lack of enthusiasm, I wasn't sure he'd even show up. I went back to my room and eventually heard a knock on my door. Phet asked if I had met with Marcus, and I answered we would meet later that day. He wanted to know if we could go touring, and I told him we would tour the next day.

"What will we do today, Edward?"

"Not much, Phet. Let's just hang out until after my meeting with Marcus."

Phet was not good at hiding his disappointment but nodded in acceptance. When we got to breakfast, I noticed there were six ducks in the cage this morning. Phet said there would be more Laotians dining at the restaurant this evening. Duck was obviously the specialty of the house. The rest of the day dragged by slowly, and Phet gave me room to contemplate my dilemma. Phet napped or worked on his computer in his room while I continued to rehearse my plea. As 1700 neared, I went to the hotel lobby and waited. Almost precisely at 1700, Marcus showed up.

"What do you want?" he asked, knowing exactly what I wanted.

"I want to talk about a dig for Dave Dinan."

He motioned toward the front porch of the hotel, and I followed him out the door and onto a concrete porch. There was no one around.

"What you got?" he queried.

I had decided that my best chance of getting Marcus to move on Dave's dig was to bring up Valerie.

"Marcus, there is more to the Dave Dinan story than I think you may be aware of. There was a woman involved."

He stared at me briefly then turned his head 90 degrees to the left and did not look at me again as I explained. I started with the day after Dave's death when Valerie came to the hooch and continued with her discharge from the Air Force and her having the baby and putting it up for adoption.

"In today's Air Force, this woman would not have lost her career or her baby and would be celebrated as a courageous woman. She is now in very poor health, and I feel our country owes it to her to recover Dave's remains before she is gone. I'm not sure she'll make it another 5 years. You are the only one who can make it happen. You have got to give it a shot!"

I stopped talking and awaited a response. For a long time, Marcus continued to look away. Finally, he turned his head, looked at me directly, and said, "There are too many 'squeaky wheels' in this process. I have a job to do, and you and folks like Ann Mills make it very difficult to get it done."

"You think I'm a squeaky wheel?" I asked.

"Damn right I do, and it makes my job very tough."

I thought about that before replying, "I suppose I might be considered a squeaky wheel, but let's not forget that if it wasn't for Ann Mills, you wouldn't even have a job."

He stewed on then conceded, "That's true."

He turned his head slightly and then went back into attack mode. Now looking directly at me, he asked, "And who is this Laotian kid you have with you? How do I know he isn't a spy for the Lao?"

Wow! I didn't see that coming. I explained my relationship with Phet and how he had demonstrated his loyalty to me over the past several years. Realizing no favorable resolution awaited me on this porch, I decided it was best to knock it off. I thanked Marcus for listening to me and we parted.

Back in my room, I tried to analyze the exchange. Bottom line was that Marcus had a great deal of pressure on him to deliver favorable results on this JFA, and it just wasn't happening. I also wondered if his statement at the League meeting about taking me to the site was causing him some consternation. Although I had been careful to never mention his declaration again, I had to believe he was having serious second thoughts about having made it. Taking me to the site would have been a bold move on his part since I was not there in any official capacity. I decided to take a low-key approach to asking him to open Dave's site again. However, first I would try to establish a good relationship with him. I maintained a high level of respect for Marcus despite this outcome.

I finally collected Phet from his room, and we wandered up to the Sengdeun Hotel for dinner. Phet could sense my frustration.

"How did it go, Edward?"

My initial response was a shake of my head. I finally answered, "I still have some work to do."

I left it at that. Phet allowed me to suffer in silence, which was exactly what I wanted. The walk up the hill was a painful hike as I tried to figure out what I could do to turn this around. Nothing magically appeared in my confused state, and I welcomed the cold Beer Lao I was about to order.

At the JFA Team's hotel restaurant, we found Marcus seated at the same table and in the same chair as we had found him the night before. He was by himself and talking to his wife on Skype. He acknowledged our presence with an unsmiling nod and continued with his call. Phet and I found a table to ourselves, and we ordered a Beer Lao and some spring rolls. I wasn't very hungry, but that Beer Lao sure tasted good, and I had a second one before the food arrived.

Off to my right, I saw Marcus finish his call, fold up his computer, and stand up as if to depart. He headed over to our table and informed me that he was on his way to a dinner with some of the local Lao officials and would be with them late. The hidden message being "…and don't wait up for me."

As we ate, Phet told me about the tours he had lined up for us the next day. Our driver was going to take us north of Ban Ban Village a few miles to a cave where 300 Laotians had been killed in 1969 (not exactly what I was in the mood to see right now) and also to some natural hot springs in the same area. My primary interest was getting some more time with Marcus to do something—I knew not what—to convince him to excavate Dave's site.

As we were finishing our meal, I noticed four members of one of the excavation teams sit at a table near us and place an order. As Phet went over to the cashier to pay the check, I wandered over to the new guys' table and greeted them with, "How's it going?" to break the ice, and I asked if I could join them. They motioned me to have a seat, and I offered to buy them a beer; two of them accepted. Phet rejoined us and sat quietly as I asked them about their work. Four of them were the management group in charge of one of the two field teams at Muang Kham. Their respective roles were team chief, archeologist, explosive ordnance disposal, and medic. These four people must be on site anytime an excavation was being conducted.

The archeologist, Curt Sedlacek, who, unlike the others, was a civilian with the Army Corps of Engineers, took on the role as a spokesperson for the group. He described their work as being slow and tedious over rough terrain. He said they had found a few items of interest but, as of yet, had made no significant finds.

"Do you know anything about the aircrew involved?" I asked, and he nodded affirmatively and said, "Yes, his name was James White and he was an F-105 pilot..." He continued to describe the circumstances of the loss of James White, but I already knew it well, and it brought back immediate recollections. James White was killed flying out of Takhli in November 1969.

In May 1969, the 34th TFS was converted from Thuds to F-4s because of the diminishing number of Thuds. I was transferred to the 44th TFS already in place at Korat. Then in late summer, the 44th was moved to Takhli where I finished out my tour. It was fun being reunited with several of my old friends from my UPT class at McConnell. We ate and drank beer together as we swapped lies, told our personal war stories, and reminisced about the "old days" as students.

During these interactions, I had met James White. He seemed like a great guy, but being in different squadrons, I did not fly with him or get to know him well. However, his story was widely known. His father was a general officer and his brother, Ed White, was killed in 1967 during an Apollo training exercise in preparation for the Moon landing, which happened in July 1969.

On a strike mission that November, White was operating in an area southwest of Ban Ban. The weather that day featured an overcast deck of clouds at medium altitude. We all hated operating beneath those cloud decks because it was easy for the gunners to see us. In those cases, we would pop above the clouds to make us "invisible"—a great tactic if you didn't need to go below the clouds again or if you could find a hole in the layer to give you visual contact with the ground so you could descend below the layer. However, if you had to descend below the layer and couldn't find a hole, it created a "Help me, Lord" moment in which you penetrated the cloud layer and hoped the last thing you would ever see was not a big jungle-covered rock.

I always considered it a big game of Rock, Paper, Scissors in which you were the scissors, the clouds were the paper, and the karsts were the rock. Rock breaks scissors every time. This is what led to the demise of James White. When we heard of his disappearance and the circumstances involved, we all assumed this had been his fate. Most of us had made the same maneuver but without this fatal outcome. We had experienced the extreme uneasiness as you punched your nose into the cloud layer and began your descent.

As Curt and his team continued to describe their site, it was clear that James White had hit a karst with a glancing blow and scattered wreckage over a wide area. Apparently, this site had been excavated several times in the past without success. Given the circumstances of the White family's loss, I could easily understand the priority of this attempt. By now the Team had finished their dinner and had another round of beer, and they decided to get some shut-eye as their day would start early the next morning. As Phet and I walked back to our hotel, I concluded that it was unlikely Marcus would pull Curt's Team off of the White case to excavate Dave's site.

Upon reaching the hotel, I noticed a few members of the Detachment 3 staff having a beer at the pub, so I told Phet to go on to bed and that I would meet him the next morning for our touring expedition. I had a few beers with them, and they asked me about my reason for being there. I told them, without mentioning my discussion with Marcus earlier that afternoon. As they explained their roles in the JFA, they often referred positively to Marcus, especially with respect to his dedication to the effort. I had hoped he might walk past on his return from his meeting with the Lao, but after a while, I decided to turn in and hoped the several beers I had consumed would allow me to sleep better than the night before.

I rested quite well and awoke to the sound of helicopters on their way to the sites. Looking out the window, I saw bright sunlight and assumed weather was not a factor for today's missions and that Marcus would probably be visiting one of the sites. I knocked on Phet's door, and after a few minutes, we met at the pub for our Lao noodle breakfast. Phet talked excitedly throughout the meal about today's tour, but my primary concern was having another conversation with Marcus, attempting, one more time, to convince him to dig Dave's site.

Our driver showed up as we were completing our meal; we hopped in the car and began our day of being tourists. Our route took us east and then north through what used to be the village of Ban Ban (it is now a collection of scattered huts with a few rice paddies). About 5 miles north of Ban Ban, we turned west on a dirt road and drove to a small gravel parking lot of a small Visitor's Center. No one was there. Our driver was familiar with the cave, so he led us up a long, rough series of stems to a large cave, which was a natural formation in the side of a tall karst. The cave was not what I had expected.

Most of the caves I had tried to destroy during the War were accessible by road, and the NVA had used them to hide supplies and trucks. This cave was not accessible from the highway, and its floor was covered with huge boulders, which would have made it difficult to store supplies there. Phet, through our driver, explained that in 1968, a large number of villagers and troops had been seeking shelter when the cave was hit by a firebomb and 300 people died. He said some of the bodies had not been removed and were still lying in some of the deeper sections of the cave. I did not express any interest in taking a look.

My guess is that the cave was the victim of a direct hit by a Bullpup missile shot from an F-105. Caves were an excellent target for the Bullpup because it could be guided by the pilot from release to impact. When we returned to the Visitor's Center, a young woman there attempted to sell us some souvenirs—I was not interested. From some of the pictures and plaques on the wall of the Center, I discovered that four "rockets" had been shot at the cave; three missed, but a fourth was right on target. It had to be Bullpups.

I felt like Phet studied me intently as I toured the cave and Visitor's Center to see if he could detect any sense of remorse. I felt nothing. I also guessed that Phet had not told our driver that I was a "Yankee Air Pilot," as he did not seem interested in my reactions to what I observed. *Thank you, Phet.*

After leaving the cave, we drove back toward Ban Ban but made another turn to the west before reaching the village and, after about 15 minutes, arrived at the site of some natural hot springs. Again, we were the only visitors, but as we toured the springs, it was obvious someone had anticipated this becoming a large tourist attraction. Work had been done to provide walkways and places to put your hands in the water. The water was very warm and carried a heavy odor of sulfur.

Upon finishing our visit to the springs, we stopped at a tourist shop, and Phet insisted I buy some scarves and material for Mary. *He will make some woman a great husband,* I thought. At the time, I did not realize that Phet had told Mary he would make sure I brought her gifts—a setup by my little friend.

Driving back to Muang Kham, I was relieved we were done touring. This whole time, my mind had been on what I might do to convince Marcus to "dig, baby dig." As we headed back, Phet had the driver stop at an intersection. "We will buy some fruit," he exclaimed, and I indifferently followed him out of the car. A vendor with several melons on a blanket was set up along the road, and Phet was soon negotiating with him. "Which one do you want, Edward?" he asked, and I pointed toward a nice-looking melon that looked a lot like the Black Diamonds I grew in my garden in Kansas. Smiling, Phet handed the vendor some kip and stated, "This will be very good fruit, Edward."

Back at Muang Kham, we ate lunch and I tried to take a nap, but my mind would not let me rest. *What will it take?* Phet also took a nap, and when I knocked on his door, I could tell he had slept much better than I.

"Are we going to stay more nights, Edward?" he asked. Without hesitation, I said, "No. Why don't you let our driver know we would like to return to Phonsavan tomorrow morning?"

"Yes, Edward," he replied, and I sensed he was not happy with this decision. "I will call him and let him know."

Once again, we made the short hike up the road to the Sengdeun Hotel for dinner. I was delighted to see Marcus at the same table, and he didn't seem to be busy, so I walked up and greeted him.

"How did it go today?" I asked.

"Better," he said without elaborating. "What did you do today?"

I explained our sightseeing journey, and he listened intently but did not respond. Then, almost as an aside, I said, "Phet and I bought a really nice watermelon on the road by Ban Ban, and we're going to slice it up and eat it at the pub next to our hotel later. Feel free to join us." He nodded but didn't commit or seem too interested.

He did not ask us to join him, and I assumed he would be on Skype shortly to talk with his family, so Phet and I found another table and ordered some Beer Lao and food. As we ate, I heard Marcus talking with his wife, and after finishing that conversation, he was joined by some other members of his Team, and they discussed plans for the next day.

With the sun about to go down and Marcus involved with his crew, I told Phet I was ready to leave. The street to our hotel was poorly lit and rough in many spots, and I didn't need a sprained ankle to add to my difficulties.

On the way, we encountered a crazed cow who was running around wildly and making a noise I recognized. Having raised cattle for many years, I had heard that sound and knew it meant she had lost her calf. Our biggest concern was if we were in her line of fire as she passed within a few feet of us twice at a high speed. "Run over by a crazed cow in

Muang Kham, Laos" was not the epitaph I wanted. For a few minutes, she continued her wild search until she heard her calf cry for its mommy and join her, and suddenly, she was another contented cow.

When Phet pointed to a booth set up on the other side of the street in front of a rundown house and exclaimed, "Edward, we must play the lottery!" I began to think I had really pissed off God but I didn't know how. Phet crossed the street with me reluctantly following, and he started talking with two Laotian women at the table. "Edward, we can become very rich. What is your lucky number?"

"I don't have one, Phet."

Phet then asked me to think of one with as many digits as I wanted, saying we would surely win and be "very rich—several million kips." I finally gave him a four-digit number, which he gave to the women. The women presented Phet a lottery ticket, and he was paying them when it began to rain gently. *Holy crap. Let me get back to the hotel and get some sleep—and get the hell out of here.*

"Hey, Ed, you still going to cut up that watermelon?"

I recognized the voice and stared into the darkness until I could make out Marcus, walking by himself back toward the hotel.

"Damn right," I exclaimed excitedly. "See you in a few minutes."

He acknowledged my reply and soon disappeared into the darkness and rain toward the hotel. Phet was still engaged in conversation with the two lottery peddlers, but I interrupted them. He bought a ticket and we continued on.

"Where is the melon?" I asked Phet.

"It is in my room."

"When we get back to the hotel, run up and get it and bring it down to the pub. Do you think they will let us borrow a knife to slice it?" I asked.

"They will slice it for us, I am sure," he replied.

As we approached the pub, I spotted Marcus at a table with two other Team members and asked if I could join them. Marcus motioned that I should have a seat.

"Phet is bringing down the melon now, and we'll get the waitress to slice it for us."

I had met the two men with Marcus the night before and knew they were involved with one of the site teams. It turns out they were involved in searching for the remains of AF personnel who were lost with the fall of Lima Site 85 (LS 85) in March 1968. The fall of LS 85 resulted in the largest ground loss of AF personnel in the Vietnam War. Chief Master Sgt. Richard Etchberger was posthumously presented the Medal of Honor for bravery as he defended the site and assisted his wounded troops in their recovery. He was subsequently killed as he was being hoisted into the helicopter. The Medal of Honor was not awarded publicly until 2010 after the War in Laos had been declassified. DOD had attempted on several occasions to recover the remains of several of those lost, but as yet, there were still 7 of 12 to be found.

So, I now knew that two of the teams were involved in recoveries that had high priority. I knew there were three teams with most JFAs, so I asked about the third team. Marcus explained they were not stationed at Muang Kham but were remotely situated at Lak Sao, further to the south, where they were searching for remains of aircrew members of a Jolly Green Giant shot down by a MiG-21 in 1970. I could now better understand Marcus's dilemma. There were three high-priority sites ahead of Dave's, and little had been achieved.

Phet had returned with the melon and was busy with the kitchen crew preparing it. He brought the melon to the table and offered it to everyone, but the two Team members declined and said they needed to get to bed. They left me and Phet alone with Marcus. I knew Marcus was weary of Phet.

"Phet, would you go get on your computer and see if we can get a hotel in Phonsavan tomorrow night and also a room for me at the Mercure the next night? You can let me know what you find out tomorrow morning."

"Yes, Edward," he answered with some confusion. I had not told him about the suspicions Marcus had expressed the previous day.

Now it was just me and Marcus and some Beer Lao and melon.

"Are you leaving tomorrow?" he asked.

"Yeah, I don't see any need to hang around here any longer."

Marcus said after a minute, "This JFA has been very tough. The weather has been such a bastard. Also, funding for blade time [helicopter usage] is not adequate."

I decided to not dwell on a dead-end topic and, instead, find out what we had in common. I soon found out that Marcus grew up in southern California with four brothers and a sister. His father ran a bakery, and Marcus planned on joining his family in California after he retired. Marcus had graduated from West Point and served several tours in Southwest Asia. Marcus is not a big man but appears extremely fit. I could easily picture him in full gear on patrol in Iraq or Afghanistan. I shared the story of Bob Carmody, which he listened to intently, giving a slow shake of his head in acknowledgment. He then shared that his brother, Matthew, had also graduated from West Point and had been killed in Afghanistan.

As I always did when I heard these kinds of revelations, I thanked God that I had chosen to be a fighter pilot. As I had lain between clean sheets in my air-conditioned hooch at Korat, I would often think about the grunts on the ground throughout the War and wonder if they would get their three square meals that day. I knew for sure I would. My forays to the front had clearly been dangerous, but I was also afforded the opportunity to retreat to a comfortable environment after being shot at. Also, there was little chance of being crippled for life. A lonely death, torture, or being a POW were the most likely outcome if the "Golden BB" found you. For me, it was more than a fair trade-off.

I communicated this perspective to Marcus, and he talked about his knowledge of aerial warfare. He stated that he spent most of his free time reading books and cited two books I knew well. *Thud Ridge* and *Going Downtown* were both written by Col. Jack Broughton and talked about his experiences as a vice wing commander at Takhli. Written from the perspective of a senior leader of an F-105 unit during the War, the books conveyed his disdain for the rules of engagement in the air war and detailed how he was court martialed by the Air Force for protecting his pilots after they had violated one of these rules.

As I listened to Marcus express his views, I concluded that this guy should become a general officer. Our country needs more leaders like

Marcus Ferrara. The only things going against him were his tenacity and honesty, and I also suspected he may be lacking in political correctness. Damn the luck! There was no doubt, however, that he is a real warrior.

As we parted that night, I said I probably wouldn't see him the next morning as Phet and I would sleep in a little and then head out. I did not ask him to do any more on Dave's case, and he did not make any promises. It was what it was.

Chapter 24

Diggin' Up Bones

I awoke Wednesday, May 18, 2016, to the sound of helicopters launching up the hill. Despite all the beer I had consumed with Marcus the previous evening, I felt pretty good. Knocking on Phet's door, I heard him rustle around before slowly opening the door. He looked like he was the one feeling ill.

"Were you able to get a hold of our driver?" I asked.

"Yes, Edward. He will be here around ten."

"Great, I'll meet you at the restaurant."

I spent a few minutes getting my clothes packed and went downstairs and across the parking lot to the restaurant. I noticed there were eight ducks packed in the cage this morning, so the restaurant would have a late night tonight. As I sat alone in the restaurant, I felt some remorse for coming away from my brief stay in Muang Kham with so little to show for my efforts. However, I was relieved that I had managed to establish a more comfortable working relationship with Marcus. As the sound of the helicopters penetrated the bright, beautiful Laotian morning, I imagined Marcus up the hill smiling, happy that the weather appeared to be giving him a break and his teams were headed to their sites.

Phet showed up shortly, bags in hand. "They are flying today. That is very good, Edward."

"Yes, it is Phet," I replied with some enthusiasm.

Phet ordered our standard bowl of Lao noodles, and after a few minutes, we were picking through our flavorful bowl trying to determine what was in today's concoction. Phet called our driver, and he showed up shortly. He must have been waiting on the street for the call. We ordered him a bowl of noodles, and I finished mine off and went back to the hotel to get my bags and check out while Phet and our driver were engaged in some Laotian banter.

The drive to Phonsavan was rather sobering. Three days in Muang Kham with little to show for it, and it might be another five years before another attempt is made. The reality of the situation grabbed me, and I became anxious to get out of Laos and retreat to Bangkok and think this reality out in friendlier confines (Wrigley Field came to mind!). I decided to return to the Novotel in Bangkok and hang out there as I waited for my flight home.

Arriving in Phonsavan, we found a room at the Hotel Phonsavan. It was a nice hotel with comfortable, air-conditioned rooms at only $27 a night. After checking in, we had our driver take us to the Xieng Khouang Airport to arrange for a flight to Vientiane the next day. Since we were changing our original reservation, Phet and some airline guy chatted for several minutes and finally got the problems resolved with no extra fee, unlike the change charge we had encountered in 2013.

I treated Phet to a celebratory massage at the hotel that afternoon. I also got a massage but had trouble enjoying it because my mind was not settled. Later we went to the restaurant across from the Mali hotel we had enjoyed in 2013 and were disappointed. The music was loud, the water level of the pond was low and with a foul odor, and the curry dish I had enjoyed in 2013 was no longer on the menu. We left and went to another restaurant that was OK, but this day was not going well. Back at the hotel, Phet got on his computer and booked a room at the Mercure for me the next night but could not change my flight on Thai Airways to get me to Bangkok the following day. As I went to bed, I was in a "this really sucks" mood.

We flew to Vientiane the next morning. I was always nervous about the one flight a day in and out of Xieng Khouang, but it had never failed me. I checked in at the Mercure, and after lunch, Phet tried again to change my Thai Airlines reservation but could not get it done on his computer. So we got a driver to take us to the Thai Airlines office, and they were able to get it done with no change fee. They were also able to book me in to the Novotel for the three nights before my departure back to the U.S. on May 23. Phet and I had dinner on the Mekong, which was a short walk from my hotel. Unlike during my earlier stay in March 2013, a lot of vendors were set up along the river in tents and makeshift huts; the smell of the food was fabulous and the many Beer Lao signs enticing.

Phet wanted to know about any future trips I might have planned, but I could only answer that it might be five years, if that. He was clearly disappointed, and I couldn't believe it myself. The food was tasty, and I downed what I suspected was my last Beer Lao—maybe forever. Maybe the only redeeming thing about Laos was its beer. We returned to the hotel where I attempted to settle up with Phet with several $100 bills, but he rejected most of them because they were not new and crisp.

"The banks will not accept them if they are soiled," he stated.

I agreed to wire him money when I returned to the U.S. Phet had to return to his village in the morning to help his family deal with his mother, who had fallen ill. I insisted we go down to the pool bar for an *adios* beer, and he agreed. Phet rarely drank beer, but he did this time, and we talked about our adventures together, not knowing if our paths would cross again. I wondered what would become of my young friend in this government that had little to offer except slogans and promises.

The next morning, I caught a taxi to the airport using my last kip, checked in for my flight, and anxiously waited to board my flight and exit Laos. The "thunk, thunk" of the main landing gear retracting into the wheel wells after takeoff gave me a comforting feeling. An hour later, I was in Bangkok, and an hour after that, I was checked in at the Novotel. I had decided to spend my three days in Bangkok writing the chapter in my book entitled "Muang Kham," working out in the fitness center, and logging lots of time at the pool and its poolside bar. Now, instead of Beer Lao, I was drinking Singha while I lounged in the beautiful pool area, reflecting on the status of my adventure and writing. I'd come so close to an excavation of Dave's site, but now, five years seemed like an eternity to this 73-year-old man.

On the evening of my second day in Bangkok, I went to the pool bar to have a beer and found myself alone with a large man seated a few feet to my left. He was talking to the bartender with a strong British accent. They were trying to get his Bluetooth system on his phone to play his music on the bar speakers. They got the problem solved, and soon we

416

were listening to hit songs from the '60s through '80s—not too loud, just right.

I broke the verbal silence. "How does that thing work?"

He went on to explain Bluetooth, and while we were talking, a lovely young Thai woman appeared opposite us from the swimming pool bar and asked for a drink and spoke to him about their schedule.

"Let's hang out here a little longer and then to our room for some rest," he answered.

She nodded and went back to swimming.

"That young lady is my daughter, and we're getting ready to catch a plane to South Africa tonight," he explained. "I'm Gavin."

I introduced myself, and soon Gavin was telling me all about the trip to South Africa, where he was going to take his daughter to a game ranch where he had secured her a position for the summer. As he was explaining their plans, she walked up and heard us talk about her new position and excitedly explained what her duties would be at the ranch. She asked Gavin a few questions about their trip and returned to the pool. Turns out Gavin was a native of South Africa but now lived on an island off the west coast of southern Thailand.

"And what brings you to this wonderful place?" he asked.

"Just left Laos and am passing through on my way back to the States."

"What were you doing in Laos?"

I gave him the short version of my quest to find my friend's remains.

He listened intently, and when I had finished, he stood up, all 6 feet and 350 pounds plus of him, and walked toward me. As he covered the 6 feet between us, all I could think was, "Oh shit, brace yourself."

He pulled me from the bar stool and wrapped his massive arms around me, giving me a hug that seemed to last 30 seconds. "Good for you, mate!"

And with that, this complete stranger had filled my tank with gas again. It's really hard for a fighter pilot to admit to even hugging another man, but to let on that it was a magic moment is even worse. But there it was, and I could feel the enthusiasm once again. We talked a bit more, and he explained his interest in the War. In fact, he pointed out a tattoo on his right arm that was dedicated to a battalion of North Vietnamese women who had fought for "Uncle Ho" during the conflict. I couldn't tell if he had chosen sides in the War, but he seemed more interested in the history of the conflict.

Shortly after, his daughter came up and said she was going to the room to pack for their trip, and he said he would join her. After shaking hands and exchanging some pleasantries, they were gone. I had another Singha or two and found my attitude about life and my quest buoyed significantly. I came by the bar again the next night hoping that Gavin's flight might have gotten canceled and I would find him at the bar again, but his bar stool was empty. I had a Singha and made a silent toast in the direction of Gavin's empty seat.

"Thank you, mate!"

The flights taking me back to Kansas were as grueling as ever, but while en route, I formulated a plan. Only a few weeks remained in this year's JFA to northern Laos, and I thought I would make one last attempt to see if I could get a team up to Dave's site. I tried to think of some argument that might get Marcus to give some attention to Dave. And then it hit me: Marcus had a great love of books and had read many of the great books about the War, including *Thud Ridge* and *Going Downtown*. Maybe this was the angle.

After a few days of thinking it through while I once again tended to my little Kansas ranch, I sent the following email to Marcus:

Marcus, have made it back to the big PX and am so glad to be here. I'm sorry I added to your stress level during my visit to Muang Kham. However, I am determined to do as much as I can to right this wrong. I was very much involved in finding Leland Sorenson and getting him to Laos two years ago, and I was very encouraged by what was discovered. I discussed this discovery with Bob Maves shortly after it took place and his comment was "dig baby, dig." Now two years later, it appears that nothing will happen on this JFA and it is unlikely anything will happen for another five—I know you, Marcus, and if this had happened in your brother's case, you would be a very squeaky wheel!

Six months ago, I began work on a book about Dave and Valerie and the attempt to return his remains to the U.S. I spent three days at the Novotel at the Bangkok Airport on my way home working on the book—mostly on the latest chapter, "Muang Kham." As you can imagine, you are the central figure in this chapter. I can't wait to read about you when the complete story is told.

On another subject, your input about funding for the recovery efforts did not go unheeded. My Representative from the 4th Congressional District of Kansas is Mike Pompeo. I have met Mike on a few occasions and have worked with his staff in D.C. and in Wichita, but I do not have a close personal relationship. If you would be interested in testifying before one of the funding Sub-Committees, I would be happy to see if I could get you on the agenda—I know you would do a great job. By the way, Mike graduated number one in his West Point Class of 1986 and seems to have a deep interest in these types of efforts.

Please let me know of any progress in the next few weeks. I also wish you the very best in your next assignment and your eventual return to CA and your dream of being near your family. Cheers, Ed Sykes (wgfp)

As I pushed "send" on this email, I wondered if Marcus would want to be portrayed as the bold hero or something else in my book. I soon received a reply:

Ed,

No worries. We haven't been able to get to our sites for most of the last week—we are currently on the fifth straight day of rain. If any team finishes their site, they will go get Dave Dinan. What I don't know is if that will happen. Bob Maves is a good man and we all agree on this one. What I can't do is predict whether we will finish our sites or not. We will do our best. Marcus

Not exactly what I wanted to hear, but a glimmer of hope remained.

Then, two days later, I got a short email from Marcus that I had to read several times to believe:

We're bringing RT3 up and will excavate 1408. I hope we have enough time. Right now, the weather is grounding us every day. Marcus

SHIT HOT!

I immediately emailed Marcus:

Marcus, in just a few hours, I will be giving a Memorial Day talk at McConnell AFB (Home of the Thud). There is a Memorial Walk at McConnell that has the names of many F-105 pilots who trained at McConnell and are now deceased and they or their families wanted them remembered on the Walk. One of the names on that Walk is 1Lt. David T. Dinan III. I had not planned on mentioning anything about my findings in Laos until now. I will

419

be happy to announce that MY COUNTRY is now fully involved in recovering the remains of one of our brothers. God Bless you, my friend—we could ask for nothing more.

God Speed, Ed Sykes (wgfp)

I sent an email to the Dinans:

Charlie and John, am back from Laos and was not happy with the news I was going to have to give you. However, looks like my efforts in discussing the situation with Marcus paid off. They are beginning the dig. I will give you the details later—they are very interesting. I am praying for good weather and good results! Please do not share this information until after the results of the dig are known. The current exercise will end on June 16. I will let you know if I find out anything. However, if they encounter success, they will probably contact you first—in that case, please let me know what you hear.

Cheers, Ed (wgfp)

Charlie responded that I would be the first to know. A few days later, we got an email from Niall informing us of what we already knew:

Gents - I've learned that a recovery team moved to Dave's site this week and began digging on Thursday!

Ed - are you still lounging around out there, drinking Laotian fire-water?!

Well, crap! I had forgotten to inform Niall of Marcus's decision. I replied:

Niall, Marcus told me on the 30th that they were going to begin digging. I am back in the U.S. after a few days' stay in Muang Kham but did manage to drink a good amount of Beer Lao and get Marcus drunk enough to decide to go to Dave's site (story to follow).

That was only a half-truth, but it sure sounded good.

Charlie sent an email to Niall letting him know he had his fingers crossed, to which Niall replied:

I also have my toes & gonads crossed Charlie, so hope I don't have to do the latter trick for too long!!!

I had to ask Niall if it hurt much to cross his gonads, and he replied:

Nah, it didn't hurt at all Ed—I learned the trick back when I used to moonlight as a Chippendale!

I also realized that I should let Leland Sorenson know about the dig:

Leland, got this from Niall yesterday. Cross your fingers. I got back from Laos a week ago and was pretty pessimistic when I left. Marcus Ferrera (I know you met him) was pretty sure they would not get to Dave's site this year despite my very serious nudging. I followed that up with a begging email and got the word last Sunday that he had decided to dig. They got the site prepped and began digging on the 2nd.

Muang Kham was just as you described it. Luckily the hotel you stayed at was full and I found a "VIP" suite at another hotel just down the road. Americans don't realize how good we have it... Cheers, Ed (wgfp)

***

Meanwhile, starting May 30, things had changed in Laos. Brandt Vollman surveyed the sky as he emerged from his hotel and noted several low clouds in the area, but it was not completely overcast. *Damn this weather,* he thought as he looked for the helicopter crews to find out what their prospects were for the day. Weather had plagued their efforts on many days during this JFA, and the chopper pilots and the Lao colonel made the decision to launch. Despite working a schedule of 13 days on and one day off since they had arrived in Lak Sao over five weeks ago, the weather had prevented them from going to the crash site over half of the days. On many days when they did make it to the site, they had been required to leave hurriedly as weather moved into the area.

Brandt had wanted to be an archeologist since his youth and, after receiving his master's degree, had landed a job with the state of Missouri working on the preservation of historical sites. In 2013, he had volunteered to be part of a DOD site excavation in Tarawa, where he assisted in the recovery of many Marines killed there during WWII. His wife, Coleen, was supportive of this effort as her family came from a long Marine tradition. His success in this adventure had encouraged him to volunteer with DPAA in 2016 to search for remains in SEA. But the success he had experienced on Tarawa was not being replicated at Lak Sao.

Brandt's team was part of the main JFA Team at Muang Kham, but in an attempt to save blade time, they were operating a good ways south of the main contingent. Marcus Ferrera came to visit the outpost from time to time, but they had a good deal of autonomy in their operation.

Like all three of the teams in this JFA, the weather was causing them significant difficulties. Plus, this particular search was very much a needle in a haystack.

The site they were working was a shoot-down of an H-53, a Jolly Green Giant (call sign JG-71) by a MiG 21 on January 28, 1970. Six crewmembers had been on board. They had been part of an effort to rescue a downed F-105 pilot in southern North Vietnam. As JG-71 and several other Jollys were preparing to refuel, a MiG 21 approached the flight from behind and fired a heat-seeker missile at JG-71. The missile entered the fuselage from the rear and ended up in the instrument panel and exploded. JG-71 immediately fell and plummeted to the karsts below and hit on the top of a peak and exploded. All six crewmembers aboard were assumed lost.

The bodies of the pilot and copilot had been found strapped in their seats by a much earlier dig, but the remaining four crewmembers had never been accounted for. Brandt's team had found some major components of the helicopter but had not found any remains of the crewmembers. It appeared that a tremendous explosion had taken place as the chopper hit the ridge of the karst, and the helicopter and its four occupants in the back had been scattered in all directions along the steep slopes of the karst.

As he emerged from the hotel, he spotted Sgt. "V" Villanueva-Lopez talking to the chopper pilots. Sgt. V was the military team leader of the Team, and as he spotted Brandt, he ended his discussion with the pilots and rushed over to talk to Brandt.

"I just got off the phone with Sammy back at Muang Kham, and he let me know we are being moved up there to dig the Alternate Site."

This was not what Brandt wanted to hear. After five and a half weeks of working the Lak Sao site, he was not happy that they had accomplished so little. Sgt. V explained that they had two days to close down their operation and move to Muang Kham. The folks there would make arrangements for their arrival. They would also begin preparing the new site for excavation. Despite his dissatisfaction, Brandt realized he had little recourse and so prepared for the move. When Sgt. V announced the decision to the Team, however, most of them were happy to make the change. After many days hanging out around Lak Sao waiting for the weather to break, they welcomed any increase in activity.

Two days later, the Team was moving into their new digs at Muang Kham and being brought up to date about their new case, 1408, Lt. David T. Dinan III. The site already had a landing zone (LZ) for the helicopter and was being cleared for the recovery operation. Three days after departing Lak Sao, the primary four members of the Team, including Brandt and Sgt. V, were transported to the site to look around. It was pointed out to them that the Lao workers had already found some aircraft wreckage.

Oops! There should not have been any wreckage in the area where Dave died. Brandt got out his GPS and compared his position with the coordinates given to him concerning the 2014 discovery of Dave's ID card. He quickly determined that they were at the wrong site. After some quick recalculations, it was discovered that, indeed, the wrong site had been cleared, and they needed to regroup. A Lao team was dispatched at once to clear an LZ at the proper site and the following day, the Team visited the new LZ and determined it was the correct site, and the Lao began clearing the entire site for excavation.

Figure 11 - Lao workers looking for remains on the hillside above "The Stream Where the American Fell," 2016. Photo courtesy of DPAA.

Brandt was amazed at how fast the Lao could clear a site out of dense jungle. Most of the smaller trees and brush were cleared using machetes. A chain saw was used for bigger trees. The search area was steep, and a rope-assist system was installed to help the diggers ascend the hill. The Lao also installed steps in some areas to assist the "fat old man" (Brandt). Two days after establishing the new site, Brandt set the datum point for the dig and marked it for an archeological dig. Then the real work began.

The Lao would dig into the soil in each section of the site and place the dirt into buckets after recording where each soil sample came from. They would then climb the hill and deposit the contents onto a screen sifter that was moved back and forth by Lao workers while Team members looked for items that were too large to pass through the screen. They found many items of interest from the start.

Figure 12 - Lao workers sifting through soil in search of remains near "The Stream Where the American Fell," 2016. Photo courtesy of DPAA.

Chapter 25

My Country Redeemed

Mary and I had rented a fabulous house in the foothills of the Rocky Mountains and had asked our kids and their families to attend our 50th anniversary celebration. All four of our children and their spouses, as well as our 13 grandchildren, were able to attend the weeklong gathering. Our children had done a great job of choosing their mates, and the group included a Protestant, Catholic, Muslim, and Jew. When my friends in Kansas asked me about my family, I would say my children had married foreigners—a Canadian, a Moroccan, a Georgian, and a delightful woman from the People's Democratic Republic of Massachusetts. We shared in a fun celebration, Mary and I taking delight in the success of our half-century relationship.

Before leaving for Colorado, I had learned that the JFA had concluded and that the Team had found a number of curious objects at the site. During the celebration, I spent a lot of time thinking about the League meeting and the briefing the Dinans were receiving in D.C., but I decided to wait until I returned to Kansas to get more information. Back on my little ranch, I didn't have to wait long.

In an email to Leland Sorensen, Charlie Dinan summed it all up:

Leland, my brother, John, and I attended the 22–25 June POW-MIA Conference in Arlington, VA, and received a briefing by the Defense POW Accounting Agency.

More than two years after finding the material evidence and David's identification card, the DPAA recovery team revisited the site and started an excavation. During the first four days, they recovered additional suspected material evidence and on days five and six recovered suspected osseous material. Unfortunately, inclement weather forced the team to temporarily shut down the site. In the meantime, the suspected osseous material is being forwarded to the lab for verification. We were not advised of the amount of the material that was recovered but were told that it was recovered over the last two days of the dig. We were also told that the excavation was started at the streambed and is progressing up the hill. It won't be completed for several additional days, when they cease to find material.

Unfortunately, we do not know when the excavation will be reopened. We were told that DPAA is meeting with Laotian officials sometime in August in an attempt to expand the recovery process, but John and I were not encouraged by the information being given to us. If the August negotiations are not successful, the earliest reopening of the site will be in 2018. I'll keep you posted.

Best Regards, Charlie Dinan

I wasn't sure what "osseous" material was, so I looked it up. Apparently the two years of Latin I had taken in high school had been a waste of time. "Oss" is the Latin word for "bone." That is good news! However, the fact that the site dig was not completed and would probably not be reopened until 2018 was bad news. The question that came to mind was if they would wait until the site is closed to analyze the items they had or would they do it sooner? I didn't have to wait long for an answer. Charlie had already put that question to Niall, and he said there was little we could do except wait.

So, what to do while we are marking time? I asked Mary if she would like to travel to Laos in the winter and meet Phet. We could travel to Xieng Khouang and she could get a feel for what I had been up to. I was a little surprised when she said she would like to go, so I promptly put her in charge of planning the trip.

I knew that Marcus was now assigned to a post in Phnom Penh, Cambodia, and sent him an email letting him know of our plans:

Marcus, hope all the dust is settling and you are on your way or settled in Cambodia. The reports we have gotten from the dig at Dave's site are pretty promising, and I thank you for your participation in this effort!

My wife, Mary, and I are beginning to plan a visit to SEA in Jan of 2017, and Cambodia will be on our list of places to visit. Any chance we could get together and we could treat you and your family to some fine Cambodian cuisine—and beer?

I'm hoping your life is much less stressful than when I last was with you in Muang Kham. Cheers, Ed (wgfp)

A few days later, I received his reply:

Ed, Life is good. Let me know when you'll be in town and hopefully we can meet up. I also agree—the results sound good and I think David Dinan may finally be home. Marcus

Great! It will be good to see him and fun for Mary to meet Phet. As Phet loved to explore, we built a trip that would include a few days of travel with him. He was very excited.

As the trip approached, I checked in with Charlie to see if he had heard anything—crickets. By now, we assumed that nothing would be done to identify Dave's remains until the case was closed in 2018 (if we were lucky). We decided we would attend the 2017 League meeting to get the standard briefing and see if we could push for a faster result.

The trip to SEA was really fun. We met Marcus and his delightful wife, Oh, for lunch in Phnom Penh. I wasn't sure if Marcus still considered me a "squeaky wheel," but the meeting could best be described as "civil." I learned that he had been selected to attend an Army Senior Officers school. This meant he would most likely be promoted to colonel and possibly more. I was elated to hear the news as I considered him to have great leadership ability.

Mary's comment after the meeting was, "He's pretty intense, isn't he? He stared at me during lunch, and he seemed to be asking me, 'Why did you marry this man?'"

I laughed it off but figured she was probably right. He's not the first guy whom I truly respect that I've also pissed off.

We met Phet and toured parts of Xieng Khouang and Luang Prabang and then took a cruise up the Mekong River. Mary and Phet hit it off at once. Phet's mother had died the year before, and he appeared to adopt Mary as a new mother. He had alluded to me being his "father" in previous discussions, so I guess this made it official.

***

By the time the 2017 League meeting arrived, we were anxious to get any new word on Dave's case. At the urging of Ann, the Dinans and I went up to the Hill and visited with all three of our U.S. representatives. My representative, Ron Estes, was friendly and promised to assist us where he could. Charlie's representative was not available, but he did meet with his military staff person, who promised to help. Finally, we met with John's representative, Rep. Rod Frelinghuysen of New Jersey. A Vietnam veteran himself, he spent nearly an hour discussing Dave's case and the mission of DPAA. In his role as chairman of the House Committee on Appropriations, he would subsequently add several million dollars to DPAA funding. We were impressed by his attention to the accounting issue.

The briefing of Dave's case was as we expected. There was no news of a special dig at Dave's site and no news that the remains found at the site were being analyzed.

"Let's go get a beer!"

John said he wanted to go to their room, so Charlie and I ventured down to the bar. As we entered, we spotted Bob Maves by himself. He came up to us as soon as he saw us and asked if we would like to come up to his room for a beer. Only one answer was acceptable, and we proceeded to his room. We grabbed a beer out of the cooler Bob always had in his room, and he got right to the point.

"Mr. Dinan, there is always a chance that they might analyze the remains that were found at Dave's site before the next investigation in northern Laos. If the identification is positive, there is a chance they would call you and let you know. They would probably also ask you if you wanted DPAA to continue digging for more remains or if you would be satisfied with what was found and allow DPAA to close this case. If you decided to let them close the case, the remains would be returned to you for burial."

After exchanging a glance, I told Charlie I would not be a part of that decision. He told Bob that he would discuss it with John and decide what their answer would be in case they got the call. Charlie asked several more questions about the process, but Bob could not elaborate any further on what might happen. The one thing he did say that had an impact was the fact that if Dave's case was closed, DPAA would be able to use its assets to move on to other cases.

We finished our beers and thanked Bob for this information and decided to find John. He was still in their room sleeping when we entered.

Charlie told John of our meeting with Bob and they discussed the matter; as they carried on their discussion, they were leaning toward closing the case and letting DPAA concentrate on other unsolved cases. It was a good conversation but, for now, was only a what-if conversation. I certainly did not make any suggestions.

A few months later, I was with my son-in-law Jeff Vaughn, returning to Wichita after a baseball adventure, when my cell phone rang. At the time, we were driving through the Flint Hills of Kansas, close to where Mary and I had spotted those Thuds crossing the highway at breakneck speed nearly a half century earlier. Every time I traverse the Flint Hills, I think of all the times I have dashed low across this beautiful landscape at well over 500 miles per hour.

The call was from an excited Charlie Dinan. He had just gotten a call from the AF Casualty Office, who informed him that the remains found at Dave's site had been positively identified. They had also asked him if the Dinan family wanted to continue to excavate the site or if they would accept the remains found as closure and have them returned for burial. Charlie answered that the family would accept the remains from the June '16 dig and allow DPAA to close the site.

Once he had given this consent to close the case, they asked him where the remains should be returned to, and he asked for a burial at Arlington National Cemetery. He told me they immediately looked into setting a date for the burial, explaining the earliest it could take place would be in April 2018. He told me he had accepted that result and would keep me posted concerning the final arrangements.

Oh, so very SHIT HOT!

My mind flashed back to that day in 2013 when I stood in a rice paddy outside the village of Xan Noy looking at the hill where the village chief had told me Dave had died. I had felt the strange spiritual sensation that Dave was with me and I had let him know, "We're coming to get you, Dave." However, I also remember that, at the time, I had had serious doubts about whether that was possible. Was it blind luck that Leland had found that ID card on the ground or had Dave somehow made it magically appear? Blind-ass luck is probably the rational answer, but as I pondered my choices, I whispered, "Thank you, Dave."

***

A few months before the ceremony at Arlington, now scheduled for April 25, I got a call from Charlie. He asked me if I would do the eulogy for Dave's ceremony.

"Of course, Charlie. I would be honored."

He proceeded to give me the ground rules that had been given to him by the folks at Arlington, mostly having to do with the length of the eulogy—five minutes maximum. I told him that would not be a problem. I did not tell him that short was better for me because I would have to memorize it, given my poor eyesight.

"Have you talked with Valerie about attending the ceremony?" I asked.

"I did call her and give her all the details, but she said it was very unlikely she would attend," he replied.

I didn't ask any further questions concerning Valerie or her daughter. I had previously let the Dinan brothers know that Valerie's inclusion in the ceremony was a family matter and I would not intrude. It didn't look like the fairy-tale ending of Valerie and her daughter being reunited at this ceremony was going to happen. The idea, however, had inspired me to write a book, and in many ways, it would still have a fairy-tale ending.

After finishing my call with Charlie, I wrote the eulogy and began practicing it that same night.

A few days before the ceremony, Mary and I departed Wichita on Saturday morning and were soon in D.C. after a brief stop in St. Louis. I had a personal celebration on the second leg of our flight by getting a Scotch on the rocks, sipping it slowly as I thought about this amazing journey. It was really happening! We grabbed a taxi to our hotel and welcomed our family over the next few days. Three of our children and 10 of our grandkids had decided to attend the ceremony, and I was so happy. However, the sense of apprehension about what could go wrong (even though not much could) was a continual burden I carried. I would wake up in the middle of the night and practice my eulogy over and over—really didn't want to fuck this one up!

On Sunday night, I was joined by my family at the Lincoln Memorial. What an inspirational sight! From there, we took the short walk over to the Vietnam Memorial and found Dave's name on the Wall. The little cross signifying "Body not recovered" had been removed and the spot had been filled in. This is where this adventure began, and now my mission was complete, nearly 9 years later. My country was redeemed, and I had helped make it right.

The arrival of the body was the big event. On a bright, windy D.C. morning, we gathered at a central location where my family, the Dinan family, and a large number of friends jumped on a waiting bus and proceeded to Washington Reagan Airport and through a special TSA gate for such events. Soon we were waiting at one of the gates in the Delta terminal. The body was due in around noon, and Mark Rockoff was due to arrive at another terminal at 1100, and somehow my family figured out a way to get him to join up with our crew.

As our plane taxied in, we were escorted down the gateway and down the steps to the ramp. I had decided to wear my uniform for the arrival, and I was asked by Dave's nephews, David and John, to join them in the official military reception line. As the casket was moving down the specially provided ramp, my knees became a little weak. *This is really happening,* I thought as the honor guard, in a most impressive fashion, picked up the casket and marched it to the waiting hearse.

You could see the passengers looking out the windows of the aircraft; the two pilots of the aircraft had exited the plane and joined the family on the ramp, smartly saluting as the casket was moved. What an impressive way to return David to the location of his eventual burial.

John Dinan Jr., an Air Force dentist, had escorted the body on the plane and told us that the crew had informed the passengers that the body of a fallen airman was on board and that their disembarkation would be delayed until the body was unloaded. He said there were no complaints. Jonesy and fellow 34th TFS comrade Al Reiter were present for the arrival. It was good to see them both.

431

Figure 13 - Dave's body arrives at Washington D.C. Leland Sorenson, forefront in suit, salutes the casket, April 24, 2018. Photo courtesy of DPAA.

We were returned to our starting point by the bus, and as we gathered in the lobby of the hotel, I exclaimed that I needed a whiskey. Niall Brannigan had the right answer. Our group went to the Liberty Tavern, and whiskey was served. As we walked in the door, I heard someone yell, "Ed Sykes!" I turned my head in the direction of the cry and standing in front of me was Bane Lyle, another Thud driver from the 34th. After arriving at the hotel, he had come over to the Liberty to get lunch and was at a table by himself. "Get your ass over here, Bane," I exclaimed, and soon Bane, Jonesy, and Al were at the end of the table spinning their lies about the War. I loved watching them in their excitement but chose not to join them, instead hanging with the Dinans and my family. The whiskey tasted as good as the euphoria of the moment. Following lunch, Mary and I joined our three kids and assorted grandkids at a house two of them had rented for the week. I mentioned how I needed a nap before dinner, and they showed me to one of the bedrooms where I lay down to get some much-needed shut-eye. I had barely drifted off when my phone rang. When I checked the number and did not recognize it, I almost let it go but decided to answer. "Yo."

"Is this Ed Sykes?" the voice on the line asked.

"Yup." Thinking it was someone wishing to sell me a cruise or life insurance, I wondered why I had answered this call.

"Hi, Ed, it's Scout Johnson, the commander of the 34th Tac Fighter Squadron. We are deployed to Kadena, Japan, but wanted to let you know there will be two F-35s to do the flyby for Lt. Dinan."

Suddenly I was wide awake as Scout explained that the Squadron would have a wake to honor Dave at the precise time as the scheduled time of the flyby—even though it would be 2300 hours in Japan. He also told me how the day before he had gathered up the lieutenants in the Squadron to talk about Dave and how much these guys had to live up to. Goose bumps! The love fest began, and I acknowledged how proud I was to have my picture on the wall of the 34th, and he professed how proud the Squadron was of its history and those who had been "Rude Rams" and "Men in Black." My nap was over. There'd be no sleeping now.

That evening, Mary and I were joined by our wonderful family at a cool pub for dinner along with Niall and Zeljka Brannigan. A minor triumph took place when Bartz announced to the family that he was no longer going to dispute my "World's Greatest Fighter Pilot" claim. So, after everything I'd gone through, this adventure accomplished more than the recovery of Dave's remains. At least my son would no longer give me shit about the "wgfp" label. Hallelujah!

Going to bed that night, I was afraid I would have trouble sleeping, but I did surprisingly well. However, I lay awake for an hour or so somewhere in the middle of the night in which I must have repeated the eulogy at least 10 times. I was feeling pretty confident I could pull it off. The next morning, I awoke shortly after daybreak and looked out the window to check the weather. It was delta sierra—foggy with a good amount of drizzle. Mary saw me at the window and asked, "How does it look?"

"I don't care if it rains, snows, we have a tornado, or an earthquake hits," I replied. "Today, it's over!"

Charlie and John and I left the hotel early and, after a temporary disorientation, found the Old Post Chapel on Ft. Myer. Walking into the Chapel, I loved the setting at once. It was a rather small sanctuary with a small altar, but the bright white walls made it look larger. As we walked down the center aisle, we saw the priest near the altar, and he welcomed us as we identified ourselves. He was an elderly man with a pleasant manner. The question I had was how long I had for the eulogy.

"How much time do you need?" he asked.

"I have a three-minute, five-minute, or more version," I replied.

"Take whatever time you need. We won't interrupt you. Anything you want to cover, funny stories or particular incidents you remember—whatever you need," he replied. I had no plans to give funny stories or even recounts of my dealings with Dave, but the extra time was welcomed because I had a couple things I wanted to tell the gathering. The priest told us we could use the Family Room next to the sanctuary. As Charlie, John, and I headed that direction, I scanned the almost-empty chapel and discovered Ann Mills-Griffiths sitting near the front. I was elated to see her there early—not like Ann, I thought. In the Family Room with the Dinan brothers, I couldn't settle myself or even sit down.

About that time, my two teenage granddaughters entered. "How are you doing, Grandpa?" asked Gracie as Abby stared intently at me, looking for any sign of my feelings. I stood there briefly before saying, "Follow me, girls, and I'll introduce you to a great woman." I led them back into the sanctuary toward Ann. "I told them I was going to introduce them to a great woman, and you're it." She stood and gave me a brief hug but quickly changed the subject.

People were drifting in, and I spotted Mary approaching me down the aisle. I motioned her toward a seat in the front, and we sat quietly for several minutes while the other attendees came in.

Eventually, the priest motioned for us to stand and the service began. After some beautiful organ music, I heard the doors at the back of the church open, and the Honor Guard brought the casket to the center aisle.

After it was placed on a roller, we stood and faced the casket as two young soldiers brought it down the aisle and rotated it to a position left of the altar. The priest conducted the service with readings from the Bible, and a short hymn was sung. The whole time, my feelings waffled between apprehension and joy. *I hope I am in the joy mode when it's my turn to speak.* As the priest began the recitation of the Lord's Prayer, I knew I was next and wasn't sure which mode I was in. When the congregation sat after the prayer, I remained standing and walked to the altar as the priest introduced me.

When I turned to face the congregation, I was amazed. The chapel was nearly full—quite a large crowd for a person who had been deceased for nearly half a century. Noticing someone arriving late and pushing an elderly woman in a wheelchair gave me a start. Could it be Valerie? I studied the woman for a few seconds and determined that it was not her. I then scanned the rest of the crowd hoping to see her, but it was not to be.

Returning to the task at hand, I wondered where to start. Then it hit me as I took a long look at Dave's casket. "David, my friend, you know this is not easy, but you have always been a trusted wingman, and I know you'll give me great cover today," I began.

Then it became easy. Knowing I had plenty of time, I first recognized Ann Mills-Griffiths as a great American patriot. I let folks know that if it hadn't been for Ann, none of this would be happening. I asked her to stand as I declared that if any woman ever had her face on our American currency, my vote was for Ann Mills-Griffiths. After a nice round of applause, she sat. I looked at her and couldn't tell if the look I was getting was "thanks, Ed" or "you asshole." I didn't care. It was a recognition that was greatly deserved.

I then briefly described my phone call from Scout, informing the crowd that halfway around the world in Kadena, Japan, the fighter pilots of the 34th Tac Fighter Squadron were gathered for a wake in honor of Dave and were drinking a toast to their departed brother. I also described the meeting Scout had with the lieutenants of the 34th, telling them they had some big shoes to fill. I ended with, "David, you have established a legacy for those young men of the 34th, and I know they will carry it forward."

OK, dumbass, let's see if you can get through this eulogy without stepping on your dick. With another look at Dave's casket, I began my formal eulogy.

"First Lt. David Thomas Dinan III was born in January of 1944 and died on St. Patrick's Day of 1969. He was barely 25 years old. Fifty years ago this spring, Dave was preparing to depart his country to participate in the last sustained, full-up air war in the history of mankind. It was a journey that would take nearly half a century to complete. And today, the journey is over.

"Today we will place Dave in a resting place in the company of so many of our nation's greatest heroes. It is an honor that he richly deserves. I'm certain that those great men and women who now rest in the beauty of Arlington will welcome him with a hearty salute and exclaim, 'Well done, brother.'

"And how would Dave want us to remember him? I suggest he would want most to be remembered as a fighter pilot. The fighter pilot fraternity that Dave lived in was an interesting group. Young men who possessed great intelligence and yet had an overwhelming desire to live on the edge. Dave was very smart, and you knew it from the moment you met him. You could also observe that he was very comfortable taking the risks that were associated with his profession. Like everyone involved in this War, he questioned the way it was being fought, but he always did his duty. He did what his country asked him to do.

"We will never know what Dave might have achieved had he not ended up on that hillside in Laos. However, we do know that as long as this country can produce warriors like Dave Dinan, our great American experiment will be sustained. Our nation thanks you, Dave, for your sacrifice. Also, thank you to the Dinan family, which sent our nation its beloved son and suffered this grievous loss."

Wow, I did it. I now called on the "Brothers' Salute."

"I would ask Dave's brothers, Charlie and John, to stand and remain standing.

"But Dave had other brothers, those men who flew their beautiful machines into combat with Dave—a true band of brothers. I would ask those who fought with Dave in the 34th Tac Fighter Squadron to stand." I watched as five elderly men stood up. Seated in a group, they rose to their feet—not as men in their 70s, but they stood briskly like young warriors. I was so proud to be associated with these great men.

"We also had silent brothers," I continued. "They were brothers we never wanted to meet. We knew they were out there to save us if we got in trouble. Would the crew of the Jolly Green Giant who bravely attempted to rescue Dave please stand?" I scanned the crowd and saw Leland jump to his feet.

"Brothers, please face the coffin.

"Brothers, ten-hut!

"Brothers, present arms."

I hesitated. "We've missed you, David. Welcome home, brother.

Figure 14 - "Brothers' Salute" to Dave led by Ed Sykes, April 25, 2018. Two men standing are John and Charlie Dinan. Photo courtesy of DPAA.

"Brothers, order arms. At ease." I then took the two steps down to Dave's casket, put my hand on it, and stated, "At rest!"

As I left the alter, the priest shook my hand, and as I sat down, Mary patted me gently on my leg, letting me know I had met with her approval—not an easy feat! A sense of relief swept over me as now I could relax and enjoy the euphoria of the moment. David's casket was removed from the chapel, and we filed out to a gentle rain and overcast sky, so I assumed the flyby would not happen. As we watched the Honor Guard place the casket on the horse-drawn caisson, I felt a tap on my shoulder and turned around to face someone I did not recognize. After staring at him for a short time, he stated, "Steve Filo."

"Fido!" I yelled, which prompted a gentle nudge from someone behind me and a comment about how I should keep it down. Steve Filo had been my weapon systems officer during my 7-year personal imprisonment in the F4 in the '80s. "Fido, thanks for coming, old friend." Another nudge let me know it was time to proceed to the cemetery. As I turned away from Fido, he gave me a great smile and a handshake, and I knew I wouldn't see him again on this day. I also knew he'd been on hospice, and this could be my final goodbye.

The walk to the graveside ceremony was a long one, and it gave me time to consider the beauty of this place and the magnitude of how many had sacrificed so much. At the graveside service, one of Charlie's sons pulled me from the crowd and sat me in a chair reserved for the family right behind Charlie and John. There was a 21-gun salute, "Taps" was played, and the band struck up a perfect rendition of "America the Beautiful." Finally, the flag was removed from the casket and folded and presented to Charlie. A second flag was presented to John, and some other dignitaries came up and offered them condolences for their loss. The priest ended the ceremony with a short prayer, and it was over for everyone except the Band of Brothers.

Al Reiter came over and grabbed me, saying it was time to "throw a nickel on the grass." The six of us gathered around the site where David would rest, and Joe Widhelm handed me a nickel when I admitted that I had forgotten mine. We stood before the site and Al began to sing,

"Hallelujah, hallelujah, throw a nickel on the grass, save a fighter pilot's ass. Hallelujah, hallelujah, throw a nickel on the grass and you'll be saved." David and I had finished our last adventure together. RIP, my friend.

Epilogue

So that's my story. It's been fun reflecting on all that has occurred over these many years. The next time I'm watching my beloved Cubs from the stands, I will once again survey the assembled crowd and think about the fact that every person there has a story to tell. I also know that I will have done something few of them ever have or will. I put my story on paper and attempted to see if anyone cares. Not that it matters much.

The act of recalling your story and putting it into words is a magnificent exercise in self-reflection and can make you think through ideas you might not have previously contemplated. It is also an interesting exercise in deciding how much of your story can be told and how much belongs to you alone. The thoughts and actions you put into words are part of the beauty of being human. Technology is rapidly beginning to be able to record all our actions and words, but if it ever becomes able to get inside our minds and view our thoughts, it won't be as much fun being human anymore.

Another interesting aspect of writing your story is the consideration of how it will be read by those whom you included in the story. For the most part, I look at life through rose-colored glasses, so I don't think I have dealt too harshly with anyone I wrote about, but, I assume, I will invariably step on some toes. I have given a good deal of editing license to my wife, Mary, when I was discussing issues that involved her or us. If I outlive her and issue a revised edition, there will probably be some "terms of endearment" and other details that she nixed. I was also careful not to offend Valerie, and I gave her an opportunity to review everything I had written concerning her. The term "victim" is thrown around way too loosely in today's culture, but in a very real sense, Valerie was, and still is, a victim of war.

So, what now? Dave is peacefully resting in Arlington, and my story is complete. However, I have new adventures on the horizon, and I'm looking forward to the challenge.

In the back of my mind is the fact that two more of my friends from the 34th TFS, Harold "Pappy" Kahler and Ron Stafford, are still listed as

MIA and their remains have not been recovered. I have asked Phet if he is ready for another adventure, and he has enthusiastically replied in the affirmative. What the hell! I'm only 76 and still have half of my vision and the financial resources to buy my way out of almost any scrape I get into.

Afterburner now!

Acknowledgments

I owe a debt of gratitude to Mary and my kids and grandkids for putting up with my craziness as I progressed through this adventure. They were great! I also found some new best friends and heroes along the way. Mark Rockoff and the Dinan brothers certainly qualify as folks I love to consider as friends. Ann Mills-Griffiths, Niall Brannigan, Leland Sorensen, Bob Maves, Dustin Roses, and Lt. Col. Marcus Ferrera have all been elevated into my personal "Hall of Heroes." To qualify for this Hall, the most important attribute is a continuing struggle to make a difference. This is especially true of Ann, and, to use some baseball jargon, she is the Most Valuable Player and deserves much higher recognition.

A final thanks to my newfound "son" Khamphet Keosiripanya, "Phet." My little Laotian buddy is the best! As we would get into scrapes along the way, I can still hear him asking, "What do we do now, Edward?" I didn't always know the answer to his query, but, invariably, Phet was part of solving the problem.

There are many others that should be given credit for their assistance during my journey and in the process of writing the book.

A special thanks to Mr. Wilfred H. (Howard) Plunkett. Any time I needed to gather historical data about the F-105 or the pilots who flew it in combat, I would contact Howard. While I am only a storyteller, Howard is a historian of the first order. For those interested in getting the "real facts" about the Thud and its pilots, see the included Bibliography for his two books, *F-105 Thunderchiefs* and *Fighting Cavaliers*.

Both the Air Force Museum at Wright-Patterson AFB and the Air Force Library at Maxwell AFB facilitated my research efforts. I visited both, and their archivists were most helpful. I also appreciate all the people who allowed me to print our email and letter correspondence in the book. Besides those mentioned in the first couple paragraphs, a big thank you goes to Bill Habeeb, Kenton Kersting, Al Teel, and Valerie Zoolakis for being so helpful and open with me on this journey.

442

The National League of POW/MIA Families was always useful as it allowed for many networking opportunities. The League families also provided great inspiration.

Despite my continual prodding, the government agencies I dealt with were generally very helpful—and patient. The Defense POW/MIA Accounting Agency (DPAA) and Stony Beach, a part of the Defense Intelligence Agency, were instrumental in the eventual recovery of Dave Dinan's remains. A special kudo to Maj. Gen. (ret.) Kelly McKeague (now the director of DPAA) for making some tough decisions on my behalf.

Finally, I would be remiss if I did not acknowledge the assistance of my author coach, Jo Lena Johnson, and my editor, Karen Tucker (the Comma Queen). They were instrumental in guiding me through a process that is well outside my abilities. I met both women through my involvement with the St. Louis Publishers Association. Ready to write your first book? These two will help you navigate the process and make you look really good. Or, as my daughter said after reading some of their work, "Dad, they really make you look smarter."

Bibliography

Hemingway, Ernest. 1944. London Fights the Robots. *Collier's,* 19 August, 17, 80–81.

Johnson, Lyndon. 1968. "Withdrawal." Speech, televised, 31 March. Voices of Democracy: The U.S. Oratory Project. Accessed April 3, 2020. https://voicesofdemocracy.umd.edu/lyndon-baines-johnson-withdrawal-speech-31-march-1968/.

Johnson, Lyndon. 1968. "Remarks on the Cessation of Bombing of North Vietnam." Speech, televised, 31 October. "Presidential Speeches: Lyndon B. Johnson Presidency," UVA Miller Center. Accessed April 3, 2020. https://millercenter.org/the-presidency/presidential-speeches/october-31-1968-remarks-cessation-bombing-north-vietnam.

Kent, Joe. 1966. *What the Captain Means*. Recorded at Cam Ranh Bay. Video accessed March 30, 2020. https://youtu.be/ft4ywx8PhAs

Plunkett, W. Howard. F-105 Thunderchiefs: A 29-Year Illustrated Operational History. Jefferson, N.C.: McFarland, 2001.

Plunkett, W. Howard and Jeff Kolln. Fighting Cavaliers: The F-105 History of the 421st Tactical Fighter Squadron 1963–1967. Scotts Valley, Calif.: CreateSpace, 2018.

Reasonable effort has been made to verify the accuracy of historical events, dates, and locations. However, some events have been re-created from the author's memory.

Key Terms

Air Force Officer Ranks
(in order from lowest to highest)

2nd Lt. (or 2Lt.)	Second Lieutenant
1st Lt. (or 1Lt.)	First Lieutenant
Lt. (or LT)	Lieutenant (1st Lt. or 2nd Lt.)
Capt.	Captain
Maj.	Major
Lt. Col. (or LTC)	Lieutenant Colonel
Col.	Colonel
Brig. Gen.	Brigadier General
Maj. Gen. (or MG)	Major General
Lt. Gen.	Lieutenant General
Gen.	General

Acronyms & Abbreviations

AAA	anti-aircraft artillery
AFB	Air Force base
AFROTC	Air Force Reserve Officers' Training Corps
ANG	Air National Guard
ARRS	Aerospace Rescue & Recovery Squadron
ATM	air turbine motor
Dash One	pilot's flight manual
Det.	Detachment
DIA	Defense Intelligence Agency
DMZ	demilitarized zone
DOD	Department of Defense
DPAA	Defense POW/MIA Accounting Agency (combined DPMO and JPAC)
DPMO	Defense Prisoner of War/Missing Personnel
EGT	exhaust gas temperature
FAC	forward air controller

IP	instructor pilot
JFA	Joint Field Activity
JPAC	Joint POW/MIA Accounting Command
KABOOM	Korat Officers Open Mess
KANG	Kansas Air National Guard
KIA-BNR	Killed in Action-Body Not Recovered
MIA/POW	Missing in Action/Prisoner of War
NKP	Nakhon Phanom (RTAFB)
NVA	North Vietnamese Army
O'Club	Officers Club
PDJ	Plaine des Jarres (or Plain of Jars)
PJ	parajumper (Air Force Pararescue)
ROTC	Reserve Officers' Training Corps
RPM	revolutions per minute
RTAFB	Royal Thai Air Force base
SAM	surface-to-air missile
SAAR	search and air rescue
SAR	search and rescue
SEA	Southeast Asia
SOW	special operations wing
TFG	tactical fighter group
TFS	tactical fighter squadron
TFW	tactical fighter wing
The War	The Vietnam War
UPT	Undergraduate Pilot Training
wgfp	world's greatest fighter pilot (Col. Ed Sykes)
WSO	weapon systems officer

About the Author

In 1973, a class of young lieutenants, undergoing training to fly the F-105 with the Kansas Air National Guard, decided to hang a label on one of their instructors. That instructor, Ed Sykes, taught them the F-105 systems in the classroom and also flew with them daily. The moniker, "World's Greatest Fighter Pilot," often shortened to "wgfp," would stay with Sykes the rest of his life.

Sykes never tried to discourage its use—not because he felt he was the world's best fighter pilot, but because he felt that being a fighter pilot of that era was a state of mind, an attitude. He will certainly claim that his fighter pilot attitude, with his acceptance of the risks and excitement of that profession, cannot be exceeded. This attitude extends well past the days when one's flying career is over.

Sykes grew up in rural western Kentucky and obtained a degree in electrical engineering from the University of Wisconsin before entering the Air Force in 1967. Following a combat tour with an F-105 squadron, he eventually ended up in the Kansas Air National Guard and spent most of his flying career with the 184th Tactical Fighter Group, including over 6 years as the unit's commander. At that time, the 184th was the largest flying unit in the Air National Guard with three squadrons of F-16s.

Following his retirement, he helped establish, along with a friend and former F-105 crew chief, his own business, which was involved in consulting and real estate investment. They sold their business in 2016. He currently manages his little "ranch" in Kansas where he raises cattle, hogs, chickens, and an assortment of other furry critters. Sykes and his wife of 53 years, Mary, have four children and 13 grandchildren (with another in the hangar). Life has been good for the World's Greatest Fighter Pilot.

Made in the USA
Columbia, SC
27 September 2020